Mulberry: Genetic Improvement in Context of Climate Change

Editors

M.K. Razdan
University of Delhi, Delhi, India

T. Dennis Thomas
Central University of Kerala, Kasaragod, India

CRC Press
Taylor & Francis Group
Boca Raton London New York

CRC Press is an imprint of the
Taylor & Francis Group, an **informa** business

A SCIENCE PUBLISHERS BOOK

First edition published 2021
by CRC Press
6000 Broken Sound Parkway NW, Suite 300, Boca Raton, FL 33487-2742

and by CRC Press
2 Park Square, Milton Park, Abingdon, Oxon, OX14 4RN

Library of Congress Cataloging-in-Publication Data

Names: Razdan, M. K., editor. | Thomas, T. Dennis, 1971- editor.
Title: Mulberry : genetic improvement in context of climate change / M. K.
    Razdan, T. Dennis Thomas.
Description: First edition. | Boca Raton, FL : CRC PRes, 2021. | Includes
    bibliographical references and index. | Summary: "Mulberry (Morus spp.)
    is an important horticultural plant in the sericulture industry. It
    belongs to the family Moraceae. The leaf of mulberry is used to feed the
    silkworm Bombyx mori L. It is also used as a fodder. Due to its economic
    and agricultural importance, mulberry is cultivated in many parts of the
    world. An estimated 60% of the total cost of silk cocoon production is
    for production and maintenance of mulberry plants. Therefore, much
    attention is needed to improve the quality and quantity of mulberry
    leaves. It is vital to increase the production of superior quality
    mulberry leaves with high nutritive value for the sericulture
    industry"-- Provided by publisher.
Identifiers: LCCN 2020046673 | ISBN 9780367024994 (hardcover)
Subjects: LCSH: Mulberry. | Genetics. | Mulberry--Genetic engineering. |
    Mulberry--Climatic factors.
Classification: LCC SF557 .M848 2021 | DDC 634/.38--dc23
LC record available at https://lccn.loc.gov/2020046673

ISBN: 978-0-367-02499-4 (hbk)
ISBN: 978-0-367-55534-4 (pbk)
ISBN: 978-0-429-39923-7 (ebk)

Typeset in Times New Roman
by Radiant Productions

Preface

Mulberry is a fast growing deciduous perennial tree or shrub belonging to the genus *Morus* and family Moraceae. It grows under various climatic conditions which range from temperate to tropics and occurs at different altitudes like from sea level to altitudes as high as 4000 m. Considering that mulberry most likely evolved in sub-Himalayan tracts, owing to its wider adaptability to divergent agro-climatic conditions, yet its growth and cultivation has spread to almost all continents including Africa, Asia, South America, Europe, North and South America. This is primarily due to the fact that mulberries are grown for different purposes around the world. Mulberry leaves are sole natural source of feeding silkworm (*Bombyx mori* L.) which is involved in production of silk, thus making mulberry a valuable economic source to sericulture industry. Besides silk production, mulberry is used in medicine, as animal fodder, and in human consumption as fruit products, e.g., jam jelly, marmalade, frozen desserts, juices, etc. Mulberry wood has equal utility in making furniture.

Mulberry being valuable economic, horticultural and industrial plant genetic resource, the threats for loss of diversity in gene pools of *Morus* spp. appear significant because of habitat destruction, deforestation, spread of invasive alien species, alterations in the land-use pattern and overall impact of ongoing climate change. Further the species that are endemic cannot grow in alternative habitats as they will be vulnerable to extinction. It is estimated that 15 to 37%, plant species will be threatened with extinction due to climate change by 2050. In this context mulberry attracts special attention since about 60% silk is produced among various continents of the world and increase in silk production is directly dependent on high silk cocoon yield. This necessitates consistent efforts for maintenance of mulberry species involved in feeding silk cocoons as well as the species of *Morus* having other valuable traits despite environmental variations. Utilization of new spontaneous or experimentally generated mulberry varieties is crucial for increase in productivity, input use efficiency, and to withstand various abiotic (drought, heat, cold, alkalinity) and biotic stresses (root-rot, root-knot, mealybug, trips, mites, whitefly) vis-à-vis the climate change. Genetic improvement thus has immense potential to cope up with unpredictable and extreme climates to sustain mulberry species.

The present book is aimed at providing overview of research carried out on various aspects in mulberry species and provides a critical appraisal of the state-of-the-art of these findings. **Section A** covers Chapters highlighting Taxonomy and Importance of Mulberry; Chapters of **Section B** describe attempts being made for genetic improvement of mulberry species using conventional and non-conventional methods; whereas other research efforts that ensure sustainable growth of mulberry are mentioned in Chapters of **Section C**. The number of recognized species in major taxonomic treatments of *Morus* has varied considerably. For example, Carl Linnaeus who established the genus *Morus* in 1753 considered the genus comprises seven species: *M. alba*, *M. indica*, *M. nigra*, *M. papyrifera*, *M. rubra*, *M. tartarica* and *M. tinctoria*. However, two of these, *M. papyrifera* and *M. tinctoria*, were subsequently moved to *Broussonetia* and *Maclura*, respectively. Whereas Linnaeus considered the characters of fruit color, leaf shape and leaf hairs as parameters in identifying these species, Bureau (1853) on the contrary treated the genus *Morus* on the characteristics of leaves and pistillate catkins, thereby recognizing five species with 19 varieties and 13 sub-varieties. Further variation in number of species within genus *Morus* has been reported from various countries (e.g., China 24, Japan 16, USA 14, Korea 6, etc.). These points to no complete agreement on the origin and number of mulberry species. The genetic diversity in important mulberry species needs revaluation of the systematic status within genus *Morus* particularly considering their occurrence in different geographical and climate conditions. This has been comprehensively revaluated in **Chapters 1** and **2.** Cultivation of mulberry is reported to have started about 5000 years back in China and it was discovered subsequently that mulberry species from different and climatic regions exist in various polyploid forms. Accordingly, the cultivation methods vary including nutrients in soil where various mulberry species grow. **Chapter 3** gives elaborative account of cultural practices of mulberry species which directly impact the productivity and yield of economic traits. Mulberry like the traditional herbs has been used in medicine for centuries. Due to its pharmacological properties mulberry fruits possess an outstanding nutritional value due to their high levels of phenolic compounds with high antioxidant activity. There are many opportunities for the food and healthcare industry to explore the health benefits of mulberry fruits as research results endorse ethnopharmacological usage of mulberry. **Chapter 4** and **Chapter 5** give an elaborative account of anti-oxidative, anti-inflammatory, anti-microbial, anti-cancer, anti-atherosclerosis, anti-Alzheimer, anti-obesity, anti-hyperlipidemic and anti-hyperglycaemic properties of mulberry fruits.

Traditionally sexual hybridisation, polyploidy and mutation breeding has contributed much to the development of high yielding mulberry varieties and other superior economic traits, particularly under the changing agro-climatic conditions created by global warming and other factors. These aspects are elucidated in **Chapter 6** and **Chapter 7.** Considering high cost of inputs like man power in conventional methods and inability of improved varieties obtained from these methods in sustaining rigors of environmental or climatic changes, the biotechnological approaches have been pursued as an alternative to genetic improvement of the existing new mulberry varieties. **Chapter 8 and Chapter 9 discusses** the applications and limitations of the plant tissue culture or *in vitro* methods in mulberry improvement, micropropagation and conservation of new genotypes; **Chapter 10** provides a general account of genetic improvement using protoplast culture and **Chapter 11** describes extent to which attempts at achieving these goals have been successful by genetic engineering.

Among other research efforts on sustainable growth of mulberry **Chapter 12** focusses on the role of various environmental factors on the growth of mulberry (*Morus alba*) under specific geographical conditions. It is apparent that impact of these factors is synergistic and with proper manipulation of these factors *Morus alba* can sustain growth when coupled with effective agro-chemical management. Similarly, the identification of tightly linked markers or candidate genes associated with trait specific QTLs have been shown to facilitate marker assisted selection (MAS) in mulberry breeding programmes for identification of stress tolerant traits resilient to climate change (**Chapter 13**). Arbuscular mycorrhizal symbiosis is another aspect that contributes significant benefits for sustainable growth and quality of mulberry plants (**Chapter 14**). A panoramic description on the development of new improved mulberry varieties by Sericultural institutes located at different climate zones and geographical regions in India including methods of their cultivation for sustainable sericulture is provided in **Chapter 15.** Furthermore this chapter mentions about protection of these varieties by management of diseases and pests in the field. Viroids, viruses and phytoplasmas are other class of most common pathogens infecting mulberry in all mulberry growing regions of the world, thus causing severe decline of the trees. The type of pathogenic symptoms induced by these agents depend on the pathogen-host combination, environment and growth conditions. There are reports of infected trees showing no symptoms of disease in some instances which can be worrisome as by the time disease shows up the damage is already done. **Chapter 16**, therefore, mentions the types of these pathogens, highlights the molecular basis of their detection, and suggests control measures for prevention of their attack on mulberry. To sustain sericulture the conservation of new mulberry genetic resources obtained through multiple research efforts is an absolute must especially when threatened under climatic change. **Chapter 17** outlines objectives and methods of germplasm conservation with reference to various mulberry species and varieties of importance.

This compendium on research findings of experts from international organizations and institutes should prove beneficial to sericulture, horticulture and pharmaceutical industry. Also it could be of use to those involved in mulberry cultivation and farming. Lastly, thanks are due to the publisher and authors of respective chapters for their sincere cooperation and support.

December 2020

MK Razdan
Delhi, India

T Dennis Thomas
Kasaragod, India

Contents

Section C: Sustainable Growth of Mulberry in Context of Climate Change

Section A

Taxonomy and Importance of Mulberry

CHAPTER 1

Systematics of the Genus *Morus* L. (Moraceae)
Taxonomy, Phylogeny and Potential Responses to Climate Change

Madhav P Nepal* and Jordan M Purintun

INTRODUCTION

The genus *Morus* L. (Tribe Moreae Dumort., Moraceae Gaudich.) is distributed in the temperate and subtropical regions of Eurasia, Africa, and the Americas, with the majority of species occurring in Asia (Berg 2005, Berg et al. 2006). The leaves of *M. alba* L., *M. australis* Poiret. and *M. nigra* L. have been valued for centuries as fodder for silkworm (*Bombyx mori* L.) larvae (Watanabe 1958). Additionally, they have been cultivated in many parts of the world for their edible fruits and as ornamental trees; they have become naturalized in several regions throughout the world. Artificial hybridization between polyploid introduced *M. nigra* and the native species was very common in East Asia for improving cultivars for the silkworm industry (Tojyo 1985). Recently, natural hybridization between the introduced *M. alba* and the native *M. rubra* in North America has been documented (Burgess and Husband 2004, Nepal 2008). Natural interspecific hybridization can pose challenges to taxonomic investigation in groups such as *Morus*. Taxonomic confusion in *Morus* is also likely in part due to the fact that species can exhibit a great deal of variation in characters, with character states overlapping among taxa. Bureau (Bureau 1873) seemed to have faced this challenge, and he "sunk" dozens of taxonomic entities previously recognized at the species level to the levels of varieties and sub-varieties within *M. alba*. There are only two genus-wide revisions since the genus *Morus* was established by Linnaeus in 1753: those of Bureau (Bureau 1873) and Koidzumi (Koidzumi 1917). Some species, however, have been treated regionally in floras more frequently (Wunderlin 1997, Berg 2001, Zhou and Gilbert 2003, Berg et al. 2006).

The number of recognized species in the major taxonomic treatments of *Morus* has varied considerably (Nepal 2008, Tikader and Vijayan 2017). Linnaeus (Linnaeus 1753) established the genus in 1753 with seven species: *M. alba*, *M. indica*, *M. nigra*, *M. papyrifera*, *M. rubra*, *M. tartarica* and *M. tinctoria*. Two of these, *M. papyrifera* and *M. tinctoria*, were later moved to *Broussonetia* and *Maclura*, respectively. Linnaeus discussed the characters of fruit color, leaf shape and leaf hairs and gave a short diagnosis of each species. The first comprehensive treatment of the genus *Morus* was primarily based on the characteristics of leaves and pistillate catkins and was given by Bureau (Bureau 1873) who recognized five species with 19 varieties and 13 sub-varieties. Varieties within *M. alba* were classified into two informal groups: one of varieties with oblong and short cylindric pistillate catkins and syncarps, and the other of varieties with long cylindric pistillate catkins and syncarps. He further divided the

Department of Biology and Microbiology, South Dakota State University, Brookings, South Dakota 57007-1898 U. S. A.
* Corresponding author: Madhav.Nepal@sdstate.edu

former group of varieties into two subgroups: one of varieties with short styles (< 1 mm) and the other of varieties with long styles (> 1 mm). The relevant names and groupings are presented in Table 1.

Bureau (1873) treated dozens of previously recognized species, including the Linnaean species *M. tatarica* and *M. indica*, as varieties of *M. alba*. The four other species he recognized were: *M. nigra* (an additional variety, var. *laciniata*), *M. rubra* (including var. *incisa* and var. *tomentosa*), *M. celtidifolia* and *M. insignis*. Greene (Greene 1910) studied *Morus* in the southwestern United States, and split *M. microphylla* sensu Buckley into 14 species (i.e., *M. albida, M arbuscula, M. betulifolia, M. canina, M. goldmanii, M. confines, M. crataegifolia, M. grisea, M. microphilyra, M. microphylla, M. pandurata, M. radulina, M. vernonii, M. vitifolia*). Schneider (1916) studied E.H. Wilson's collection of *Morus* from China housed at the Arnold Arboretum of Harvard University. He described one new species, *M. notabilis*, upgraded Bureau's *M. alba* var. *mongolica* to the species level, and enumerated *M. alba, M. acidosa* and *M. cathayana* based on this collection. Koidzumi (1917) presented the most recent genus-wide treatment and recognized 24 species in two sections: *Dolichostylae* Koidz. and *Macromorus* Koidz. These sections were recognized based on the character of the style length: the former with the long style and the latter section with the short style. In his classification, Koidzumi promoted some of Bureau's varieties to the level of species. Leroy (Leroy 1949) classified *Morus* into three subgenera: *Eumorus* J.F. Leroy, including all Asian and North as well as Central American *Morus*; *Gomphomorus* J.F. Leroy, including South American species; and *Afromorus* A. Chev., which includes African species.

In addition to traditionally employed taxonomic characters in *Morus*, several attempts have been made to identify additional taxonomically useful characters that would enable workers to better distinguish the species. For example, Hotta (Hotta 1954) studied variation in shape and position of leaf cystolith cells in *M. alba, M. australis* and *M. mongolica*, Katsumata (Katsumata 1971) studied size and shape variation of leaf ideoblast and used this character to classify several races of *M. alba* and *M. australis*. However, their results showed that these characters were variable and often transcended the species boundaries. Venkataramana (Venkataraman 1972), in his review article on wood phenolics of the family Moraceae, examined three types of bark flavonoids present in *M. rubra* were absent in Asian species (*M. alba, M. serrata, M. laevigata* and *M. indica*). He further showed that some other flavonoids in the bark such as mulberrin, mulberrochromene, cyclomulberrin, cyclomulberrochromene and three other unnamed pigments present in Asian species were absent in *M. rubra*. Further five other wood phenolics found to be present in different concentrations in Asian species were absent in *M. rubra*. If phenolic differences are consistent within species (infraspecific variation, if assessed, was not discussed), it may be useful to explore these chemical characters as potential taxonomic characters.

Recent taxonomic work on *Morus* includes description of new species such as: *M. deqinsis, M. liboensis* and *M. jimpinensis* and *M. barkamensis* by Chang (1984), *M. mongolica* var. *hopeiensis* and *M. australis* var. *incisa* by Wu and Chang (Zhengy and Xiushi 1989), and *M. gongshanensis* and *M. mongolica* var. *longicaudata* by Cao (Cao 1991). Revision and lectotypfication have been accomplished for *M. nigra* L. (Bhopal and Chaudhri 1977); *M. alba* L., *M. indica* L. and *M. tatarica* L. (Rao and Jarvis 1986); *M. insignis* (Berg 1998a, b) and *M. rubra* (Reveal 2007). Regional taxonomic work on *Morus* also includes revision of *Morus* in Africa by Berg (1988a); in North America by Wunderlin (1997); in the Neotropics by Berg (2001), and in China by Zhou and Gilbert (Zhou and Gilbert 2003). Reveal (2007) recognized 12 species in China alone; most of these species had been recognized by Koidzumi (Koidzumi 1917) and included new species of Chang (1984), Wu and Chang (1989), and Cao (1991). Berg (2001) studied the genus *Morus* in the Neotropics, and recognized *M. insignis* and *M. celtidifolia* from Central and South America. Berg (2005) estimated the number of species worldwide as ca. 12 (eight in Asia, one in Africa and three in the New World), and later estimated the worldwide number at 10–15 (Berg et al. 2006); however, he did not provide a list of all recognized species in either of these publications. Currently, the International Plant Names Index (IPNI, https://www.ipni.org/accessed through the Global Biodiversity Information Facility [GBIF] data portal, http://data.gbif.org/datasets/resource/40, August 14, 2019) endeavors to list all validly published names. It lists 234 names in *Morus*, most of which are either synonyms or names of infrataxa.

Species including *M. alba, M. australis*, and *M. nigra* have long been subjects of domestication and artificial selection. Their introduction to several countries in the world, domestication, escape from place of cultivation with seeds in the millions dispersed by birds and bats (see Tang et al. 2008) and establishment in the natural habitat of other species may have contributed to taxonomic confusion that may present challenges for conservation of the

Table 1. History of major taxonomic treatments of Morus.

Linnaeus 1753	Bureau 1873	koidz umi 1917 Section Dolic hostyleae koidz. = I and Section Macromorus koidz. = II	Leroy's list 1949, Subgenus *Eumorus* = E, *Gemphomorus* = G and *Afromorus* (A)	Flora of China, Zhou and Gilbert 2003
M. alba L.	*M. alba* Var. *vulgaris* Subvar. *tenuifolia* Subvar. *rosea* *colombasa* Subvar. *colombassetta* Subvar. *rebalaira* Subvar. *romana* Subvar. *macrophylla* Subvar. *tokwa* Subvar. *tatarica* Var. *italica* Var. *pyrimidalis* Var. *bungeana* Var. *venosa* Var. *constantinopolitana* Subvar. *pumila* Var. *mongolica* Var. *serrata* Var. *nigriformis* Var. *indica* Var. *cuspidata* Var. *stylosa* Var. *arabica* Var. *atropurpurea* Var. *latifoila* Var. *laevisgata* Subvar. *laeviegata* Subvar. *viridis*	*M. alba* (II) *M. multicaulis* Perr. (II) *M. rotundiloba* Koidz. (I) *M. tiliaefolia* Makino. (II) *M. mongolica* (I) *M. serrata* Roxb. (II) *M. boninensis* Koidz. (II) *M. nigriformis* Koidz. (I) *M. arabica* Koidz. (I) *M. bombycis* Koidz. (I) *M. acidosa* Griff. (I) *M. kagayameae* Koidz. (I) *M. atropurpurea* Roxb. (II) *M. macroura* Miq. (II) *M. laevigata* Wall. (II)	 *M. tilaefolia* (E) *M. mongolica* (E) *M. serrata* (E) *M. australis* (E) *M. laevigata* (E) *M. macroura* (E) *M. wittiorum* (E)	 *M. mongolica* *M. serrata* *M. australis* *M. trilobata* *M. macroura* *M. wittiorum* *M. liboensis*
M. tatarica L.	*M. abla* Var. *vulgaris* Subvar. *tatarica*			
M. nigra L.	*M. nigra* Var. *laciniata*	*M. nigra* L. (II)	*M. nigra* (E)	*M. nigra*
	M. insignis	*M. insignis* Bur. (II)	*M. insigins* (G) *M. trianae* (G)	
M. rubra L.	*M. rubra* Var. *tomentosa* Var. *incisa*	*M. ruba* L. (II)	*M. rubra* (E)	
M. indica L.	*M. alba* Var. *indica*	Synonym of *M. ocidosa*		
M. tinctoria L.	=*Maclura tinctoria*			
M. Papyrifera L.	=*Broussonetia papyrifera*			
	M. celtidifolia Kunth.	*M. celtidifolia* Kunth. (II) *M. mollis* Rusby. (II)	*M. celtidifolia* (E)	
		M. mesozygia Stapf. (II) *M. microphylla* Buckl. (II)	*M. mesozygia* (A) *M. lacteal* (A) *M. microphylla* (E) *M. koordersinia* (E)	
		M. cathayana Hemsl. (II)	*M. cathayana* (E)	*M. cathayana*
		M. notobilis C.K. SCHN. (I)	*M. notabilis* (E)	*M. notabilis*

native species. Hybridization between two species, the native North American *M. rubra* and the introduced *M. alba*, has been well documented in Canada (Burgess and Husband 2004). Interspecific hybridizations of *M. alba* with *M. australis* and *M. serrata* have also been shown to produce a high percentage (> 80%) of fertile seeds (whereas a cross between *M. alba* and *M. macroura* produced no fertile seeds (Das 1965). According to Tojyo (Tojyo 1985), *M. nigra* (2n = 308), introduced in Japan from western Asia, had been hybridized with the native species to produce varieties exhibiting several ploidy levels. Many introduced *Morus* species including *M. alba* are easily and vegetatively propagated and can be opportunistically apomictic (Griggs and Iwakiri 1973), which may increase their adaptability in the novel habitats. Considering these facts, clarification on taxonomy of the native species is an important step toward protection of native species. Diversity of the genus *Morus* in remote areas of Asia and tropical montane forests might have hindered the rigorous taxonomic treatment. The major objectives, therefore, of this study is to re-evaluate the taxonomy of the genus *Morus* worldwide and summarize phylogenetic relationships based on our previous study (Nepal and Ferguson 2012).

Materials Examined

In addition to the field observations of *Morus* species, over 2000 specimens from the major herbaria worldwide were examined. The herbaria acronyms follow Index Herbariorum (Holmgren and Holmgren 1992): A: Arnold Arboretum; FLAS: University of Florida Herbarium, GH: Gray Herbarium, Harvard University; MO: Missouri Botanical Garden; KSC: Kansas State University; NY: New York Botanical Garden; P: Herbier National de Paris (National Herbarium of Paris); LINN: Herbarium of the Linnaean Society of London; B: Berlin Herbarium. We used morphological characters of the bud (size, bud-scale banding), the leaf (base, petiole, hair distribution, size, venation, margin, apex), the inflorescence (shape, size, number and type of unisexual flowers), the style (length) and the infructescence (shape, size, color) for species identification and description. Herbarium label data on collection times were observed as a proxy to the flowering and fruiting times the specimens were collected. *Morus alba* and the native *M. rubra* co-occur in many plant communities in the Flint Hills region. Field observation of 13 populations of *M. alba* and nine populations of *M. rubra* complemented herbarium study for these species. All collected specimens were deposited at KSC. For all materials studied, morphological characters of the stem, stipule, bud, bark, branch, leaf, inflorescence, style, and fruit were thoroughly examined. Flowering and fruiting times were obtained from the specimen labels. Observations on some features that were not present in the studied specimens were taken from previous literature on *Morus*; these included observations on habit and bark (Bureau 1873, Berg 2001, Zhou and Gilbert 2003), bud scales (Wunderlin 1997, Zhou and Gilbert 2003), and chromosome numbers.

Floras from regions around the world, personal communication with experts, and herbarium databases all facilitated this study. Photographs or drawings of some type specimens were directly downloaded from the websites of various herbaria (US, http://www.nmnh.si.edu/botany; MO, http://www.mobot.org; NY, http://sweetgum.nybg.org/vh/specimen_list.php, P, http://www.mnh n.fr; LINN, http://www.linnean-online.org/view/plants_alpha/morus.html) or were obtained directly from herbarium personnel. All examined images and drawings were deposited at KSC. DNA sequences (nuclear Internal Transcribed Spacer [ITS] and chloroplast *trn*L-*trn*F spacer regions) developed by the first author Nepal (Nepal and Ferguson 2012) were reanalyzed to provide an overview of phylogenetic relationships within the genus. The species recognition in this study was based on the literature review and study of herbarium specimens available mostly in the western herbaria. An intensive study of Asian taxa in the future will be necessary to complement the findings of the present study.

Taxonomic Revision of the Genus *Morus*

Thirteen species of the genus *Morus* are recognized. A brief taxonomic description of the genus, a key to the species, and a short description of each species are presented below.

Morus Linnaeus in *Species Plantarum* 2: 986 (1753); Bureau in DC, *Prodromus* 17: 237 (1873). Koidzumi, *Bulletin of Sericultural Experimental Station* (Tokyo) 3:1 (1917). Nakai, *Journal of Arnold Arboretum* 8:234 (1927). Berg, *Flora Neotropica Monograph* 83: 25 (2001). *Flora Malesiana* 17:23 (2006) [Type species: *M. nigra* L. Lectotype: designated by Britton and Brown, *Flora of the Northern United States*. Second ed., 1: 631 (1913); by Bhopal and Chaudhri, *Pakistan Systematics* 1(2): 29 (1977)].

Dioecious, subdioecious, or monoecious shrubs to trees with milky sap. Terminal bud in the stem dies. Buds conical or ovoid, the outer bud scales pubescent with dark, brown, or white apical margins. Stipules caducous to semi-persistent, pubescent. Leaves alternate, stipulate, cauducous. Leaf blades ovate to lanceolate, unlobed or lobed, the margins serrate to dentate to crenate; primary veins usually 3, secondary veins pinnate. Inflorescences pedunculate, axillary, with 1–5 catkins produced per bud; staminate catkins cylindric; pistillate catkins oblong to capitate. Flowers unisexual on the same or different plants. Staminate flowers with four perianth parts; imbricate; stamens inflexed in bud; anthers dithecous, introse and dorsifixed; pistillodes present. Pistillate flowers with four perianth parts; imbricate; ovary superior; style long/short/absent; stigma 2-branched. Infructescence oblong or cylindric. Fruits sub-drupaceous, each enclosed by enlarged succulent perianth. Base chromosome number (x) = 14, 2n varies from 28, 42, 56, 84 to 308 (Ammal 1948, Azizian and Sonboli 2001).

Distribution: Broad in temperate and subtropical regions, montane forests in the tropics and subtropics and lowlands of tropical Africa.

Key to the species of Morus

Note: The number in front of the species name in the dichotomous key refers to the order of accompanying taxonomic description.

1. Leaf ovate to orbicular, secondary venation (from the mid-rib) less prominent and scalariform except two to three pairs toward the leaf apex. Peduncle longer than inflorescence...........................**7. *M. mesozygia***

1. Leaf ovate to lanceolate, secondary venation prominent and not scalariform. Peduncle shorter or equal to inflorescence (2)

2(1). Pistillate flowers with distinctly long style (> 1 mm) (3)

2. Pistillate flowers with no or short style (< 1 mm) (5)

3(2). Leaf margin with acute dentation characterized with a short to long seta.........................**9. *M. mongolica***

3. Leaf margin without seta as mentioned above (4)

4(3). Infructescence elongated, < 2 cm (excluding peduncle), leaf shape variable........................**2. *M. australis***

4. Infructescence cyndric, 2–4 cm, leaf broadly ovate with cordate base..............................**11. *M. notabilis***

5(2). Infructescence longer than 2 cm (excluding peduncle) (6)

5. Infructescence shorter than 2 cm (8)

6(5). Infructescence 2–5 cm..**3. *M. cathayana***

6. Infructescence 5–16 cm or longer (7)

7(6). Axillary bud minute, petiole 1–2.5 cm, leaf blade usually lanceolate to elliptic, margin minutely serrate to subentire, peduncle < 0.5 cm..**5. *M. insignis***

7. Axillary bud larger, petiole 2.5–6 cm, leaf blade ovate to broadly ovate, margin sub-entire to minutely serrate, peduncle > 0.5 cm..**6. *M. macroura***

8(5). Leaf blade usually bright green, adaxially usually glabrous, abaxially sparse pubescent along the veins, leaf margin irregularly dentate, leaf apex usually obtuse...**1. *M. alba***

8. Leaf blade usually dull green, adaxially slightly scabrous, abaxially pubescence all over, leaf margin with acute serrations, leaf apex acute to subcaudate (9)

9(8). Leaf margin with regularly spaced triangular teeth, bud scales and stipules semi-persistent....**13. *M. serrata***

9.   Leaf margin not as above, bud scales and stipules immediately cauducous (10)

10(9).   Leaves broadly cordate at base, glabrous adaxially or slightly scabrous, sparsely pubescent along the veins abaxially, the leaf margin with wider teeth. Infructescence oblong, 1.5–2.5 cm wide, up to 2.5 cm; stigma long pubescent...**10. *M. nigra***

10.   Leaves deeply cordate at base, densely pubescent along the veins adaxially, sparsely pubescent in the interveinal areas. Infructescence cylindric, 0.5–2 cm wide, up to 2 cm; stigma short pubescent (11)

11(10).   Leaf blade abaxially pubescent, adaxially usually scabrous. Stem branches are horizontally spread in a characteristic pattern. Fruits compactly arranged in a fleshy cylindrical infructescence...........**12. *M. rubra***

11.   Leaf blade adaxially slight to harsely scabrous, fruits loosely arranged, globose or capitate, not as fleshy as in *M. rubra* (12)

12(11).   Shrub to small tree, mature leaf blade less than 6 cm, ovate to ovato-lanceolate, abaxially scabrous or pubescent, infructescence small (ca. 0.5 cm; excluding the peduncle) adaxially harshly scabrous..**8. *M. microphylla***

12.   Small to big tree, mature leaf up to 4–20 cm, abaxially harshly pubescent to scabrous, oblong to lanceolate, base usually unequal to cordate, adaxially glabrous to slightly scabrous. Infructescence 1–2 cm sometime longer..***M. celtidifolia***

Species Description

1. ***Morus alba*** Linnaeus in *Species Plantarum* 2: 986 (1753). [Bureau in DC, *Prodromus* 17:238 (1873), Koidzumi, *Bulletin of Sericultural Experimental Station* (Tokyo) 3: 1 (1917). Nakai, *Journal of Arnold Arboretum* 8:234 (1927). Berg in *Flora Neotropica Monograph* 83: 25 (2001), Zhou and Gilbert in *Flora of China* (2003). Berg in *Flora Malesiana* 17:23 (2006).] (Lectotype: Herb. Linn. No. 1112.1 (LINN), upper left specimen by Rao and Jarvis, *Taxon* 35: 705 [1986]).

Synonyms. *M. pendula* Sudw., *M. alba* Sudw., *M. atropurpurea* Roxb., *M. alba* var. *vulgaris* Bur., *M. arabica* Koidz., *M. bullata* Balb. ex Loud., *M. byzantina* Sieber ex Steud., *M. colombassa* Hort. ex Dippel, *M. constantinopolitana* Hort. ex Poir., *M. cucullata* Bonaf., *M. dulcis* Royle, *M. fastigiata* Hort. ex dippel, *M. furcata* Hort. ex Steud., *M. guzziola* Hort. ex Steud., *M. heterophylla* Loud., *M. hispanica* Hort. ex Loud., *M. italica* Poir. ex Lam., *M. japonica* Audib. ex Ser., *M. kaki* Hort. ex Lavallee, *M. laciniata* Audib. ex Loisel., *M. latifolia* Hort. ex Spach, *M. levasseurei* Hort. ex Lavallee, *M. lhou* Koidz., *M. lucida* Hort. ex Loud., *M. macrophylla* Hort. ex Steud., *M. mariettii* Hort. ex Steud., *M. mauritiana* Jacq., *M. membranacea* Hort. ex Steud., *M. moretti* Audib. ex Bur., *M. multicaulis* Raf., *M. nana* Audib. ex Loisel., *M. nigriformis* Koidz., *M. patavia* Audib. ex Dippel, *M. patavina* Hort. ex Spach., *M. pumila* Balb., *M. romana* Lodd. ex Spach, *M. rubra* Lour., *M. serotina* Mart. ex Bur., *M. sinensis* Hort. ex Loud., *M. stylosa* Ser., *M. subalba* Hort. ex Steud., *M. venassaini* Hort. ex Steud., *M. venosa* Delile ex Spach.

Description. Subdioecious trees to 12 m. Bark: gray, smooth or shallowly furrowed with wide ridges colored reddish tan or yellow. Branches: irregular or diffused, shorter or slightly condensed in bushy trees, finely to densely pubescent when young. Buds: terminal buds on the stem and branches caducous. Axillary buds reddish brown, 1–4 mm long, ovoid, greenish white to creamy with white bands on the apical margins, pubescent. Leaves: leaf scars slightly raised and nearly circular. Stipules lanceolate to linear, 2–3.5 cm long, greenish white to whitish brown, caducous, and thin pubescent. Petioles densely pubescent, 0.5–5 cm long. Leaves tri-nerved with two lateral veins extending through one- to two-thirds of the lamina, the secondary veins angled acutely (30–60°) to the mid-veins. Blades ovate, unlobed or lobed with 1–5 sinuses, 2–20 cm × 1.5–18 cm, the bases rounded to cordate. Margins crenate, serrate, or dentate, the teeth often irregularly tapering to obtuse tiny lobe, dentate to doubly dentate, sometime spacely serrate, teeth blunt. Leaf apices usually obtuse but sometimes acute. Adaxial surfaces glabrous and bright green, abaxial surfaces dull green with conspicuous pubescent along the veins. Inflorescence: staminate catkins 1–7 per node, each with 5–30 pedunculate flowers, 1–5 cm long; the peduncles 0.5–2 cm long, pendulous, and pubescent with white hairs. Pistillate catkins 1–5 per node, each with 3–20 pedunculate flowers, 1–3 cm long; the peduncles 1–3 cm long, pubescent. Flowers: staminate flowers whitish or greenish yellow, the perianths pale green and broadly elliptic; filaments inflexed in bud, the anthers globose to reniform. Pistillate flowers greenish white, the perianths ovoid to oblong; styles 0.0–0.7 mm long, stigmas branched and longer in shrubby specimens with lobed leaves. Fruits: infructescences oblong, their component fruits loose to tightly arranged; fruits ovoid to ellipsoid, 1–3 cm long (excluding the peduncles), white, dark red, pink, purple, or black when mature. Flowering: March–May. Fruiting: April–July.

Distribution. Native to South and Central China, this species is now cultivated nearly worldwide in a variety of tropical and temperate environments. Floras Asian countries report recent spotting in many areas, where there was no prior report of occurrence. This perhaps attributes to the increased breeding practices in the regions mainly for silkworm industries.

Economic importance. The primary economic value of this species is the use of its leaves as the primary fodder for silkworm larvae. It is also used as fodder for livestock, the bark fibers are used for textiles, paper and as medicine, and the fruits are edible.

Specimens examined. CHINA. –Nanking: Chekiang, *Ho 985* (GH), April 22, 1932. Shantung: *Cheo and Yen 175* (GH), 1936. Western Hupei: *Wilson 3308* (GH), May 1907. –Hong Kong: *Hu 5015* (GH), March 9, 1968. –Hupei: without date and number (GH). –Kwanf Tung: Honam Island, *Levine 329* (GH), February 10, 1917. –Southern Shansi: *Tang 680* (GH), May 3, 1929. –Szechuan: *Chengtu 5927* (GH), April 21, 1937. –Tibet: Tali range, *Bulley* et al. *4732* (GH), July 1907. –Yunnan: Likiang, *Rock 8489* (GH), 1923. –Yunnan: *Tsiang and Wang 16221* (GH), May 1939. –Yunnan: *Ten 49* (GH), March 10, 1916. USA. –Kansas: KPBS (39.1036 °N and 96.597 °W), *Nepal 642* (KSC), April 19, 2005. Images examined. LINN 1112.1 (LINN).

Both artificial and natural selections have contributed to the development of *Morus alba* cultivars worldwide. The economic importance in silk production, along with its utility as a fast-growing tree hardy across a wide range of environments, has led to its introduction to almost every country in the world. This species may outcompete native vegetation (GEPP Council 2006), and there is evidence that it is threatening the species integrity of the native congener *M. rubra* through asymmetric, introgressive hybridization and genetic swamping in the United States (Nepal 2008) and in Canada (Burgess et al. 2005). Trees escaped from cultivation in North America appear to have evolved freezing tolerance within the last several decades, and their range has shifted north to much higher latitudes. The Russians/Hutterites apparently brought some pretty cold hardy cultivars with them to the United States during the late 1800s/early 1900s. Herbarium label data reveals that the collection time of recently collected flowering and fruiting vouchers in North America generally precedes those collected 50 years or earlier.

2. *Morus australis* Poir. in *Encycl. (Lamarck)* 4: 380. 1797. [Type: P]

Synonyms. *M. acidosa* Griffith, *M. alba* L. var. *indica* Bur., *M. alba* var. *nigriformis* Bur., *M. alba* var. *stylosa* Bur., *M. atropurpurea* Roxb., *M. australis* var. *hachijoensis* (T. Hotta) S. Kitamura, *M. australis* var. *hastifolia* (Z. Y. Cao) Z. Y. Cao, *M. australis* var. *incisa* C. Y. Wu, *M. australis* var. *inusitata,* (Levl.) C. Y. Wu, *M. australis* var. *lineuripartita* Z. Y. Cao, *M. australis* var. *oblongifolia* Z. Y. Cao, *M. australis* var. *trilobata* S. S. Chang, *M. bombycis* Koidz., *M. bombycis* Koidz. var. *flexuosa* Hotta, *M. bombycis* Koidz. var. *ikuchuensis* Hotta, *M. bombycis* Koidz. var. *mikamiana* Hotta, *M. bombycis* Koidz. var. *saishuensis* Hotta, *M. bombycis* var. *tiliifolia,* Koidz., *M. formosensis* Hotta, *M. hachijoensis* Hotta, *M. hastifolia* Wang & Tang ex Z. Y. Cao, *M. inusitata* H. Lev., *M. kagayamae* Koidz., *M. longistylus* Diels, *M. tiliaefolia* Makino, *M. trilobata* (S. S. Chang) Z. Y. Cao, *M. wallichiana* Koidz.

Description. Dioecious shrubs or small trees to 7 m. Bark: gray or brown. Lenticels round to elliptic. Branches: gray or brown, glabrous or rarely pubescent, fewer and with longer internodes than in *M. alba*. Buds: conical to cylindric with broader bases tapering to tip, 0.5–2 mm, usually incurved, bud scale brown with white band on the apical margin. Leaves: leaf scars shallow, circular, and not as distinct as in *M. alba*. Stipules lanceolate to linear-lanceolate, 0.5–1 cm long, pubescent. Petioles pubescent, 0.5–1.5 cm long. Blades lanceolate to broadly ovate, simple or lobed with 3–5 sinuses, 1–7 cm × 0.4–4 cm, the bases cordate or rounded. Margins usually regularly serrated, apex subcaudate to acuminate. Abaxial surfaces pubescent all over but especially along the veins, the adaxial surfaces slightly to strongly scabrous. Leaves tri-nerved, the lateral veins extending through one- to two-thirds of the lamina, the secondary veins fewer than in other species and angled at 30–60° to the mid-vein. Abaxial veins are more prominent than axial. Inflorescence: staminate catkins 1 per node, each with 5–20 flower, 1–1.5 cm long; the peduncles 0.2–0.9 cm long. Pistillate catkins 1 per node, each with 3–10 flower, 0.5–1 cm long, the peduncles 0.2–0.5 cm long and shorter than the catkin they subtend. Flowers: staminate flowers with green, ovate perianths; anthers yellow. Pistillate flowers with dark, oblong perianths; styles long (0.2–0.8 cm) with bifid stigma. Fruits: infructescences globose to oblong, 0.5–1.5 cm × 0.5 cm, green, white, red, or dark purple at maturity. Achenes yellowish brown. Flowering: March–April. Fruiting: April–May.

Distribution. China (Hupei, Pingtang), Bhutan, India, Japan, Korea, Myanmar and Nepal.

Economic importance. In China, the bark fibers are used for making paper and the fruit are edible (Zhou and Gilbert 2003).

Specimens examined. CHINA. –Hupei, *Wilson 3301* (GH), May–June, 1907. –Schechuan: *Yu 622* (GH), May 3, 1932. –Chekiang: *Chen 1240* (GH), April 30, 1933. –Kwangsi: *Chung 81558* (GH), May 12, 1936. –Tsingchen: *Teng 90376* (GH), May 28, 1936. NEPAL. –Eastern Nepal: Dharapani-Sanguri Bhanjyang, *Kanai et al.* (without number) (GH), without date. –Namochee: *Rukma 1626* (GH), October 5, 1965. TAIWAN. –Hsiang: *Liu 104* (GH), March 25, 1993. JAPAN. –Okinawa: Kadena, *Moran 4965* (GH), March 5, 1955. –Tohoku, *Yateishi et al. 15613* (GH), May 6, 1991.

3. ***Morus cathayana*** Hemsl. in *J. Linn. Soc., Bot.* 26: 456 (1894). [Type: K, GH]

Synonyms. *M. cathayana* var. *gongshanensis* (Z. Y. Cao) Z. Y. Cao, *M. cathayana* var. *japonica* (Makino) Koidz., *M. chinlingensis* C. L. Min, *M. gongshanensis* Z. Y. Cao, *M. rubra* L. var. *japonica* Makino, *M. tilaefolia* Makino.

Description. Dioecious or monoecious shrubs or small trees to 7 m. Bark: gray to dark brown. Branches: young branches pubescent with oblong lenticels, internodes longer than in *M. alba*. Buds: oblong to subglobose, to 4 mm long, brown, the bud scales with white bands on the apical margins, acute apex, narrowly supported. Buds differ from *M. alba* with relatively larger size, and dark-brown apical margin in the budscale. Leaves: stipules lanceolate, to 1.5 cm long. Petioles 1–3.5 cm long, pubescent. Blades broadly ovate to orbicular, sometimes lobed with 2–3 sinuses, 8–20 cm × 6–15 cm, bases cordate to truncate. Margins serrate to serrate-crenate, the serrations spaced out. Leaf apices acute to acuminate. Abaxial surfaces pubescent with dense white hairs. Adaxial surfaces scabrous, the interveinal areas pubescent. Lateral veins extending through up to ½ of the leaf, the secondary veins at 45–75° to the mid-rib. Inflorescence: staminate catkins 1–2 per node, each with 35–40 flower, 3–6 cm long, the peduncles 1.5–3.5 cm long. Pistillate catkins 1 per node, each with 15 to 35 flower, 2.5–5 cm long, the pedunicules 1–1.5 cm long. Flowers: staminate flowers with ovate perianths; stamens 4, a small pistillode present. Pistillate flowers with obovate perianths; styles short with bifid stigmas. Fruits: infructescences 2.5–4 cm long, red, dark purple, or white at maturity. Flowering: March–May. Fruiting: April–June.

Distribution. China, Japan, and Korea.

Specimens examined. CHINA. – Huangshan: Tan Krou, *Deng and Yao 79132* (GH), May 15, 1979. –Hupei: *Chow 231* (GH), 1934; Dong 686 (MO), May 10, 1995; *Henry 6378* (GH); 1885-88; *Wilson 365* (GH), April, 1900; *Xia and Ren* 123 (GH), June 7, 1999. –Szechuan: *Wang 20742* (GH), May 10, 1930; *Wilson 9310* (GH), May 22, 1908. –Tschen-Keou-Tin: *Farges 1477* (GH), May 1899. JAPAN – Hongsu: *Dequichi amd Tsugaru 3779* (GH), May 19, 1983. – Tsushima, *Ohashi* and *Sohma 10666* (GH), June 26, 1968.

Notes. In the original description of *M. cathayana*, Hemsley listed A. Henry's collection numbers 5543, 5860, and 6378 as the type specimens. He also mentioned four other specimens (Fortune 35, and collection numbers 1409, 5435, and 5487 of A. Henry from Hupei, China) which he suggested may be attributable to a different species. I examined *Henry 6378* and *1409* (GH) and found all of these specimens to be indistinct from the other specimens listed by Hemsley in the original description.

Specimens collected from Yunnan by H. T. Tsai in 1933 and from Hong Kong by Hu and Hu in 1969 possess certain characters, particularly leaf margins, venation patterns, and style lengths, that are intermediate between those typical of *M. alba* and *M. australis*. Hybridization between these two species should be assessed.

4. *Morus celtidifolia* Kunth in Humboldt and Bonpland, *Nov. Gen. Sp.* 2: 27 (1817). [Type: P]

Synonyms. *M. corylifolia* Kunth., *M. mexicana* Benth.

Description. Dioecious or monoecious trees to 10 m. Bark: grayish white, lenticels circular to elliptic. Branches: gray or dark brown, glabrous or sparsely pubescent, usually with longer internodes than *M. microphylla*. Buds: to 1.2 cm long, conical or ovoid, the bud scales white with brown bands or brown with white bands on the apical margins. Leaves: leaf scars circular, deeply concave, and almost circular. Stipules lanceolate, to 1 cm long, pubescent. Petioles 1–4 cm long, pubescent. Blade lanceolate to oblong elliptic, unlobed or rarely lobed with 3–5 sinuses, 4–25 cm × 1–8 cm, bases truncate to cordate. Margins serrate. Apex usually acuminate, sometimes sub-caudate. Lateral veins extend through up to two-thirds of the leaf blade and often converging towards the apex, the secondary veins angled at 45° to the mid-rib. Adaxial surfaces usually glabrous with few hairs. Abaxial surface pubescent along the veins, the veins are more prominent than on the adaxial surface and are often yellow. Inflorescences: catkins unisexual or cosexual. Staminate catkins 2 per node, each with 10–30 flowers, 2.5–3 cm long, the peduncles 0.5–2.5 cm long. Pistillate catkins 1 per node, each with 15–20 flowers, 1–4.5 cm long, the peduncles 0.5–2 cm long. Flowers: Staminate flowers widely spaced, sometimes retaining bud scales, the perianths ovate to elliptic; pistillodes present. Pistillate flowers also widely spaced, the perianths ovoid; styles very short or absent, stigmas branched. Fruits: infructescence 1–4 cm long, oblong with loosely arranged fruits, these dark red or brown at maturity and appearing drier than in other species. Flowering: March–April. Fruiting: April–July.

Distribution. Mexico to Honduras. The occurrence of this species in the Andes may be due to the introduction of trees for fruit production (Berg 2001).

Notes. Berg (Berg 2001) studied the type specimens of *Morus* from the southwestern United States, including the types for *M. albida*, *M. arbuscula*, *M. betulifolia*, *M. canina*, *M. confinis*, *M. crataegifolia*, *M. goldmanii*, *M. grisea*, *M. microphylla*, *M. pandurata*, *M. radulina*, *M. vernonii*, and *M. vitifoila*, all sensu Greene (Greene 1910) and concluded that these names are synonyms of *M. celtidifolia*. They are recognized here as synonyms of *M. microphylla*.

Specimens examined. GUATEMALA. *Steyermark 50515a* (GH), August 13, 1942. MEXICO. –Chiapas: *Breedlove 50802* (GH) from, April 11, 1981. –Nanchitita: *GBH 3627* (GH), March 20, 1933. –Other localities: *Caranza 1560* (GH), 1989. *Dodge 44* (GH), July 1891. *GBH 1336* (GH), July 16, 1898. *GBH 3843* (GH), March 20, 1933. *Lundell 5471* (GH), July–August, 1934. *Matuda 1384* (GH), May 1937. *Matuda 30675* (GH), April 11–12, 1945. *Palmer 158* (GH), Feb 1–April, 1907. *Palmer 563* (GH), April 11, 1905. *Pringle 6791* (GH), April 23, 1898. *Sargent 5 &8* (GH), 1887. *Sharp 45308* (GH), 1945. –Sonora: *Gentry 3641* (GH), April 14, 1938.

5. *Morus insignis* Bureau in De Candolle, *Prodromus* 17: 247 (1873). [Type: P, BM, F, G, K]

Synonyms. *Morus peruviana* Planch. ex Koidz., *M. trianae* Leroy, *M. marmollii* Legname.

Description. Dioecious trees to 10 m. Bark: brown or gray with elliptical lenticels. Branches: coppery red or brown, sparsely pubescent. Buds: small, 0.2–0.5 cm × 0.2 cm, ovoid, and brown; bud scales brown with dark bands on the apical margins. Leaves: stipules lanceolate, 0.5–1 cm × 0.2–0.5 cm, brown and pubescent. Petioles 0.5–2.5 cm long, pubescent. Blades elliptic to oblong, 5–20 cm × 3–15 cm, the bases unequal, obtuse. Margins shallowly serrate to dentate. Leaf apices acuminate. Veins not as prominent as in other *Morus* species; the lateral veins extending through one-half to two-thirds of the lamina. Adaxial surfaces glabrous to scabrous. Abaxial surface pubescent with short white hairs. Inflorescences: staminate catkins 2 per node, each with 30–100 flowers, 6–12 cm long, the peduncles 0.2–0.5 cm long. Pistillate catkins 1–2 per node, each with 20–60 flowers, 3–15 cm long, the peduncles 0.1–0.3 cm long. Flowers: staminate flowers with bud scale still intact, the perianths ovoid. Pistillate flowers with ovoid perianths; styles less than 1 mm long, the stigmas branched. Fruits: infructescences 5–15 cm long with fruits loosely arranged. Flowering: February–April (Central America); March–May (South America). Fruiting: May (Central America); September–November (South America).

Distribution. Cloud forests from Southern Mexico to Central America, Ecuador, Colombia, Argentina, and Venezuela.

Specimens examined. COLOMBIA. –Antiqua: *MacDoughal and Roldan 3563* (MO), January 29, 1989. COSTA RICA. –Las Nubes San Jose: *Standley 38808* (GH), March 20–22, 1924. –San Jose de la Montana: *Fosberg et al. 47800* (MO), May 17, 1996. ECUADOR. –Bolivar: *Gray and Tippas 12497* (MO), July 19, 1993. *Zak and Jarmillo 2705* (GH), September 3, 1987. –Napo: *Palacious 5982* (MO), October 1990. –Other: *Palacious 6228* (MO), October 12, 1990. PERU. –Bangara: *Smith and Vasquez 4900* (MO), September 2, 1983. –Oxapama: *Smith and Pretel 8027* (MO), July 25, 1984. –Piura: *Gentry et al. 74981* (MO). September 22, 199. VENEZUELLA. – Bocono: *Dorr and Barnett 7523* (MO), October 31, 1990. –Other: *Weberbouer 7040* (GH), August 1914.

We see a slight shift in flowering and fruiting times of *M. insignis* in subtropical and temperate habitats from the tropical habitats (often vary with altitudes). Further investigation involving field studies in Central and South Americas would help us understand the change in phenology better.

6. *Morus macroura* Miq. in *Pl. Jungh.* 1: 42 (1851). [Type: B?]

Synonyms. *M. alba* var. *laevigata* Wall. ex Bur., *M. laevigata* Wall. ex Brand., *M. liboensis* S. S. Chang, *M. wallichiana* Koidz., *M. wittiorum* Handel-Mazzetti, *M. wittiorum* var. *mawu* Koidz., *M. jinpingensis* S. S. Chang.

Description. Dioecious trees to 20 m. Bark: dark brown. Branches: young branches pubescent at nodes. Buds: ovoid, 0.2–1 cm × 0.5 cm, brown or white and pubescent, the bud scales with white bands on the apical margins. Leaves: leaf scars slightly raised, semi-circular and extrose, stipules mostly linear, 0.5–3 cm long, pubescent. Petioles ca. 1 cm long, pubescent. Blades broadly ovate, unlobed, 5–20 cm × 3–7 cm, the bases rounded, rarely cordate. Margins usually minutely serrate to almost entire. Leaf apices acute to shortly acuminate. Lateral veins extend through one-half to two-thrids of the lamina, the secondary veins angled at 45–75° to the mid-rib. Adaxial surfaces glabrous to sub-scabrous. Abaxial surfaces subglabrous with sparse hairs along the veins. Inflorescences: staminate catkins 2 per node, each with 30–150 flowers, 4–16 cm long, the peduncles 1–2.5 cm long. Pistillate catkins 2 per node, each with 20–80 flowers, 6–16 cm long, the peduncles ca. 1 cm long. Flowers: staminate flowers with ovate perianths, these pubescent; anthers globose. Pistillate flowers with oblong perianths oblong; styles absent, the stigma branched and pubescent. Fruits: infructescences 6–16 cm long, yellowish white at maturity, the fruits often dry and sparsely arranged. Flowering: March–April. Fruiting: April–May.

Distribution. Tropical montane forests of China, Nepal, Bhutan, Indochina, Malaysia, Myanmar, India, Thailand, and Indonesia.

Economic importance. Bark fiber is used for papermaking, while the wood and leaves are used to make dyes.

Notes. *M. wittiorum* and *M. liboensis* are recognized as separate species in the Flora of China (Zhou and Gilbert 2003) but are treated here within *M. macroura*. Zhou and Gilbert (2003) recognized these species using only a limited number of specimens, and the postulated taxa possessed overlapping distributions while being differentiated from *M. macroura* and one another by only minor differences in leaf morphology. Future taxonomic research using more specimens, field observations and molecular data will be needed to resolve the diversity of this group.

Specimens examined. BURMA. –North Burma: *Kingdom-Ward 476* (GH), April 2, 1939.–Tangkhul: *Kingdom-Ward 22049* (GH), March 21, 1953. BHUTAN. –Gaylegphung: *Grierson and Long 4106* (GH), March 19, 1982. CHINA. –Hainan: *Lau 25825* (GH), March 21, 1936. –*Kaifu 2426* (MO), May 16, 1980. –Kwansi: Ping Nan, *Wang 39218* (GH), May 21, 1936. –Kwangtung: Chaapung village, *Taam 693* (GH), May 1–24, 1938. –Kwangtung: *Tsang 26442* (GH), May 21–30, 1936. –Yunan: *Henry 12019A* (GH), (without date). Thailand. –Chiang Mai: *Maxwell 90-272* (GH), February 28, 1990. –Lampoon: *Maxwell 94–296* (GH), March 2, 1994. – Other: *Dickason 7118* (GH), March 1938. –Sutep, Siam, *"Native" S436* (GH), April 11, 1949.

7. *Morus mesozygia* Stapf ex A. Chev. in *Vegt. Ut. Afr. Trop. Fr.* 5 : 263 (1909). [Type : P, K]

Synonyms. *Celtis lactea* Sim., *Morus lactea* (Sim.) Mildbr., *M. mesozygia* var. *lactea* (Sim.) A. Chev.

Description. Dioecious or monoecious trees reported to be up to 35 m. Bark: gray or brown with densely distributed short elongated lenticels. Branches: gray, glabrous. Buds: small, brown, and ovoid, the bud scales with gray bands on the apical margins. Leaves: stipules lanceolate, 1 cm × 0.5 cm, sparsely pubescent. Petioles 0.5–2.5 cm long. Blade elliptic, oblong, lanceolate, or suborbicular, 6–12 cm × 5–8 cm, acuminate to caudate, the bases truncate or cordate. Margins crenate to serrate. Abaxial surfaces pubescent in the axils of the veins. Adaxial surfaces glabrous. Leaves tri-nervate with 2–5 distinct secondary veins orientated at 60–75° to the mid-rib, scalariform, and orientated at almost 90° to the mid-rib, secondary veins are > 60° with the lateral veins towards the margin, the tertiary veins less contrasting or often inconspicuous. Inflorescences: staminate catkins 2 per node, each with 15–20 flowers, the peduncles 1–2.5 cm long. Pistillate catkins subglobose, 2 per node, each with 5–10 flowers, 0.5–1 cm long, the peduncles 1.5–2.5 cm long. Flowers: staminate flowers crowded. Pistillate flowers with ovoid perianths; the styles short (< 1 mm) with branched stigmas. Fruits: infructescence, subglobose to globose, 0.5–2.5 cm × 0.5 cm. Flowering: April–September. Fruiting: March–August.

Distribution. Native to tropical Africa, Senegal, Nigeria, Congo, North-western Angola, south-western Ethiopia, and the Republic of South Africa.

Economic importance. The fruits are edible, and the wood is used as timber in Africa (Berg 1977).

Selected Specimens examined. MOZAMBIQUE. *Safala 3163* (MO), July 23 1941. NIGERIA.–Ibadan: *Okafor OL455* (MO), August 7, 1967. –Oyo: *Daramola 330* (MO), September 27, 1993. UGANDA. *Egglig 1183* (MO), without date. SIERRA LEONE. *Georges 22886* (MO), December 30, 1965. SOUTH AFRICA. 0.2°22'N and 16° 09' E, *Gentry and Harris 62776* (MO), without date. ZAIRE. 1° 26' N, 28° 33' E, *Hart 1342* (MO), August 1, 1966. –Other. Guine Portugusa: *Espirito 1961* (MO), April 25, 1945. Reserve de Lamto, *Par and Gautier-Begum 1105* (MO), March 2, 1989.

8. *Morus microphylla* Buckley in *Proc. Acad. Nat. Sci. Philadelphia.*14: 8 (1862). [Type: PH]

Synonyms. *M. albida* Greene, *M. arbuscula* Greene, *M. betulifolia* Greene, *M. canina* Greene, *M. cavaleriei* H. Lev., *M. confinis* Greene, *M. crataegifolia* Greene, *M. goldmanii* Greene, *M. grisea* Greene, *M. microphilyra* Greene, *M. mollis* Rusby, *M. pandurata* Greene, *M. radulina* Greene, *M. vernonii* Greene, *M. vitifolia* Greene.

Description. Dioecious or monoecious shrubs or trees to 5 m. Bark: gray or yellowish gray with densely distributed elongated lenticels. Branches: gray, pubescent, the internodes of flowering branches relatively more condensed compared to *M. celtidifolia*. Buds: elliptic or ovoid with acute apices, 1–3 mm long, the bud scales brown or white with dark bands on the apical margins. Leaves: leaf scars nearly circular, deeply concave, upwards facing. Stipules linear-lanceolate, 3–5 mm long, pubescent. Petioles 0.2–1 cm, pubescent. Blades lanceolate to ovate, unlobed or lobed with 3–5 sinuses, 1–6 cm × 0.6–3 cm, the bases usually cordate, sometimes round. Margins usually regularly serrated. Leaf apices acuminate, sometimes subcaudate. Adaxial surfaces harshly scabrous. Abaxial surfaces pubescent and scabrous. Inflorescences: staminate catkins 1 per node, each with 5–10 flowers, 0.5–0.8 cm long, the peduncles 0.2–0.5 cm long. Pistillate catkins each with 3–8 flowers, 0.5–0.8 cm long, the peduncles ca 0.5 cm long, pubescent. Flowers: staminate flowers with ovoid perianths, compactly arranged. Pistillate flowers with ovoid perianths; the styles absent with 2 stigmas. Fruits: infructescences short cylindric or globose, 0.5–1 cm long, red or deep purple to black and fleshy to dry at maturity. Flowering: March–April. Fruiting: April–June.

Distribution. USA (Arizona, Texas, and New Mexico), Mexico, and Costa Rica.

Specimens examined. USA. –Arizona: Graham Co, *Maguire 10531* (GH), April 6, 1935; *McKelvery 11999* (GH), June 1, 1929; Tanque verde Canyon, *Brass 14281* (GH), April 6, 1940. –Texas: Calahan Co, *Palmer 13681* (GH), May 26, 1918; Edwards Co, *Cory 18983* (GH), May 20, 1936; Elpaso Co, *Franks 6630* (GH), August 20 1996; New Braunnfels, Comanche Spring, *Lindheimer 1166* (GH), May 1849; San Antonio, *Schulz 64* (GH), April 17, 1922. –New Mexico: near Silver City, *Eastwood 8448* (GH), May 6, 1999.

Notes. Berg (2001) recognized several *Morus* species sensu Green (1910) from the southwestern United States (Texas, Arizona and New Mexico) as *Morus celtidifolia*. The present study recognizes them as synonyms of *M. microphylla*. There are many specimens from northern and eastern Mexico that have characters intermediate between *M. microphylla* and *M. celtidifolia*. Further assessment of natural hybridization between these two species may resolve this taxonomic ambiguity.

9. *Morus mongolica* (Bureau) C. K. Schneid. in Sargent, *Pl. Wilson.* 3: 296 (1916). [Type: P]

Synonyms. *M. alba* var. *mongolica* Bur., *M. barkamensis* S. S. Chang, *M. deqinensis* S. S. Chang, *M. mongolica* var. *barkamensis* (S. S. Chang) C. Y. Wu & Z. Y. Cao, *M. mongolica* var. *diabolica* Koidz., *M. mongolica* var. *hopeiensis* S. S. Chang & Y. P. Wu, *M. mongolica* var. *longicaudata* Z. Y. Cao, *M. mongolica* var. *pubescens* S. C. Li & X. M. Liu, *M. mongolica* var. *rotundifolia* Y. B. Wu, *M. mongolica* var. *vestita* Rehder, *M. mongolica* var. *yunnanensis* (Koidz.) C. Y. Wu & Z. Y. Cao, *M. yunnanensis* Koidz.

Description. Dioecious shrubs or small trees to 6 m. Bark: gray, brown, or brown-black, the lenticels light colored and elliptic. Branches: gray, reddish (coppery) to brown black. Buds: often small and conical, sometime ovoid, 1–3 mm long, the bud scales grayish brown with dark bands on the apical margins. Leaves: stipules usually linear to lanceolate, 1.5–4 cm long, pubescent. Petioles 1–3.5 cm long, pubescent. Blades lanceolate to broadly ovate, unlobed or lobed with 3–5 sinuses, 6–15 cm × 5–8 cm, the bases usually cordate. Margins usually regularly serrate with each tooth tapering into a hair like seta 0.8–3 mm long. Leaf apices acuminate. Adaxial surfaces glabrous. Abaxial surfaces sparsely pubescent along veins. Veins in the mature leaves reddish brown or whitish yellow, the lateral veins extending through one-third to one-half of the lamina, the secondary veins orientated at 45–60° to the midrib. Inflorescences: staminate catkins 1 per node, each with 15–30 flowers, 2–5 cm long, the peduncles 1–2 cm long. Pistillate catkins 1 per node, each with 10-20 flowers, 1–2 cm long, the peduncles 1.5–3 cm long and shorter than those of the male catkins. Flowers: staminate flowers with ovate perianths. Pistillate flowers with oblong perianths; the styles long (2–6 mm) with bifid stigmas. Fruits: infructescences 1–1.5 cm long, red, purple, or white at maturity. Flowering: March–April. Fruiting: April–July.

Distribution. China, Japan, Korea, and Mongolia.

Notes. Schneider (1916) listed seven specimens collected in 1907 by E. H. Wilson (collection numbers 8ᵃ-8ᶠ) from Western Hupeh, China. There are also other specimens mentioned such as those collected by A. Von Rosthorn (without number and date), J. G Jack in 1905, and specimens discussed by E. E. Maire in his description of *M. mongolica*. Several of these reference specimens are at GH and were observed as part of the present study.

Specimens examined. CHINA. –Beijing: *Wang et al. 21* (MO), April 4, 1986.–Chihly: *Smith 958* (MO), 1928. – Chihly: *Hers 2218* (GH), without date. –Guizhou: Dancheen, *Liu 20538* (MO), May 15, 1996. –Hubei: Dang Yang, *Jiang and Tao 117* (MO), April 26, 1982. –Kunming: *Fanan 1537* (MO), Guilin, May 27, 1980. –Pu Kecheng: *Liu and Zheng 1473* (MO), June 28, 1984. –Shaansi: *Smith 6417* (GH), without date. –Shaanxi: *Zhang 19360* (MO), June 12, 1983. –Shansi: *Smith 6329* (MO), July 1924. –Sichuan: Longkon, *Dai 100715* (MO), May 30, 1958. –Hupei: *Wilson (8ᵃ-8ᶠ)* (GH), June 1907–1908. –Yunnan: *Chu 1217* (MO), May 14, 1988. –Yunnan: *Rock 2861* (GH), without date. –Yunnan: *Maire 527* (GH), without date.

10. *Morus nigra* L. in *Species Plantarum* 2: 986 (1753). [Type: LINN]

Synonyms. *M. atrata* Raf., *M. cretica* Raf., *M. laciniata* Mill., *M. petiolaris* Raf., *M. scabra* Moretti (Moretti 1842), *M. siciliana* Mill.

Description. Subdioecious trees to 10 m. Bark: gray or dark brown. Branches: brown, pubescent. Buds: grayish brown, ovoid, 0.5 cm × 0.2 cm, the bud scales brown with white apical margins. Leaves: leaf scars slightly raised, cupoid. Stipules mostly lanceolate, 2–3 cm long, pubescent. Petioles 2–6 cm long, pubescent. Blades broadly ovate, unlobed or lobed with 2–3 sinuses, 4–22 cm × 3–14 cm, the bases usually cordate. Margins usually regularly serrate. Leaf apices acute to shortly acuminate. Adaxial surfaces glabrous to slightly scabrous. Abaxial surfaces pubescent along the veins. Inflorescence: staminate catkins 1 per node, each with 15–25 flowers, the peduncles 0.5 cm long, pubescent. Pistillate catkins 1 per node, each with 15–35 flowers, the peduncles 0.5 cm long. Flowers:

staminate flowers with ovate perianths. Pistillate flowers with oblong perianths, these with ciliate margin; the styles very short (< 1 mm) with branched stigmas. Fruits: infructescences elliptic to ovoid, 1–2.5 cm × 1–2 cm, purple to black at maturity. Flowering: April–May. Fruiting: May–June.

Distribution. Native to western Iran, introduced to many countries worldwide.

Specimens examined. BERMUDA. *Manuel 652* (GH), January 2, 1964. CHINA. Yunnan, *Yu 5271* (GH), without date. COSTA RICA. *Condeez 17451* (GH), 1909. CUBA. *Shafer 8816* (GH), Ferbruary 16, 1911; Harvard Tropical Garden Soledad, *Jack 5004* (GH), March 26, 1927. EL SALVADOR. *Cardenson 288* (GH), November 1921. SOUTHWEST AFRICA. *Rodin 9215* (GH), without date. USA. Hawaii. *Neal 1236* (GH), May 8, 1944. WEST INDIES. *Procter 18232* (GH), April 4-June12, 1958. **Images examined**. LINN 1112.3 (LINN).

11. *Morus notabilis* C. K. Schneid. in Sargent, *Pl. Wilson.* 3: 293 (1916). [Type: MO]

Description. Dioecious trees to 15 m. Bark: grayish to dark brown, the lenticels round or sometime elliptic and light colored. Branches: white or dark brown and spreading with long internodes, glabrous or pubescent with short, sparse hairs. Buds: conical or ovoid, 1–6 mm long, grayish brown, the bud scales with dark banding on the apical margins. Leaves: leaf scars nearly or wholly circular, shallow on the surface. Stipules linear to lanceolate, 1.5–3 cm long, pubescent. Petioles 3–5 cm long, pubescent. Blades broadly ovate to orbicular, usually unlobed, 7–25 cm × 6–20 cm, the bases mostly cordate. Margins usually regularly (spaciously) dentate with acute tips. Leaf apices acuminate to obtuse. Lateral veins extend through up to two-thirds of the lamina, the secondary veins converging together by the margin and orientated at 45–75° to the mid-rib. Adaxial surfaces glabrous. Abaxial surfaces sparsely pubescent. Inflorescences: staminate catkins usually 2 per node, each with 30–50 flowers, 3–7 cm long, the peduncles 2–3 cm long. Pistillate catkins 1 per node, each with 25–45 flowers, 3–5 cm long, the peduncles 3–4 cm long. Flowers: staminate flowers with ovate perianths; pistallodes present. Pistillate flowers with oblong perianths; the styles long (0.5–1 cm) with bifid stigmas. Fruits: infructescences 2.5–4 cm long, white or purple red at maturity. Flowering: April–June. Fruiting: May–August.

Distribution. Sichuan and Yunnan of China.

Specimens examined. CHINA. –Jing Chikeng-Shan: Anhiu, *Wang Shilong A178* (MO), without date. –Sichuan: Emei, *Honggui 5380* (MO), June 13, 1995. *Honggui 5708* (MO), without date. –Szechuan: Ma-pien Hsian, *Wang 23074* (GH), May 29, 1931. *Wilson 919* (MO), 1907. –Tibet: 28°24' N and 98°28' E, *Forrest 20241* (GH), September 1921. –Yunnan: Shang-pa, *Tsai 58727* (MO) without date. –Zhejiang: Anji Longwang, *Fang and Deng 975071* (MO), May 8, 1997.

12. *M. rubra* L. in *Species Plantarum* 2: 986 (1753). [Type: LINN]

Synonyms. *M. canadensis* Poir., *M. caroliniana* Hort. ex Moretti, *M. riparia* Rafinesque, *M. rubra* var. *tomentosa* (Rafinesque) Bureau, *M. missouriensis* Audib. ex Moretti, *M. pensylvanica* Nois. ex Loud., *M. reticulata* Raf., *M. scabra* Willdenow, *M. tomentosa* Raf., *M. virginica* Duham. ex Dippel. *Morus murrayana* Saar & Galla.

Description. Subdioecious tree to 15 m. Bark: usually gray with orange tint, relatively thin compared to *M. alba* with flattened ridges and shallow furrows. Branches: gray to reddish brown, the young branches pubescent becoming glabrous at maturity; mature trees with characteristic pattern of branch spreading branching. Buds: ovoid to conical with pointed apices, the bud scales 0.5–0.8 cm long and gray with dark or brown bands on the apical margins. Leaves: leaf scars circular, slightly raised. Stipules linear, ca. 1 cm long, and densely pubescent. Petioles 1–3 cm long, pubescent. Blade broadly ovate, sometimes irregularly lobed with 3–5 sinuses, 5–30 cm × 3–22 cm, the bases usually cordate, sometimes unequal to truncate. Margins regularly serrated. Leaf apices acuminate, sometimes subcaudate. Abaxial surfaces sparsely to densely pubescent. Adaxial surfaces slightly scabrous. Inflorescence: catkins unisexual or cosexual. Staminate catkins 1–5 per node, each with 10–35 flowers, to 2.5 cm long, the peduncule 1–1.5 cm long. Pistillate catkins 1 per node, 1–1.5 cm long, the penducles to 2 cm long. Flowers: staminate flowers with ovoid, purplish green perianths; pollen 20–25 × 17–20 μm. Pistillate flowers with very short (< 0.5 mm) styles and branched stigmas. Fruits: infructescences 1–2.5 cm long, red or deep purple to black at maturity. Achenes compactly arranged. Flowering: March–May. Fruiting: April–July.

Distribution. Eastern North America up to the eastern margin of the Great Plains and as far north as southern Ontario, Canada. Distribution range and center of diversity is expected to shift in response to climate change (see Prasad et al. 2007, Iverson et al. 2008).

Economic importance. The stem bark, root, and tree sap were used medicinally by Native Americans (Wunderlin 1997). This species is threatened in Connecticut and Massachusetts and is endangered in Vermont and Michigan (USDA PLANTS 2007).

Specimens examined. USA.–Georgia: *McDuffie, Bartlett 2647* (KSC), April, 1884. –Illinois: Peoria, *Chase 14403*, May 9, 1957. –Kansas: Osage Co., *Hitchcock* (without collection number), 1895; Decatur Co, *Gates 18045*, May 22, 1935; Riley: Konza, 39.10521° N and 96.6021° W, *Nepal 706*, May 25, 2005; Saline Co, *Hancin 60*, February 3, 1893. –Kentucky, *Kellerman* (without number), Fayette, May 19, 1882. –Oklahoma *Goodman 7130*, May 13, 1961; Payne, *Waugh* (without number), Jan 30, 1935. –Pennsylvania: Franklin Co, *Smith 38*, July 5, 1936 (all these are KSC specimens). **Images examined**. LINN 1112.6 (LINN).

13. *Morus serrata* Roxburgh in *Fl. Ind., ed.* 3: 596 (1832). [Type: P]

Synonyms. *Morus alba* var. *serrata* (Roxb.) Bur., *M. altissima* Miq., *M. gyirongensis* S. S. Chang, *M. pabularia* Decne., *M. vicorum* Jacquem.

Description. Dioecious trees to 15 m. Bark: dark brown with almost circular lenticels. Branches: young branches are densely pubescent. Buds: ellipsoid to ovoid, 0.2–1 cm × 0.1–0.5 cm, the bud scales persistent, brown to chocolate with white apical margins, pubescent. Leaves: leaf scars slightly raised, semi-circular. Stipules linear-lanceolate, 0.5–2 cm × 0.2–0.5 cm, pubescent. Petioles densely pubescent, 4–6 cm long. Blades broadly ovate, unlobed or lobed with 1–3 sinuses, 5–20 cm × 3–14 cm, the bases usually cordate. Margins dentate to doubly dentate, often regularly serrate. Leaf apex acute to shortly caudate. Lateral veins extend through up to the half of the lamina. Adaxial surfaces glabrous to slightly scabrous. Abaxial surfaces densely pubescent, yellowish veins, more prominent than on the adaxial surfaces. Inflorescence: staminate catkins 3–6 per node, each with 10–30 flowers, 2–6 cm long, the peduncule 0.5–1 cm, long. Pistillate catkins 1–4 per node, each with 5–20 flowers, 1–2.5 cm long, the peduncules 0.5–1 cm long. Flowers: staminate flowers with ovate perianths. Pistillate flowers with oblong perianths; styles absent, bifid stigmas branched and pubescent. Fruits: infructescence 1.5–2 cm long, red, white, or pink at maturity. Flowering: April–May. Fruiting: May–June.

Distribution. Montane forests of China, Nepal, and India.

Economic importance. The leaves of this species are used as fodder in Nepal, and the fruits are edible.

Specimens examined. INDIA. –Dehra Dun: *Beli 115* (GH), June 16, 1922. *Singh 91* (GH), November 5, 1912. –Kumaon: *Strachey and Winterbottom 4* (GH), without date. –Punjab: Kulu, *Koelz 4788* (GH), June 1–5, 1933. *Stewart 1958* (MO), May 23, 1917.–Uttar Pradesh: Musoorie, *Fleming 417* (GH), April 30, 1948. NEPAL. –Dhaman: *Bisram 195* (GH), April 1, 1929. –Doti, *Bisram 297* (GH), April 23, 1929. PAKISTAN. –Kashmir: *Stewart and Hasiu 25506* (GH), April 25, 1953; *Stewart 27326*, April 21, 1954. –Punjab: *Parker 2850* (GH), May 17, 1928.

Phylogeny of the Genus Morus

Phylogeny based on the nuclear internal transcribed spacer (ITS) region and the chloroplast *trnL-trnF* intergenic spacer region suggests that the majority of *Morus* species form a monophyletic group corresponding to the *Morus* subgenus *Morus sensu* (Leroy 1949) and are divided between two clades: one of eight species native to Asia, and another of three species native to North America (Nepal and Ferguson 2012). The same analyses indicate that the genus *Trophis sensu stricto* may form a sister group to these core *Morus* clades, while the African *M. mesozygia* and the Neotropical *M. insignis* are more distantly related to both, making the genus as it is currently circumscribed non-monophyletic.

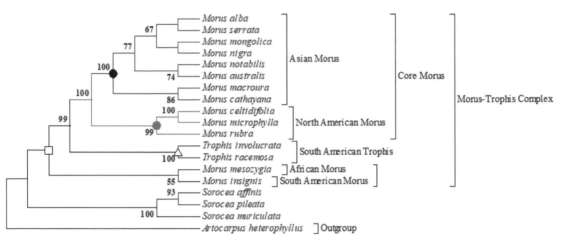

Figure 1. Phylogeny of the genus Morus based on the combined ITS/trnL-trnF data set using a K81uf + G evolutionary model. Numbers above the branches are ML bootstrap values based on 3,000 replicates. Bootstrap values less than 50% are not shown. For details, see Nepal and Ferguson (2012).

The relationships among taxa in Asian clade remain poorly resolved. The species *M. australis*, *M. mongolica*, and *M. notabilis* are distinguished from the rest of the genus by the development of long (> 1 mm) styles, and these species were recognized as comprising Section *Dolichostylae* by Koidzumi (1917). Zhekun and Gilbert (2003) followed this approach in their treatment of the genus for *Flora of China*, and the molecular analyses of Nepal and Ferguson (2012) show a weak support to the hypothesis that these long-styled species form a monophyletic group within the Asian clade.

Nepal and Ferguson (2012) also found weak molecular evidence for a sister relationship between *M. cathyana* and *M. macroura*, which are distinguished from the other species in the Asian clade by their long (> 2 cm) catkins and larger (0.2–1.0 cm × 0.5 cm) axillary buds (Nepal 2008). The North American clade is comprised of three species, *M. rubra*, *M. microphylla*, and *M. celtidifolia*, and is sister to the Asian clade discussed above. *M. microphylla* and *M. celtidifolia* are sister species and are native to southwestern North America and Central America, respectively, while *M. rubra* is widespread in the deciduous forests of eastern North America. The introduced *M. alba*, a member of the Asian clade, forms natural hybrids with *M. rubra* and in some areas may threaten it via genetic swamping (Burgess et al. 2005, 2008).

The phylogenetic analyses also indicate that the African *M. mesozygia* and the Neotropical *M. insignis* are less closely related to the Asian and North American *Morus* clades than is the genus *Trophis sensu stricto*, such that the inclusion of these species in *Morus* makes the genus non-monophyletic. Other molecular studies on the Moraceae have reached similar conclusions on the close relationship of *Trophis* and *Morus* and have also indicated that *Trophis* itself is non-monophyletic (Datwyler and Weiblen 2004, Zerega et al. 2005, Clement and Weiblen 2009). Further taxonomic work to resolve the systematics of *Trophis* will also need to incorporate the genus *Morus*.

Morus Response to Climate Change

Majority of the cultivated *Morus* species are rapidly growing trees with high photosynthetic activities, and therefore can reduce atmospheric carbon dioxide thereby lowering the greenhouse effects. *Morus* species are key to silkworm industry because they provide food for silkworms at an industrial level. *M. alba* cultivars are preferred for windbreak and for controlling soil erosion by landscape industries and the United States Army Corps of Engineers (USACE). There is a growing consensus that mulberries are multi-purpose plants that can be used in sustainable agriculture because of their roles in sequestering carbon in the soil; they are source of fodder for cattle, produce delicious edible fruits, beautiful shade trees, have medicinal values and serve as energy source for heating and lighting homes (e.g., fire wood for cooking or lighting the houses) in rural communities at the foothills of the Himalayas, where cooking gas and electricity are not available. Therefore in Asia, *Morus* species (*M. alba*, *M. austalis*, *M. cathayana*, *M. macroura*, *M. nigra*, *M. notabilis*, *M. serrata*) are considered multi-purpose plants.

As we have briefly outlined above in the taxonomic revision section, herbarium data shows a slight shift in flowering and fruiting times of *Morus* species. Naturalized *M. alba* is rapidly invading the natural habitats of native tree species, endangering the integrity of the native *M. rubra* (by genetic swamping) and expanding its range in the United States and in Canada. How much of this *M. alba* invasion is attributable to global change or climate change, is yet to be investigated although there are reports on the effects of climate change on species diversity, richness and adaptability. We are unaware of research data specific to the effects of climate change on Asian and African *Morus* species. Herbarium data also shows a phenological variation in *M. insignis* in subtropical and temperate habitats from the tropical habitats (often varying with altitudes). Further investigation involving field studies and historical data in Central and South Americas could yield insight into the change in phenology (Nepal 2008). Below we describe an example of projected shift in distribution range of *M. rubra* in the United States in response to climate change event.

Iverson et al. (Prasad et al. 2007, Iverson et al. 2008) assessed potential climate change response of 133 tree species including *M. rubra* in nine states of the northeastern United States. Their goal was to assess current and projected future distributions in response to simulated expected climate change parameters. Potential impacts of climate change on suitable habitat of trees including *M. rubra*, under the high emissions trajectory were discussed. They have predicted "Importance Value (IV)" based on the relative number of stems and the relative basal area of stems in each plot (Iverson and Prasad 1998) examined. The IVs shown in Figure 2 represent average plot data for each 20 × 20 km cell for *Morus rubra*. Increased fluctuation in temperatures during the growing season in the United States is expected to make plant species more vulnerable to herbivores (de Sassi and Tylianakis 2012), insect pest and disease (Williams et al. 2000), and fires. Fluctuations in precipitation patterns and increased CO_2 levels are expected to lead to modifications in both natural and modified forests (Kirilenko and Sedjo 2007). The modeling results showed that *M. rubra* may gain substantial areas of suitable habitat in the state of New York by outcompeting other native species, including *Juglans nigra*, *Picea mariana*, *Quercus palustris*, and *Ulmus americana*.

We recently reviewed herbarium specimens collected from the eastern United States and found many instances of misidentification among specimens of *M. rubra* from *M. alba*. The problem of misidentification between *M. rubra* and *M. alba* is perhaps further complicated by rampant bidirectional hybridization between *M. rubra* and introduced *M. alba*. *M. alba* and *M. alba x rubra* hybrids are abundant in the natural habitats of the native *M. rubra*, thus threatening its genetic integrity. This is partly because of much higher magnitude of introgression toward

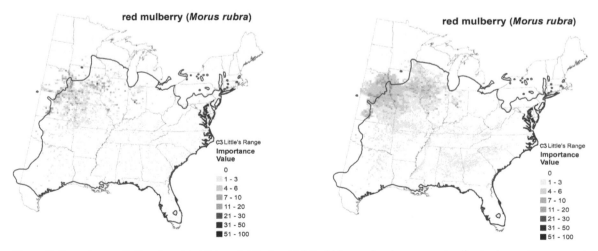

Figure 2. Current range (left) and modeled future distribution (right) of *Morus rubra* in response to climate change. (Figure panels were adapted from Prasad et al. (2007) and Peters et al. 2019: Climate Change Tree Atlas accessed at https://www.fs.fed.us/nrs/atlas/tree/682; June 16, 2019.)

M. alba maternal parents than that toward *M. rubra* as maternal parents (Burgess et al. 2005). This particular example underscores the need of clear understanding of species delineation and application of the necessary knowledge for species management. We infer that any taxonomic work is valuable for addressing both local and global issues.

Conclusions

Although there have been previous attempts to study the phylogeny of *Morus*, those efforts were limited by incomplete sampling of the genus. Here we recognize 13 species in the genus *Morus* with species distributed across temperate and tropical regions worldwide. Phylogenies based on the nuclear internal transcribed spacer (ITS) region and the chloroplast *trnL-trnF* intergenic spacer region suggest that *Morus* is non-monophyletic: the subgenus *Morus* sensu Leroy is resolved as a monophyletic group of Asian and North American *Morus* species. Our previous study has further shown that two species of the genus *Trophis*, including the type species, formed a clade sister to the Asian and North American species. The same analyses placed *M. mesozygia* and *M. insignis* outside the *Morus-Trophis* clade, making the genus non-monophyletic. Asian clade of *Morus* accounts for the majority of species diversity within the genus and includes most economically important species. It still remains unclear if the long-styled species of Koidzumi's Section *Dolichostylae* form a monophyletic clade, or if the species *M. macroura* and *M. cathyana* are sister species. Further extended taxonomic work on Asian *Morus* is needed to confirm our species recognition. More work is also needed in the taxonomic study of the genus *Trophis*—the close relationship of certain species of *Trophis* with *Morus* suggests the need of a comprehensive analysis of both genera. Finally, molecular evidence indicates that the genus *Morus*, as currently circumscribed, is non-monophyletic, and additional taxonomic studies in *Morus* and related genera will ultimately necessitate a taxonomic revision that reflects this information. Phenological inferences drawn from the herbarium specimens indicate slight variation in flowering and fruiting times in several *Morus* species and rapid range expansion of naturalized *M. alba* in Americas perhaps in response to global climate change.

Acknowledgements

We would like to acknowledge the curators of the Missouri Botanical Garden Herbarium (MO) and the Harvard University Herbarium (GH) for loaning us specimens for this research. We appreciate valuable suggestions given by C. C. Berg of Linden Austria, C. Y. Wu of the Herbarium of Kunming Institute of Botany (KUN), and Michael Gilbert of the Kew Botanical Garden Herbarium. Drs. Carolyn Ferguson and Mark Mayfield from Kansas State University Herbarium supervised part of this work (as part of MPN's PhD dissertation). They contributed valuable discussion and provided valuable resources. We benefited from online access to the digital images of type specimens from the herbarium of the Linnaean Society of London (LINN), Harvard University Herbarium (GH), Missouri Botanical Garden Herbarium (MO), and National Herbarium of Paris (P). This project was supported by South Dakota State University, partly by Kansas State University Herbarium (KSC) and partly by the United States Department of Agriculture hatch projects (SD00H469-13 and SD00H659-18) to M.P.N.

References

Ammal, E. 1948. The origin of black mulberry. Journal of the Royal Horticultural Society 73.

Azizian, D. and A. Sonboli. 2001. Chromosome counts for five species of Moraceae from Iran.

Berg, C.C. 1977. Revisions of African Moraceae (excl. Dorstenia, Ficus, Musanga and Myrianthus). Bulletin du Jardin Botanique National de Belgique 47: 267–407.

Berg, C.C. 1998a. The genera Trophis and Streblus (Moraceae) remodeled. Proceedings of the Koninklijke Nederlandse Akademie van Wetenschappen, Amsterdam, the Netherlands, Series C 91: 345–362.

Berg, C.C. 1998b. Moraceae. pp. 1–128. *In*: Harling, B. and L. Andersson (eds.). Flora of Ecuador Vol. 60. Arlov, Sweden: Elanders Berlings.

Berg, C.C. 2001. Moreae, Artocarpeae, and Dorstenia (Moraceae): With introductions to the family and Ficus and with additions and corrections to Flora Neotropica Monograph 7. Organization for Flora Neotropica.

Berg, C.C. 2005. Moraceae diversity in a global perspective. Biologiska Skrifter 55: 423–440.

Berg, C.C., E.J.H. Corner and F. Jarrett. 2006. Flora Malesiana. Series I, Seed plants. Volume 17, Part 1: Moraceae-genera other than Ficus. Nationaal Herbarium Nederland.

Bhopal, F. and M. Chaudhri. 1977. Flora of Pothohar and adjoining areas. Part-II. Casuarinaceae to Polygonaceae. Pakistan Systematics 1: 1–98.

Bureau, E. 1873. Moraceae. Prodromus Systematis Naturalis Regni Vegetabilis 17: 211–279.

Burgess, K.S. and B.C. Husband. 2004. Maternal and paternal contributions to the fitness of hybrids between red and white mulberry (*Morus*, Moraceae). American Journal of Botany 91: 1802–1808.

Burgess, K., M. Morgan, L. Deverno and B. Husband. 2005. Asymmetrical introgression between two *Morus* species (*M. alba*, *M. rubra*) that differ in abundance. Molecular Ecology 14: 3471–3483.

Burgess, K.S., M. Morgan and B.C. Husband. 2008. Inter-specific seed discounting and the fertility cost of hybridization in red mulberry (*Morus rubra* L.). The New Phytologist. 177: 276–84.

Cao, Z.Y. 1991. New taxa of Morus (Moraceae) from China. Acta Phytotaxonomica Sinica 29: 264–267.

Chang, S. 1984. New taxa of Moraceae from China and Vietnam [*Cudrania*, *Morus*, *Ficus*]. Chih wu fen lei hsueh pao = Acta Phytotaxonomica Sinica 22(1): 64–76.

Clement, W.L. and G.D. Weiblen. 2009. Morphological evolution in the mulberry family (Moraceae). Systematic Botany 34: 530–552.

Datwyler, S.L. and G.D. Weiblen. 2004. On the origin of the fig: phylogenetic relationships of Moraceae from *ndh*F sequences. American Journal of Botany 91: 767–777.

Das, B. 1965. Some observations on inter-specific hybridization in mulberry. Indian J. Seric. 4: 1–8.

de Sassi, C. and J.M. Tylianakis. 2012. Climate change disproportionately increases herbivore over plant or parasitoid biomass. PLOS ONE 7: e40557.

GEPP. Council. 2006. List of non-native invasive plants in Georgia. Retrieved from URL http://www. gaeppc. org/list. cfm.

Greene, E. 1910. Some southwestern mulberries. Leaflets of Botanical Observation and Criticism 2: 112–121.

Griggs, W.H. and B.T. Iwakiri. 1973. Development of seeded and parthenocarpic fruits in mulberry (*Morus rubra* L.). Journal of Horticultural Science 48: 83–97.

Holmgren, P.K. and N.H. Holmgren. 1992. Plant Specialists Index: Index to Specialists in the Systematics of Plants and Fungi, Based on Data From Index Herbariorum (Herbaria), edition 8. Koeltz Scientific Books.

Hotta, T. 1954. Fundamentals of *Morus* plants classification. Kinugasa Sanpo 390: 13–21.

Iverson, L.R. and A.M. Prasad. 1998. Predicting abundance of 80 tree species following climate change in the eastern United States. Ecological Monographs 68: 465–485.

Iverson, L., A. Prasad and S. Matthews. 2008. Modeling potential climate change impacts on the trees of the northeastern United States. Mitigation and Adaptation Strategies for Global Change 13: 487–516.

Katsumata, F. 1971. Shape of idioblasts in mulberry leaves with special reference to the classification of mulberry trees. The Journal of Sericultural Science of Japan 40: 313–322.

Kirilenko, A.P. and R.A. Sedjo. 2007. Climate change impacts on forestry. Proceedings of the National Academy of Sciences 104: 19697–19702.

Koidzumi, G. 1917. Taxonomy and phytogeography of the genus Morus. Bull. Seric. Exp. Station, Tokyo (Japan) 3: 1–62.

Leroy, J.F. 1949. Les Muriers sauvages et cultives. La sericiculture sous les tropiques. Journal d'agriculture traditionnelle et de botanique appliquée 29: 481–496.

Linnaeus, C. 1753. *Morus*. Species Plantarum 2: 968.

Moretti, G. 1842. Prodromo di una monografia delle specie del genere Morus (letto nella sedata dell'IR instituto di scienze, lettere ed arti del giorno 4 feb. 1841).

Nepal, M. 2008. Systematics and reproductive biology of the genus *Morus* L. (Moraceae) [dissertation]. Retrieved from ProQuest Dissertations and eses.

Nepal, M.P. and C.J. Ferguson. 2012. Phylogenetics of *Morus* (Moraceae) Inferred from ITS and trnL-trnF Sequence Data. Systematic Botany 37: 442–450.

Peters, M.P., L.R. Iverson, A.M. Prasad and S.N. Matthews. 2019. Utilizing the density of inventory samples to define a hybrid lattice for species distribution models: DISTRIB-II for 135 eastern US trees. Ecology and Evolution 9: 8876–8899.

Prasad, A., L. Iverson, S. Matthews and M. Peters. 2007. A climate change atlas for 134 forest tree species of the eastern United States [database]. Retrieved from US Department of Agriculture, Forest Service website: http://www. nrs. fs. fed. us/atlas/tree.

Rao, C.K. and C. Jarvis. 1986. Lectotypification, taxonomy and nomenclature of *Morus alba*, *M. tatarica* and *M. indica* (Moraceae). Taxon: 705–708.

Reveal, J. 2007. *Morus rubra* L. P. 683 in Order out of chaos. Index of Linnean plant names and specimens. Charlie Jarvis (eds.).

Schneider, C.K. 1916. *Morus notabilis* C. K. Schneider In Sargent, Pl. Wilson 3.

Tang, Z.H., M. Cao, L.X. Sheng, X.F. Ma, A. Walsh et al. 2008. Seed dispersal of *Morus macroura* (Moraceae) by two frugivorous bats in Xishuangbanna, SW China. Biotropica 40: 127–131.

Tikader, A. and K. Vijayan. 2017. Mulberry (*Morus* Spp.) genetic diversity, conservation and management, pp. 95–127 in Biodiversity and Conservation of Woody Plants. Springer.

Tojyo, I. 1985. Research of polyploidy and its application in *Morus*. JARQ.

USDA PLANTS. 2007. *Morus rubra* L. Accessed at https://plants.usda.gov/core/profile?symbol=moru2, March 15, 2007.

Venkataraman, K. 1972. Wood phenolics in the chemotaxonomy of the moraceae. Phytochemistry 11: 1571–1586.

Watanabe, T. 1958. Substances in mulberry leaves which attract silkworm larvae (Bombyx mori). Nature 182: 325.

Williams, D.W., R.P. Long, P.M. Wargo and A.M. Liebhold. 2000. Effects of climate change on forest insect and disease outbreaks, pp. 455–494 in Responses of Northern US Forests to Environmental Change. Springer.

Wunderlin, R. 1997. *Morus* [Moraceae] in Flora of North America north of Mexico, vol. 3. Oxford University Press, New York, NY: 390–392.

Zerega, N.J., W.L. Clement, S.L. Datwyler and G.D. Weiblen. 2005. Biogeography and divergence times in the mulberry family (Moraceae). Molecular Phylogenetics and Evolution 37: 402–416.

Zhekun, Z. and M.G. Gilbert. 2003. Flora of China. vol. 5. St. Louis: Missouri Botanical Garden Press. p. 21–73. Moraceae Flora of China Editorial Committee.

Zhengy, W. and Z. Xiushi. 1989. Taxa nova nonnulla Moracearum sinensium. Acta Botanica Yunnanica 11: 24–34.

Zhou, Z. and M. Gilbert. 2003. Moraceae. Flora of China, pp. Science Press, Beijing.

CHAPTER 2

Genetic Diversity in Mulberry Genotypes

Muzaffer İpek,* Şeyma Arikan, Ahmet Eşitken and *Lütfi Pırlak*

INTRODUCTION

Mulberry is a fast-growing deciduous and perennial tree or shrub in genus *Morus* of the family *Moraceae* (Linnaeus 1753). Mulberries are grown for different purpose around the world. It is widely grown to feed silkworm (*Bombyx mori* L.) for sericulture. Mulberry leaves are natural, single food for silkworm. As a result mulberry cultivation has gained importance in the sericulture industry. In terms of production of silkworm cocoon (reelable), China, Japan, India, Thailand, and Viet Nam are prominent countries of the far east, followed by Iran Islamic Republic and Uzbekistan in middle Asia. In the production of a silkworm cocoon, the production cost of leaf covers approximately 60% of all expenditure in India (Das 1965). China has produced 398.212 tons of reelable silkworm cocoon, an approximate 67% of the world cocoon production (Anonymous 2017). India being the second biggest producer after China has produced 155.315 tons of reelable cocoon. China is the biggest raw silk producer which is approximately 75% (126.001 tons) of the world total silk production (Anonymous 2014). China as a leader in both the production of the cocoon and raw silk is also the leader exporting the raw silk and unreelable cocoons worth 316 million USD and 35 million USD, respectively (Anonymous 2016). Because of these reasons, mulberry provides essential input to the sericulture industry that provides employment to a great number of people in countries such as China, India, Japan, Iran Islam Republic, Uzbekistan, and other Asian countries. However, in Turkey, mulberry is widely grown for market fruit products with a yield of 67.000 tons from nearly 2.7 million trees growing in 2200 hectares in 2018 (Anonymous 2019).

Besides being food of silkworm, mulberry has other uses especially in consumption of the mulberry fruit produced from *M. alba* L., *M. laevigata* Wall., *M. nigra* L., *M. rubra* L., which have high phenolic acids and flavonoid content (Arfan et al. 2012), and in processed fruit (Pekmez, Pestil, Köme, dried fruit, jam, jelly, marmalade, pulp, juice, paste, ice cream and wine) industry (İpek et al. 2012). Mulberry is also useful in production of high-quality musical instruments and furniture obtained from the lumber of *M. laevigata* Wall. and *M. serrata* Roxb. Further, mulberry plants have been used as a resource for some pharmaceuticals in Asian countries (Vijayan et al. 2014, Sheet et al. 2018) such as some antioxidants (Yen et al. 1996) and hypoglycaemic compounds (Kelkar et al. 1996). Its fruits also are used to cure dysentery, constipation, and hypoglycemia (Lee et al. 2011). Furthermore, mulberry trees (e.g., *M. alba* L., and *M. alba* var. pendula Dipp.) also offer some advantages like the protection of soil against erosion, maintenance of water table due to its low water consumption, providing of food to birds, providing shelter to shade-loving plants, and using in landscape architecture.

The University of Selçuk, Faculty of Agriculture, Horticulture Department, Konya-Turkey.
Emails: arikan@selcuk.edu.tr; aesitken@selcuk.edu.tr; pirlak@selcuk.edu.tr
* Corresponding author: mipek@selcuk.edu.tr

Origin and Distribution

The Himalayan foothills are accepted as an origin of the *Morus* spp. (Koidzumi 1917) and later mulberries spread into the southern hemisphere. The major cultivated areas of mulberry are tropical regions. Today, mulberries exist in between 50° North and 10° South latitude (Yokoyama 1962) covering the Europe, South-East Asia, North-East Asia, Middle East, South Africa, South and North America including Mexico (Figure 1) from sea level to high altitude (~ 4000 m) (Machii 1999, Sharma et al. 2000, Vijayan et al. 2004a) on earth. According to Watt (1873) certain *Morus* species were truly wild in India, however, Vavilov (1926) reported that primarily center of origin for mulberry could be China, Korea, and Japan. Mulberries are widely present in China, Japan, India, Pakistan, and Bangladesh and other some Asian countries. Due to the fact that the mulberry fruit is eaten by birds, seeds of these fruits are transported over long distances and into areas where the mulberry does not grow naturally; therefore, there is not yet a complete agreement on the origin of mulberry species.

The largest genetic diversity of mulberry occurs in China with 24 *Morus* species of which 17 species are endemic, followed by Japan (19 species), USA (14 species) and Korea (6 species) (Table 1). In Turkey, three mulberry species, such as *M. alba* L., *M. nigra* L., and *M. rubra* L. are both grown for their fruit and feeding the silkworm (İpek et al. 2012).

High genetic diversity in *Morus*. wide geographical distribution, the morphological plasticity, and high natural hybridization make taxonomy of the genus *Morus* more complex and disputed (Vijayan et al. 2011a). According to Figure 1 reproduced from Sharma et al. (2000), five species are present in Continental America; one of five *Morus* species, *M. insignis*, occurs in South America and other four *M. celtidifolia* Kunth., *M. microphylla* Bickl., *M. mollis*, and *M. rubra* L., are in North America. In Asian continent 16 *Morus* species are reported but only three species, *M. alba* L., *M. multicaulis* Perr., *M. atropurpurea* Roxb. are cultivated for sericulture industry and other species are considered wild species. In the Middle East, *M. arabica* Koidz. and *M. nigra* L. are cultivated. *M. nigra* L. is named "black mulberry", *M. alba* L. "white mulberry", and *M. rubra* L. "red mulberry". The *M. mesozygia* Stapf., which grows in a semiarid, sub-humid and humid area, has been reported in South Africa and called "African mulberry". Similarly, *M. tartarica* is named "Russian mulberry", *M. serrata* Roxb. "Himalayan mulberry", *M. celtidifolia* Kunth. "Mexican mulberry", *M. microphylla* Bickl. "Texas mulberry", and *M. australis* Poir. "Chinese mulberry".

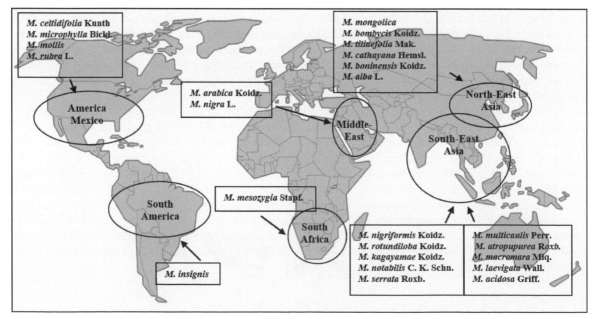

Figure 1. Geographical distribution of *Morus* species (Reproduced from Sharma et al. 2000).

Table 1. Geographical distribution of *Morus* sp. (Singh and Saratchandra 2004).

Countries	Total Number of *Morus* sp.	The Number of Endemic *Morus* sp.
China	24	17
Japan	19	14
USA	14	9
Korea	6	1
India	4	1
Taiwan	4	1
Indonesia	3	2
Colombia	3	1
Mexico	3	2
Turkey	3	-
Thailand	2	2
Argentina	1	1
Peru	1	1

Taxonomy of *Morus* sp.

The higher adaptation to different ecological conditions, easy spontaneous and manual hybridization, and large genetic diversity make mulberry genetic structure rather complicated and highly heterozygous (Dandin 1987). This makes difficult to classify mulberry (Figure 2). Although there is a detailed chapter on mulberry taxonomy in this book yet brief mention its given here being co-related to genetic diversity of mulberry genotypes.

Mulberry taxonomy was started by Linnaeus (1753) and he reported seven *Morus* species (Table 2). About a century later, Seringe (1855) noticed eight *Morus* species. In India, Brandis (1874) described four *Morus* species including *M. alba* L., *M. indica* L., *M. laevigata* Wall., and *M. serrata* Roxb. and he divided *Morus* genus into two sections. Hooker (1885) classified mulberries in the genus *Morus* in the tribe *Morae* of the family of *Moraceae* under *Urticales* order. Since the early 1900s, the taxonomy of the genus *Morus* has been extensively studied. Koidzumi (1917), classified the genus *Morus* into two sections (dolichostyle and macromorus) comprising 24 species based on stigma nature in male and style length in the female flower. In

Figure 2. Differences in fruit characteristics (Krishna et al. 2018).

Table 2. Some taxonomic studies of the genus *Morus* (1753–1979).

Linnaeus (1753)	reported 7 species in *Morus*
Seringe (1855)	reported 8 species in *Morus*
Brandis (1874)	described 4 species and classified the *Morus* genus into two sections
Hooker (1885)	recognized 4 species of *Morus* in India
Koidzumi (1917)	reported 24 species in *Morus* and divided the genus *Morus* into two sections
Koidzumi (1923)	reported 25 species in *Morus*
Leroy (1949)	divided the genus *Morus* into three subgenera (Eumorus, Gomphomorus, and Afromorus)
Hotta (1954)	divided the genus *Morus* into two sections; Dolychocystolithiae and Brachycystolithiae
Iyer (1954)	reported *M. alba* L., *M. indica* L., *M. nigra* L., *M. chinensis*, and *M. multicaulis* Perr.
Gururajan (1960)	suggested *M. alba*, *M. bombycis* Koidz., and *M. latifolia* Poir. in India
Shah and Kachroo (1979)	classified the genus *Morus* into two sections (Section 1: *Morus nigra* L. and Section 2: *Morus alba* L., *Morus bombycis* koidz., and *Morus altifolia* P.) by leaf anatomy and wood characters
Hirano (1982)	suggested *M. alba* L., *M. bombycis* Koidz., and *M. latifolia* Poir. together under one heading
Sanjappa (1989)	reported 68 species in genus *Morus*

the second report Koidzumi (1923) divided the genus into two sections (*Dolichostyle* and *Macromorus*) on the basis of style length, and each section was further divided into two subsections (*Pubescentae* and *Papillosae*) considering stigma hairiness. Leroy (1949) divided the genus *Morus* into *Eumorus*, *Gomphomorus*, and *Afromorus*, but this division has not been widely accepted. Hotta (1954) described 35 species *Morus* in Japan under two headings *Dolychocystolithiae* and the *Brachycystolithiae* mainly on characteristics of cystolith cells in the leaf. Iyer (1954) reported five *Morus* species, *M. alba* L., *M. indica* L., *M. nigra* L., *M. chinensis*, and *M. multicaulis* Perr., from India. Later, Gururajan (1960) suggested that there were two more species (*M. bombycis* Koidz., and *M. latifolia* Poir.) in addition to five *Morus* species reported by Iyer (1954) in India. Shah and Kachroo (1979) divided the genus *Morus* into two sections; which include *M. nigra* L., *M. alba* L., *M. bombycis* Koidz., and *M. latifolia* Poir. using leaf anatomy and wooden characteristics of mulberries. Hirano (1982) supported Shah and Kachroo (1979) based on results which he obtained from analyzing leaf proteins of *M. alba* L., *M. bombycis* Koidz., and *M. latifolia* Poir. Sanjappa (1989) reported the existence of 68 mulberry species.

However, of the more than 150 species cited in Index Kewensis, most of them especially in *M. alba* L., are either synonyms or cultivars. In this context, the classification of Koidzumi (1917) seemed widely acceptable with *Morus* comprising 24 species (Table 3) which is supported by chromosome studies. The Table 3 shows chromosome number of *Morus* species varying from $2n = 2x = 28$ to $2n = 22x = 308$. These studies reported different ploidy status of some species: *M. bombycis* Koidz. (diploid, triploid and tetraploid), *M. latifolia* Poir. (diploid, triploid and mixoploid), *M. laevigata* Wall. (triploid and tetraploid), *M. rotundiloba* Koidz. (diploid and triploid), *M. cathayana* Hemsl. (hexaploid and octoploid) and *M. notabilis* C. K. Schn. (haploid, diploid and triploid) (Janaki 1948, Vijayan et al. 2011b, Yamanouchi et al. 2017). More studies on classification or taxonomy of *Morus* are detailed in Chapter 1 and Chapter 3 of this book.

Genetic Diversity in *Morus* spp.

As mentioned previously genetic diversity in mulberry is quite complex and varied due to its high adaptability to different climatic and soil conditions, natural (open pollination) or easy manual hybridization, and vegetative or generative propagation. As a result mulberry parts such as a leaf, fruit, timber, etc., develop having wide economic use for different purposes. One of the reasons for genetic complexity is the variation in flower forms like monoecious, dioecious or hermaphrodite which result in diversity of genotypes within mulberry species. The diversity in genotypes of mulberry have therefore produced synonym of mulberry species among different genres. The agronomic and morphologic features of the mulberry species being highly heterogeneous are polygenic in inheritance, each controlled by two or more genes. Thus mulberry genotypes easily adapt to different environmental conditions and agricultural practices making it difficult to identify genotypes or cultivars of mulberry simply

Table 3. Chromosome studies of *Morus* species (see Koidzumi 1917).

M. acidosa Griff. (2n = 2x = 28)	*M. formosensis* Hotta	*M. multicaulis* Perr. (2n = 2x = 28)
M. alba L. (2n = 2x = 28)	*M. indica* L. (2n = 2x = 28)	*M. nigra* L. (2n = 22x = 308)
M. arabica Koidz.	*M. kagayamae* Koidz. (2n = 2x = 28)	*M. nigriformis* Koidz.
M. atropurpurea Roxb. (2n = 2x = 28)	*M. laevigata* Wall. (2n = 6x = 84)	*M. notabilis* C. K. Schn. (n = x = 14)
M. boninensis Koidz. (2n = 6x = 84)	*M. latifolia* Poir. (2n = 2x = 28)	*M. serrata* Roxb. (2n = 8x = 112)
M. bombycis Koidz. (2n = 4x = 56)	*M. microphylla* Bickl.	*M. rotundiloba* Koidz. (2n = 2x = 28)
M. cathayana Hemsl. (2n = 6x = 84)	*M. macroura* Miq.	*M. rubra* L.
M. celtidifolia Kunth (2n = 8x = 112)	*M. mesozygia* Stapf.	*M. tiliaefolia* Mak. (2n = 8x = 112)

by morphological characteristics. Many approaches have been reported to identify *Morus* genetic diversity. Morphological differences in mulberry species were firstly commonly used for identification of genetic diversity. Some researchers have subsequently attempted biochemical, such as leaf proteins (Hirano 1982), isozymes (Hirano and Naganuma 1979, Venkateswarlu et al. 1994), cytological (Katsumata 1979), besides morphological (Katsumata 1972, Machii et al. 1997, Biasiolo et al. 2004, Yilmaz et al. 2012, Türemiş et al. 2017), and pomological (Boubaya et al. 2009, Yilmaz et al. 2012) approaches for identifying genetic diversity in *Morus*. However, a description of the *Morus* genus comprising 24 species reported by Koidzumi (1917) has been widely accepted for many years. With beginning of this millennium, genetic studies containing molecular characterization, DNA fingerprinting, and application of DNA markers have significantly contributed to the understanding of genetic diversity and phylogenetic relationships among mulberry species. In order to solve complex genetic diversity in mulberry, use of more reliable and rapid molecular techniques such as Random Amplified Polymorphic DNA (RAPD), Inter-Simple Sequence Repeat (ISSR), Simple Sequence Repeat (SSR), Amplified Fragment Length Polymorphism (AFLP), Sequence-Related Amplified Polymorphism (SRAP), and Sequence Characterized Amplified Region (SCAR) DNA markers have become important. Application of RAPD as molecular technique for genetic characterization of mulberry was first report published by Xiang et al. (1995), followed by Feng et al. (1996), Lou et al. (1996), Feng et al. (1997), Lou et al. (1998), and Zhang et al. (1998). RAPD has been used to identify genetic characterization of mulberry by many researchers. In addition, the ISSR is also the most used marker system by researchers to work out the genetic diversity of mulberry genotypes (Table 4). ISSR has also been used together with RAPD. Further, AFLP, SSR and SRAP markers can be used to characterize wild and cultivated species, varieties of mulberry (Table 5).

Most of the existing genetic studies were carried out by the Indian researchers, followed by Chinese, Turkish, Korean, Japanese and Italian researchers. The reason for many studies conducted on mulberry in India and China has been because both countries have a wide range of genetic resource collections (Table 5). The largest number (2600) of mulberry genetic resources, which include cultivars, genotypes, and accessions are in China, followed by Japan (1375), India (1271), South Korea (208), Bulgaria (140), France (70), Italy (50), USA (23), and others (23) (Vijayan et al. 2018). Turkey has 4 cultivars and 95 mulberry selected genotypes, such as *M. alba* L. (64), *M. nigra* L. (29), and *M. rubra* L. (2). Mulberry genotypes are selected and conserved in Apricot Research Institute in Malatya-Turkey.

The most genetic studies in the genus *Morus* focused on cultivars, genotypes or accessions of *M. alba* L., *M. lhou* Koidz., *M. multicaulis* Perr., *M. indica* L., *M. latifolia* Poir. and other *Morus* species. Some researchers attempted to divide *Morus* species into some groups according to the genetic distance, while others clustered by UPGMA in the same group. Previous genetic studies reported that dendrogram generated by RAPD and ISSR markers on *Morus* species, *M. alba* L. and *M. lhou* Koidz., showed the close genetic relationship (Kalpana et al. 2012). Sung et al. (2001) also reported genetic similarity between both species was 88%. Awasthi et al. (2004) too determined the high genetic similarity existed between *M. alba* L. and *M. lhou* Koidz. and clustered these species in the same group, while having found high genetic distance between *M. alba* L. and *M. tiliaefolia* Mak. Whereas, *M. alba* L. and *M. lhou* Koidz. seemed to have more genetic similarity compared with other wild and cultivated species (Awasthi et al. 2004), paradoxically Sheet et al. (2018) reported that there was a high genetic distance between *M. alba* L. and *M. lhou* Koidz. unlike other scientific reports. Importantly several studies on mulberry genetic diversity have

Table 4. Some selected references of molecular markers used in study of genetic diversity in mulberry.

Species or Varieties	DNA Markers	References
M. alba L., *M. australis* Poir., *M. bombycis* Koidz., *M. cathayana* Hemsl., *M. indica* L., *M. laevigata* Wall., *M. latifolia* Poir., *M. lhou* (ser) Koidz., *M. multicaulis* Perr., *M. nigra* L., *M. rotundiloba* Koidz., *M. rubra* L., *M. serrata* Roxb., *M. sinensis* Hort., *M. tiliaefolia* Mak.	*ISSR-RAPD*	Awasthi et al. (2004)
M. acidosa Griff., *M. alba* L., *M. bombycis* Koidz., *M. indica* L., *M. latifolia* Poir.	*ISSR-RAPD*	Vijayan (2004)
M. alba L., *M. indica* L., *M. indica* L. *x M. latifolia* Poir., *M. nigra* L. *x M. multicaulis* Perr., *M. nigra* L. *x M. multicaulis* Perr. *x M. indica* L.	*ISSR-RAPD*	Vijayan et al. (2004a)
M. serrata Roxb.	*ISSR*	Vijayan et al. (2004b)
M. alba L., *M. laevigata* Wall., *M. bombycis* Koidz., *M. indica* L., *M. latifolia* Poir.	*ISSR-RAPD*	Vijayan et al. (2004c)
Limoncina, Schinichinose, Kattaneo, Obawasa, Rangoon, China white, China black, Canton china, Almora local, Punjab local, Sujanpur-2	*ISSR-RAPD*	Srivastava et al. (2004)
M. alba L., *M. australis* Poir., *M. indica* L., *M. laevigata* Wall., *M. latifolia* Poir.	*ISSR*	Vijayan et al. (2005)
Xiansang 305, Beisangyihao, Nongsang 8, Huangluxuan, Jihu 4, Dazhonghua, Xinyiyuan, Nongsang 14, Yu 237, Xuanqiu 1, 7307, Husang 32, Xiang 7920, Canzhuan 4, Huamingsang, 7946, Yu 2, Shigu 11-6, Xuan 792, Yu 711, Yu 151, Hongxin 5, Lunjiao 40, Wan 7707	*ISSR*	Weiguo et al. (2006a)
M. alba L., *M. alba* var. *macrophylla* Loud., *M. alba* var. *pendula* Dipp., *M. alba* var. *venose* Delile., *M. atropurpurea* Roxb., *M. australis* Poir., *M. bombycis* Koidz., *M. cathayana* Hemsl., *M. laevigata* Wall., *M. mongolica* Schneid., *M. multicaulis* Perr., *M. nigra* L., *M. rotundiloba* Koidz., *M. wittiorum* Hand-Mazz.	*ISSR*	Weiguo et al. (2006b)
M. alba L., *M. macroura* Miq., *M. indica* L., *M. serrata* Roxb.	*ISSR*	Vijayan et al. (2006a)
M. indica L.	*ISSR*	Vijayan et al. (2006b)
M. alba L., *M. alba* var. macrophylla Loud., *M. alba* var. pendula Dipp. *M. alba* var. venose Delile., *M. atropurpurea* Roxb., *M. australis* Poir., *M. bombycis* Koidz., *M. cathayana* Hemsl., *M. laevigata* Wall., *M. mongolica* Schneid., M. multicaulis Perr., *M. nigra* L., *M. rotundiloba* Koidz., *M. wittiorum* Hand-Mazz.	*ISSR-SSR*	Weiguo et al. (2007b)
66 local mulberry varieties	*ISSR*	Weiguo et al. (2007a)
M. alba L., *M. indica* L., *M. laevigata* Wall.	*ISSR*	Kar et al. (2008)
DD, Chinapeaking, Karanahalli local, *M. macroura* Mig., *M. lhou* Koidz., M5, MR2, Mysore local, Newvar20, R127, R175, Srinagar, S1, S13, S34, S36, S145, S146, S1635, V1	*ISSR-RAPD*	Chikkaswamy et al. (2012)
M. alba L., *M. nigra* L.	*ISSR-RAPD*	İpek et al. (2012)
M. alba L., *M. lhou* Koidz.	*ISSR-RAPD*	Kalpana et al. (2012)
M. laevigata Wall., *M. serrata* Roxb.	*ISSR-RAPD*	Naik et al. (2015)
China White, Kanwa2, Mandalay, S1531, S146, S34, S13, Triploid-8, Triploid-10	*RAPD- DAMD*	Bhattacharya and Ranade (2001)
Mysore Local, V-1	*RAPD*	Naik et al. (2002)
M. alba L.	*RAPD*	Orhan et al. (2007)
47 mulberry genotypes	*RAPD*	Ozrenk et al. (2010)
M. alba L., *M. lhou* Koidz.	*RAPD*	Sheet et al. (2018)
M. acidosa Griff., *M. alba* L., *M. atropurpurea* Roxb., *M. bombycis* Koidz., *M. boninensis* Koidz., *M. cathayana* Hemsl., *M. celtidifolia* Kunth, *M. indica* L., *M. kagayamae* Koidz., *M. laevigata* Wall., *M. latifolia* Poir. *M. macroura* Miq., *M. mesozygia* Stapf., *M. microphylla* Bickl., *M. multicaulis* Perr., *M. nigriformis* Koidz., *M. notabilis* C. K. Schn., *M. rotundiloba* Koidz., *M. rubra* L., *M. serrata* Roxb., *M. tiliaefolia* Mak.	*AFLP*	Sharma et al. (2000)

Table 4 contd. ...

...Table 4 contd.

Species or Varieties	DNA Markers	References
M. alba L., *M. bombycis* Koidz., *M. latifolia* Poir.	*AFLP*	Botton et al. (2005)
M. alba L., *M. nigra* L., *M. rubra* L.	*AFLP*	Kafkas et al. (2008)
M. alba L., *M. alba* var. macrophylla Loud., *M. alba* var. venose Delile., *M. alba* var. pendula Dipp., *M. atropurpurea* Roxb., *M. bombycis* Koidz., *M. cathayana* Hemsl., *M. laevigata* Wall., *M. mongolica* Schneid., *M. multicaulis* Perr., *M. nigra* L., *M. rotundiloba* Koidz., *M. wittiorum* Hand-Mazz.	*SRAP*	Weiguo et al. (2009)
M. alba L.	*SRAP*	Bajpai et al. (2014)

Table 5. List of *ex situ* genetic resources in some Asian countries (Vijayan et al. 2018).

Morus Species	Japan	China	India	Korea
M. acidosa Griff.	44	-	-	1
M. alba L.	259	762	93	105
M. atropurpurea Roxb.	3	120	-	-
M. australis Poir.	-	37	2	-
M. bombycis Koidz.	583	22	15	97
M. boninensis Koidz.	11	-	-	-
M. cathayana Hemsl.	1	65	1	-
M. celtifolia Kunth.	1	-	-	-
M. formosensis Hotta.	2	-	-	-
M. indica L.	30	-	350	5
M. kagayamae Koidz.	23	-	-	1
M. laevigata wall.	3	19	32	1
M. latifolia Poir.	349	750	19	128
M. macroura Miq.	1	-	-	-
M. mesozygia Stapf.	1	-	-	-
M. microphylla Bickl.	1	-	-	-
M. mizuho Hotta	-	17	-	-
M. mongolica Schneider	55	-	-	-
M. multicaulis Perr. Perr.	-	-	15	-
M. nigra L.	2	1	2	3
M. nigriformis Koidz.	3	-	-	-
M. notabilis C.K.Schn.	14	-	-	-
M. rotundiloba Koidz.	24	4	2	-
M. rubra L.	1	-	1	-
M. serrata Roxb.	3	-	18	-
M. tiliaefolia Mak.	1	-	1	14
M. wittiorum Hand-Mazz.	-	8	-	-
Morus spp. (unknown)	15	-	106	259

shown that *M. alba* L. belongs to same group comprising *M. multicaulis* Perr., *M. laevigata* Wall., *M. nigra* L, *M. wittiorum* Hand-Mazz, *M. cathayana*, *M. indica* L., *M. australis* Poir., *M. rotundiloba* Koidz., *M. atropurpurea* Roxb., and *M. bombycis* Koidz. based on cluster analysis (Botton et al. 2005, Vijayan et al. 2005, Vijayan et al. 2006b, Weiguo et al. 2006b, Weiguo et al. 2007b, Kar et al. 2008, Weiguo et al. 2009). On the contrary, it has also been reported *M. alba* L. genotype (24Ke10) was outside the main group which consisted of genotypes of *M. alba* L. and *M. nigra* L. whereas black mulberry genotype (01KaD2), was in the main group using both marker systems

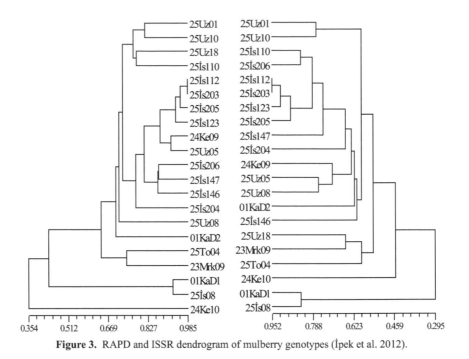

Figure 3. RAPD and ISSR dendrogram of mulberry genotypes (İpek et al. 2012).

(Figure 3). This suggests (*M. alba* L.) genotype 24Ke10 and *M. nigra* L. genotype 01KaD2 could be a hybrid between *M. alba* L. and *M. nigra* L. or vice versa (İpek et al. 2012).

A study that both supported and opposed the result of İpek et al. (2012) stating that *M. nigra* L. was genetically distant from other *Morus* species, forming a single cluster in the phylogenetic tree, could be explained due to the high chromosome numbers (2n = 308) preventing *M. nigra* L. to hybridize with other *Morus* species easily (Weiguo et al. 2009). Consequently, this ensues difficulty in speciation such as in *M. alba* L., *M. latifolia* Poir., and *M. bombycis* Koidz. because these species could be hybrids or introgressants being a result of higher adaptation capability, monoculture cultivation, easy natural hybridization, environmental stress, and natural selection among various mulberry species (Botton et al. 2005). Vijayan et al. (2004c) suggested that *M. alba* L., *M. latifolia* Poir. and *M. bombycis* Koidz. should be joined together under one species because of forming an internal group in all dendrograms of Hirano (1977, 1982). Evidence from isozymes profiles generated on the basis of geographical distribution by Srivastava et al. (2004), Orhan et al. (2007) and Bajpai et al. (2014) found high genetic similarity among the *M. alba* L. genotypes collected from different regions and it was concluded that there was a weak correlation between geographic distance and genetic similarity. However, Sharma et al. (2000) noticed that geographic origin was though helpful for identification of genotypes but no clear correlations were found between geographic origin and genetic relationship. Nevertheless, some studies report that genetic diversity in the same species or between different species is entirely related to geographic distance (Weiguo et al. 2007a, Naik et al. 2015). The genetic diversity in mulberry seems negatively affected by cultivation or monoculture, in comparison to the wild *Morus* species (Weiguo et al. 2007a). All studies therefore point out that genetic diversity of mulberry genotypes is too high and complex.

Conclusion

The culture of the mulberry cultivation dates back to approximately 5000 years because of its economic importance. Various mulberry species are used for different purposes such as in silk industry, processed fruit industry, landscape architecture, etc. It is well known that the mulberry species have more heterozygous genetic structure and as a result adapt to changed climatic conditions in different regions far away from their center of origin. The migration of mulberries from their origin to different parts of the world has caused some problems, due to misleading local nomenclature, incorrect germplasm recordkeeping, variations in phenotype under different ecological conditions, easy natural and manual hybridization among species, varieties, and/or genotypes. As a consequence, these problems may have contributed over the years to an erroneous or confused classification of mulberry species. The complex genetic structure and confusing classification leads to extension of the process of genetic diversity. In the last two decades, remarkable results have been obtained in genetic research of the genus *Morus* and new mulberry varieties have been introduced via biotechnological methods. The transgenic cultivars have ability to show tolerance to abiotic stress such as cold, salinity, and drought. With on set of the global warming or climate change, abiotic stress conditions continue to increase day by day. In order to reduce harmful effects of the climate change on mulberry growing areas, new mulberry cultivars having wider adaptation capability and higher tolerance level to stress conditions can be introduced by exploring diversity in mulberry genotypes and through biotechnological methods that include plant tissue culture, DNA marker techniques, Marker Assisted Selection (MAS), mutation breeding techniques and gene transfer technology.

References

Anonymous. 2014. http://www.fao.org/faostat/en/#data/QP (Access Date: 11.06.2019).

Anonymous. 2016. http://www.fao.org/faostat/en/#data/TP. (Access Date: 11.06.2019).

Anonymous. 2017. http://www.fao.org/faostat/en/#data/QL. (Access Date: 11.06.2019).

Anonymous. 2019. https://biruni.tuik.gov.tr/medas/?kn=92&locale=tr (Access Date: 11.06.2019).

Arfan, M., R. Khan, A. Rybarczyk and R. Amarowicz. 2012. Antioxidant activity of mulberry fruit extracts. International Journal of Molecular Sciences 13(2): 2472–2480.

Awasthi, A.K., G. Nagaraja, G. Naik, S. Kanginakudru, K. Thangavelu and J. Nagaraju. 2004. Genetic diversity and relationships in mulberry (genus *Morus*) as revealed by RAPD and ISSR marker assays. BMC Genetics 5(1): 1(1–9).

Bajpai, P.K., A.R. Warghat, R.K. Sharma, A. Yadav, A.K. Thakur, R.B. Srivastava and T. Stobdan. 2014. Structure and genetic diversity of natural populations of *Morus alba* in the Trans-Himalayan Ladakh Region. Biochemical Genetics 52(3-4): 137–152.

Bhattacharya, E. and S.A. Ranade. 2001. Molecular distinction amongst varieties of mulberry using RAPD and DAMD profiles. BMC Plant Biology 1(1): 3(1–8).

Biasiolo, M., M.T. Da Canal and N. Tornadore. 2004. Micromorphological characterization of ten mulberry cultivars (*Morus* spp.). Economic Botany 58(4): 639–646.

Botton, A., G. Barcaccia, S. Cappellozza, R. Da Tos, C. Bonghi and A. Ramina. 2005. DNA fingerprinting sheds light on the origin of introduced mulberry (*Morus* spp.) accessions in Italy. Genetic Resources and Crop Evolution 52(2): 181–192.

Boubaya, A., M. Ben Salah, N. Marzougui and A. Ferchichi. 2009. Pomological characterization of the mulberry tree (*Morus* spp.) in the South of Tunisia. J. Arid Land Stud. 19(1): 157–159.

Brandis, D. 1874. The Forest flora of north-west and central India: A Handbook of the Indigenous Trees and Shrubs of Those Countries. Atlas, WH Allen, London, UK.

Chikkaswamy, B.K., R.C. Paramanik, A. Debnath and M. Sadana. 2012. Evaluation of genetic diversity in mulberry varieties using molecular markers. Nature Sci. 10: 45–60.

Dandin, S. 1987. Cross ability studies in mulberry. Indian J. Seric. 26: 11–14.

Das, B.C. 1965. Some observations on inter-specific hybridization in mulberry. Indian J. Seric. 4: 1–8.

Feng, L., G. Yang, M. Yu, Y. Ke, C. Jing and Z. Xiang. 1996. Studies on the genetic identities and relationships of mulberry cultivated species (*Morus* L.) via a random amplified polymorphic DNA assay. Acta Sericologica Sinica 22(3): 135–139.

Feng, L., G. Yang, M. Yu, X. Zhang and Z. Xiang. 1997. Study of relationships among species in *Morus* L. using random amplified polymorphic DNA (RAPD). Zhongguo Nongye Kexue 30(1): 52–56.

Gururajan, M. 1960. Varieties of mulberry-a classification. Indian Silk J 1: 12–15.

Hirano, H. 1977. Evaluation of affinities in mulberry and its relatives by peroxidase isozyme technique. Jpn. Agric. Res. 11: 228–233.

Hirano, H. and K. Naganuma. 1979. Inheritance of peroxidase isozymes in mulberry (*Morus* spp.). Euphytica 28(1): 73–79.

Hirano, H. 1982. Varietal differences of leaf protein profiles in mulberry. Phytochemistry 21(7): 1513–1518.

Hooker, J.D. 1885. Flora of British India, Vol-4, L. Reeve & Co., London, UK.

Hotta, T. 1954. Taxonomical study on cultivated mulberry in Japan. Botanical Institute, Faculty of Textile Fibres, Kyoto University of Industrial Arts and Textile Fibres, Kyoto, Japan.

İpek, M., L. Pirlak and S. Kafkas. 2012. Molecular characterization of mulberry (*Morus* spp.) genotypes via RAPD and ISSR. Journal of the Science of Food and Agriculture 92(8): 1633–1637.

Iyer, Y. 1954. Field crops of India with reference to Mysore. The Bangalore Printing and Publishing Co. Ltd., Mysore Road, Bangalore: 619.

Janaki, A.E.K. 1948. The origin of the black mulberry. Journal of the Royal horticultural Society 73: 113–120.

Kafkas, S., M. Özgen, Y. Doğan, B. Özcan, S. Ercişli and S. Serçe. 2008. Molecular characterization of mulberry accessions in Turkey by AFLP markers. Journal of the American Society for Horticultural Science 133(4): 593–597.

Kalpana, D., S.H. Choi, T.K. Choi, K. Senthil and Y.S. Lee. 2012. Assessment of genetic diversity among varieties of mulberry using RAPD and ISSR fingerprinting. Scientia Horticulturae 134: 79–87.

Kar, P.K., P.P. Srivastava, A.K. Awasthi and S.R. Urs. 2008. Genetic variability and association of ISSR markers with some biochemical traits in mulberry (*Morus* spp.) genetic resources available in India. Tree Genetics & Genomes 4(1): 75–83.

Katsumata, F. 1972. Relationship between the length of styles and the shape of idioblasts in mulberry leaves, with special reference to the classification of mulberry trees. The Journal of Sericultural Science of Japan 41(5): 387–395.

Katsumata, F. 1979. Chromosomes of *Morus nigra* L. from Java and hybridization affinity between this species and some mulberry species in Japan. The Journal of Sericultural Science of Japan 48(5): 418–422.

Kelkar, S.M., V.A. Bapat, T.R. Ganapathi, G.S. Kaklij, P.S. Rao and M.R. Heble. 1996. *Morus indica* L. shoot cultures: detection of hypoglycemic activity. Curr. Sci. 71: 71–72.

Koidzumi, G. 1917. Taxonomy and phytogeography of the genus *Morus*. Bull. Seric. Exp. Station, Tokyo 3: 1–62.

Koidzumi, G. 1923. Synopsis specierum generis. *Morus*. Bull. Seric. Exp. Station 11: 1–50.

Krishna, H., D. Singh, R.S. Singh, L. Kumar, B.D. Sharma and P.L. Saroj. 2018. Morphological and antioxidant characteristics of mulberry (*Morus* spp.) genotypes. Journal of the Saudi Society of Agricultural Sciences (DOI: j.jssas.2018.08.002) (*In Press*).

Lee, Y., D.E. Lee, H.S. Lee, S.K. Kim, W.S. Lee, S.H. Kim and M.W. Kim. 2011. Influence of auxins, cytokinins, and nitrogen on production of rutin from callus and adventitious roots of the white mulberry tree (*Morus alba* L.). Plant Cell, Tissue and Organ Culture 105(1): 9–19.

Leroy, J. 1949. Contribution a l'etude des Monochlamydees: documents nouveaus sur des plantes de Madagascar, de Sumatra et de Colombie. Bull. Mus. Hist. Nat. Ser. 221: 725–732.

Linnaeus, C. 1753. Hortus cliffortianus. Amstelaedami, Amsterdam, Holland.

Lou, C., Y. Zhang and Y. Zhang. 1996. Studies on RAPD in mulberry. Acta Agriculturae Universitatis Chekianensis 22(2): 149–151.

Lou, C., Y. Zhang and J. Zhou. 1998. Polymorphisms of genomic DNA in parents and their resulting hybrids in mulberry *Morus*. Sericologia 38(3): 437–449.

Machii, H., A. Koyama and H. Yamanouchi. 1997. Manual for the characterization and evaluation of mulberry genetic resources Misc. Publ. Natl. Inst. Seric. Entomol. (22): 105–124.

Machii, H. 1999. A list of genetic mulberry resources maintained at National Institute of Sericulture and Entomological Science. Misc. Publ. Natl. Seric. Entomol. Sci. (Japan) 26: 1–77.

Naik, V.G., A. Sarkar and N. Sathyanarayana. 2002. DNA fingerprinting of Mysore Local and V-1 cultivars of mulberry (*Morus* spp.) with RAPD markers. The Indian Journal of Genetics and Plant Breeding 62(3): 193–196.

Naik, V.G., S. Dandin, A. Tikader and M.V. Pinto. 2015. Molecular diversity of wild mulberry (*Morus* spp.) of Indian subcontinent. NISCAIR-CSIR 14: 334–343.

Orhan, E., S. Ercisli, N. Yildirim and G. Agar. 2007. Genetic variations among mulberry genotypes (*Morus alba*) as revealed by random amplified polymorphic DNA (RAPD) markers. Plant Systematics and Evolution 265 (3-4): 251–258.

Ozrenk, K., S.R. Gazioglu, C. Erdinc, M. Guleryuz and A. Aykanat. 2010. Molecular characterization of mulberry germplasm from Eastern Anatolia. African Journal of Biotechnology 9(1): 1–6.

Sanjappa, M. 1989. Geographical distribution and exploration of the genus *Morus* L. (Moraceae). Genetic resources of mulberry and utilization (Edited K Sengupta and SB Dandin), Central Sericultural Research and Training Institute (CSRTI), Mysore, India.

Seringe, N.C. 1855. Description culture et taille des muriers: leurs espèces et leurs variétés, V. Masson, Paris, France.

Shah, A. and P. Kachroo. 1979. The structure of wood in some species of *Morus*. Recent research in plant science. Edited by SS Bir. Kalyani Publishers, New Delhi, India.

Sharma, A., R. Sharma and H. Machii. 2000. Assessment of genetic diversity in a *Morus* germplasm collection using fluorescence-based AFLP markers. Theoretical and Applied Genetics 101(7): 1049–1055.

Sheet, S., K. Ghosh, S. Acharya, K.P. Kim and Y. Lee. 2018. Estimating genetic conformism of Korean mulberry cultivars using random amplified polymorphic DNA and inter-simple sequence repeat profiling. Plants 7(1): 21.

Singh, T. and B. Saratchandra. 2004. Principles and techniques of silkworm seed production. Discovery Publishing House New Delphi, India.

Srivastava, P.P., K. Vijayan, A.K. Awasthi and B. Saratchandra. 2004. Genetic analysis of *Morus alba* through RAPD and ISSR markers. Indian Journal of Biotechnology 3: 527–532.

Sung, G., K. Ryu, H. Kim, H. Nam, T. Goo and S. Cho. 2001. Analysis of ITS nucleotide sequences in ribosomal DNA of *Morus* species. Korean J. Seric. Sci. 43(1): 1–8.

Türemiş, N., L. Pirlak, A. Eşitken, Ü. Erdoğan, A. Tümer, B. İmrak and A. Burğut. 2017. A field survey of promising mulberry (*Morus* spp.) genotypes from Turkey. Erwerbs-Obstbau 59(2): 101–107.

Vavilov, N.I. 1926. Studies on the origin of cultivated plants. Leningrad Press, Petersburg, Russia.

Venkateswarlu, M., B. Susheelamma, A. Sarkar and R. Datta. 1994. Isozyme studies in mulberry germplasm introduced from Rajasthan. Indian J. Sericult. 33: 98–99.

Vijayan, K. 2004. Genetic relationships of Japanese and Indian mulberry (*Morus* spp.) genotypes revealed by DNA fingerprinting. Plant Systematics and Evolution 243(3-4): 221–232.

Vijayan, K., A. Awasthi, P. Srivastava and B. Saratchandra. 2004a. Genetic analysis of Indian mulberry varieties through molecular markers. Hereditas 141(1): 8–14.

Vijayan, K., P. Kar, A. Tikader, P. Srivastava, A. Awasthi, K. Thangavelu and B. Saratchandra. 2004b. Molecular evaluation of genetic variability in wild populations of mulberry (*Morus serrata* Roxb.). Plant Breeding 123(6): 568–572.

Vijayan, K., P. Srivastava and A. Awasthi. 2004c. Analysis of phylogenetic relationship among five mulberry (*Morus*) species using molecular markers. Genome 47(3): 439–448.

Vijayan, K., C. Nair and S. Chatterjee. 2005. Molecular characterization of mulberry genetic resources indigenous to India. Genetic Resources and Crop Evolution 52(1): 77–86.

Vijayan, K., A. Tikader, P. Kar, P. Srivastava, A. Awasthi, K. Thangavelu and B. Saratchandra. 2006a. Assessment of genetic relationships between wild and cultivated mulberry (*Morus*) species using PCR based markers. Genetic Resources and Crop Evolution 53(5): 873–882.

Vijayan, K., P. Srivastava, C. Nair, A. Awasthi, A. Tikader, B. Sreenivasa and S.R. Urs. 2006b. Molecular characterization and identification of markers associated with yield traits in mulberry using ISSR markers. Plant Breeding 125(3): 298–301.

Vijayan, K., B. Saratchandra and J.A. Teixeira da Silva. 2011a. Germplasm conservation in mulberry (*Morus* spp.). Scientia Horticulturae 128(4): 371–379.

Vijayan, K., A. Tikader, Z. Weiguo, C.V. Nair, S. Ercisli and C.H. Tsou. 2011b. *Morus*. In wild crop relatives: genomic and breeding resources. Springer (p: 75–95), Berlin, Heidelberg.

Vijayan, K., P.J. Raju, A. Tikader and B. Saratchnadra. 2014. Biotechnology of mulberry (*Morus* L.). A review. Emirates Journal of Food and Agriculture 26(6): 472.

Vijayan, K., G. Ravikumar and A. Tikader. 2018. Mulberry (*Morus* spp.) breeding for higher Fruit production. Advances in plant breeding strategies: Fruits (p: 89–130), Springer, Cham.

Watt, G. 1873. A dictionary of economic products of India. Delhi, India, Periodical Experts.

Weiguo, Z., M. Xue-Xia, Z. Bo, L. Zhang, P. Yi-Le and Y.-P. Huang. 2006a. Construction of finger printing and genetic diversity of mulberry cultivars in China by ISSR markers. Acta Genetica Sinica 33(9): 851–860.

Weiguo, Z., Z. Zhihua, M. Xuexia, W. Sibao, Z. Lin, P. Yile and H. Yongping. 2006b. Genetic relatedness among cultivated and wild mulberry (Moraceae: *Morus*) as revealed by inter-simple sequence repeat analysis in China. Canadian journal of plant science 86(1): 251–257.

Weiguo, Z., Y. Wang, T. Chen, G. Jia, X. Wang, J. Qi, Y. Pang, S. Wang, Z. Li and Y. Huang. 2007a. Genetic structure of mulberry from different ecotypes revealed by ISSRs in China: An implication for conservation of local mulberry varieties. Scientia Horticulturae 115(1): 47–55.

Weiguo, Z., Z. Zhihua, M. Xuexia, Z. Yong, W. Sibao, H. Jianhua, X. Hui, P. Yile and H. Yongping. 2007b. A comparison of genetic variation among wild and cultivated Morus species (Moraceae: *Morus*) as revealed by ISSR and SSR markers. Biodiversity and Conservation 16(2): 275–290.

Weiguo, Z., R. Fang, Y. Pan, Y. Yang, J. Chung, I. Chung and Y. Park. 2009. Analysis of genetic relationships of mulberry (*Morus* L.) germplasm using sequence-related amplified polymorphism (SRAP) markers. African Journal of Biotechnology 8(11): 2604–2610.

Xiang, Z., X. Zhang and Y. Maode. 1995. A Preliminary Report on the Application of RAPD in Systematics of *Morus*. Acta Sericologica Sinica 4.

Yamanouchi, H., A. Koyama and H. Machii. 2017. Nuclear DNA Amounts of Mulberries (*Morus* spp.) and Related Species. Japan Agricultural Research Quarterly 51(4): 299–307.

Yen, G.C., S.C. Wu and P.D. Duh. 1996. Extraction and identification of antioxidant components from the leaves of mulberry (*Morus alba* L.). Journal of Agricultural and Food Chemistry 44(7): 1687–1690.

Yilmaz, K., Y. Zengin, S. Ercisli, M. Demirtas, T. Kan and A. Nazli. 2012. Morphological diversity on fruit characteristics among some selected mulberry genotypes from Turkey. J. Anim. Plant. Sci. 22: 211–214.

Yokoyama, T. 1962. Synthesized Science of Sericulture. Central Silk Board: 39–46.

Zhang, Y., L. Chengfu, Z. Jinmei, Z. Hongzi and X. Xiaoming. 1998. Polymorphism studies on genomic DNA of diploids and polyploids in mulberry. Journal of Zhejiang Agricultural University 24: 79–81.

CHAPTER 3

Cultivation of Mulberry
An Important Genetic and Economic Source

Kunjupillai Vijayan,[1,*] *Prashanth Sangannavar,*[1] *Soumen Chattopadhyay,*[2]
Daniel Dezmirean[3] *and Sezai Ercisli*[4]

INTRODUCTION

Mulberry is a fast growing deciduous tree cultivated under various climatic conditions ranging from temperate to tropics (Yokoyama 1962, Vijayan et al. 2011a), from sea level to altitudes as high as 4000 m (Machii et al. 1999, Tutin 1996), in various forms such as low bush, high bush and tree. Although mulberry is thought to have evolved in sub-Himalayan tracts, owing to its wider adaptability to divergent agro-climatic conditions, it is seen at present in almost all continents including Africa, Asia, South America, Europe, North and South America (Le Houerou 1980, Rodríguez et al. 1994). Mulberry is amenable to both sexual and vegetative propagation. Thus, diverse levels of polyploidization in the genus are reported with wide variation of chromosome numbers, such as $2n = x = 14$ chromosomes in *M. notabilis*, $2n = 2x = 28$ in *M alba* and *M. indica*, $2n = 3x = 42$ in *M. bombysis*, $2n = 4x = 56$ in *M. laevigata*, *M. cathayana*, and *M. boninensis*, $2n = 6x = 84$ in *M. serrata* and *M. tiliaefolia*, and $2n = 22x = 308$ in *M. nigra* (Maode et al. 1996, Basavaiah et al. 1989). Though it has been argued that the basic chromosome number of the genus *Morus* is $x = 14$ (Azaizan and Sonboli 2001) but, recent chromosome karyotyping study (He et al. 2013) confirmed that the previous prediction of Dutta (1954) and Das (1961) that the basic chromosome number of the genus is seven. Although the cytological status is pretty well delineated, the species demarcation of the genus *Morus* is still a matter of great debate and disputes (Vijayan et al. 2004). One of the major reason for such ambiguity in the species differentiation is the high success rate it enjoys in cross hybridization among the so called species (Vijayan 2010). Thus, a large number of interspecific hybrids are present in nature, which interfere the process of species identification (Tikader and Dandin 2007). Hence, their true taxonomic identity is still a matter of great controversies (Berg 2001). Presence of 150 species of mulberry have been cited in the Index Kewensis, though a majority of them were later treated either as synonyms or as varieties rather than species and transferred to allied genera (Wang and Tanksley 1989). Currently, the family Moraceae is placed into Rosales by the Angiosperm

[1] Research Coordination Section, Central Silk Board, BTM Layout, Madiwala, Bengaluru-560068, Karnataka, India.
  Emails: dr.sangannavar@gmail.com
[2] Central Sericultural Research and Training Institute, Berhampore-742101, West Bengal, India, Email: soumenchatto@rediffmail.com
[3] Faculty of Animal Science and Biotechnology, University of Agricultural Science and Veterinary Medicine ClujNapoca, 3-5 Mănăşturstr-400372 ClujNapoca, Romania, Email: ddezmirean@usamvcluj.ro
[4] Ataturk University, Faculty of Agriculture, Department of Horticulture, Erzurum-25240, Turkey, Email: sercisli@gmail.com
* Corresponding author: kvijayan01@yahoo.com

Phylogeny Group III (Chase et al. 2009) with 37 genera. The type genus *Morus* consists of about 11–13 species with ~ 1100 varieties (Nepal and Ferguson 2012). Yet species number in the genus *Morus* is still a matter of debate and more recently number of species has been further reduced to 8, e.g., *M. alba*, *M. nigra*, *M. notabilis*, *M. serrata*, *M. celtidifolia*, *M. insignis*, *M. rubra*, and *M. mesozygia*. (Zeng et al. 2015).* However, only a small fraction of these species is cultivated on commercial purposes and the rest of the species are still growing in the wilderness.

Cultivation of Mulberry

Cultivation of mulberry started ~ 5000 years before in China for feeding the silkworms *Bombyx mori* L. (FAO 1990, Barber 1991). Diploid and triploid forms of *M. alba*, *M. indica*, *M. bombycis*, *M. latifolia* and *M. multicaulis* are cultivated for sericulture purpose and tetraploids like *M. laevigata* and *M. nigra* for fruits which are bigger and sweeter.

Planting Materials and Methods

Mulberry is moderately resistant to biotic and abiotic stresses, hence, it can be grown in a wide range of places. Luxurious growth of mulberry need soils which are flat, deep, fertile, well drained, loamy to clayey, porous with good moisture holding capacity and the pH ranging from 6.5 to 6.8. Soil with higher or lower pH can be ameliorated by applying adequate quantity of gypsum or lime as indicated in Tables 1 & 2. The ideal temperature ranges from 24°C to 28°C and the annual rainfall from 600 mm to 2500 mm.

To initiate the plantation, the land is ploughed to a depth of 30 to 35 cm, leveled properly with a basal dose of well decomposed farm yard manure (FYM), containing approximately 0.5% N, 0.2% P_2O_5 and 0.5% K_2O, at the rate of 20 MT/ha. The most common method of mulberry propagation is through stem cuttings as it generally gives 60–80% survival rate. Cuttings of 8–10 months old shoots (15–18 cm long and, 50 mm thick with 3 to 4 healthy buds) are ideal planting materials. The success in rooting of stem cuttings depends on the genotype, environmental factors, and physiological state of the cuttings (Tikader et al. 1995, 1996). In areas where irrigation facilities are limited, mulberry is planted using saplings raised in nursery. Six to eight month old saplings are used for plantations. In a single pit only one sapling is planted. In general, planting is done during the rainy seasons in India and spring season in European countries.

However, in temperate countries where the plant cannot be propagated through cutting, because of poor rooting, the grafting is the usual practice. Depending on the type of material used, e.g., nature, age and portion of the stock, grafting's are mainly of five types: wedge and crown grafting, whip, root and bud grafting. However, Spring bud grafting is the most common method, where in a T-cut is made on the rootstock and a smooth, sloping cut is made on the lower end of the scion. The scion is then inserted into the T-cut and wrapped and sealed. The successful graft union is evident by the close matching of the callus producing tissues near the cambium layers, whereas, the incompatible matching depends on the season and physiological condition of the materials. It is observed that T-buddings give the highest (40.6%) success in comparison to that of patch buddings (34.6%) and

Table 1. Amendments of alkali soils for mulberry cultivation (Datta 2000).

pH range (To bring pH to 6.8)	Quantity of gypsum/ha*
7.4 to 7.8	2.0 MT**
7.9 to 8.4	5.0 MT
8.5 to 9.0	9.0 MT
9.1 to above	14.0 MT

* ha = hectare; ** MT = Metric Tonnes.

---

* A comprehensive study on taxonomic evaluation of mulberry (*Morus* sp.) is detailed in Chapters 1, 2.

Table 2. Amendments of acidic soils for mulberry cultivation (Datta 2000).

pH range	Quantity of lime/ha to bring pH to 6.8		
	Plain	Hilly areas	Soil type
5.5 to 6.5	1.25 MT	2.5 MT	Sandy
	2.50 MT	5.0 MT	Sandy loamy
	5.0 MT	7.5 MT	Loamy
	7.5 MT	8.75 MT	Clay loamy

crown grafts (21.7%). Similarly, buddings made in May month are reported to give better success rate (43.2%), than buds made in June (31.8%) and August (16.3%) (Vural et al. 2008).

After the branch development the bushes are given specific production cuts that are trade marks for the intensive mulberry farming. *Morus alba* ssp. has a specific particularity as regards the buds excitability. Shoots formed from these buds exhibit a very vigorous growth, being long and without ramifications; this aspect being very advantageous for the harvest of leaves.

Application of Manures and Fertilizers

Mulberry is known to be a nutrient hungry plant requiring high doses of fertilizer because of its fast growing habit to producing huge quantity of leaves. Depending on the variety, cultural conditions, type of plantations, leaf harvesting frequency and soil conditions, the fertilizer requirement of mulberry varies considerably in different parts of India. In southern and eastern parts of India, the mulberry shoot is harvested or leaf plucked 5 to 6 or 4 to 5 times in a year, respectively, in order to rear the silkworm at 2 to 3 months' interval under irrigated conditions. In Rainfed gardens of northeastern states, leaf harvest is done 3 to 4 times per year. Hence, the fertilizer requirement for the mulberry in irrigated areas is much higher compared to other regions where shoot harvesting is restricted to 2–4 times in a year. According to Craiciu (1972) mulberry plantation consumes a lot of nutrients, for obtaining high yields of mulberry leaves. It utilizes about 24 kg of nitrogen, 12 kg of phosphorus and 15 kg of potassium to produce one ton of leaf. Nitrogen is one of most important macronutrient which plays a very important role in the vegetative growth of the mulberry. Similarly, phosphorus determines the development of roots, ripening of the wood, and level of resistance to the frost. Likewise, potassium intensifies the process of formation of chlorophyll and starch, and accelerates the processes of lignifications of shoots. Therefore, for a well maintained mulberry garden which yields 50–60 MT of foliage per hector needs approximately 350 MT of chemical fertilizer to be added annually. Further, to reduce the loss of nitrogen fertilizers through leeching, it may be applied in split doses after every crop like other fertilizers that are applied in split doses (Tables 3 and 4). However, in temperate regions where mulberry has a restricted growth the proportionate application of fertilizer, such as 10:10:10 NPK, annually will maintain satisfactory growth (Rathore et al. 2011). In addition to these chemical fertilizers, organic manures such as FYM Compost, Vermi compost, poultry manure, biofertilizers such as *Azatobacterium, Azospirillum,* phosphobacterium, etc., are also added at the rate of 10–20 MT/ha/yr to make the soil healthier. Further macronutrients and micronutrients also need to be supplemented according to the requirement based on soil analysis and appearance of deficiency symptoms. Micronutrients can be applied either directly to the soil or sprayed on the leaf as aqueous solutions of required concentrations.

Pruning of Plants

Depending on climatic conditions, objective of the plantation, and the level of mechanization, mulberry is pruned in different forms as trees, high bushes and low bushes. One of the basic purposes of pruning mulberry is to sustain the growth and leaf quality as well as easy harvest of the foliage for silkworm rearing throughout the year. Mulberry is pruned 4–5 times annually after every silkworm crop. Pruning is generally done by sharp sickle or pruning saw at a height of 25–30 cm from the ground in most of the places (Jolly 1987). However, high bush or

Table 3. Recommendation for fertilizer and manure applications in mulberry garden of different regions of India.*

Fertilizer & manure application	Garden condition	Irrigated			Rainfed			Remarks
		N	P	K	N	P	K	
I. *Southern India*								
Fertilizer recommendation (kg/ha/y)	Chawki garden	260	140	140	-	-	-	In 8 split doses as straight fertilizers
	Late age garden	350	140	140	-	-	-	In 5–6 split doses as shown in Table 4
Manure recommendation (MT/a/y)	40 MT and 20MT of FYM in 2 split doses for Chawki and late age gardens respectively.							
II. *Eastern & north eastern India*								
Fertilizer recommendation (kg/ha/y)	Chawki and late age garden	336	180	112	150	50	50	*Irrigated:* In 4–5 split doses as straight fertilizers Rainfed: 2–3 split doses as straight fertilizers
Manure recommendation (MT/a/y)	40 MT and 20MT of FYM in 2 split doses for Chawki and late age gardens respectively.							
III. *Northern India*								
Fertilizer recommendation (kg/ha/y)	Sub-tropical (Rainfed)	150	100	100	-	-	-	Nitrogen in 2 split doses
	Temperate	250	120	120				Nitrogen in 2 split doses

* based on the Technology descriptor of CSR&TI, Mysore, Berhampore and Pampore.

Table 4. Fertilizer application for mulberry in Southern India* (Source: Krishnaswamy (1986)).

Fertilizer application	Irrigated condition						Rain fed condition		
	Row system			Pit system					
	N	P	K	N	P	K	N	P	K
Recommendation (kg/ha/yr)	300	120	120	280	120	120	100	50	50
Split doses (Kg/crop)									
First crop	60	60	60	60	60	60	50	50	50
Second crop	60	-	-	40	-	-	50	-	-
Third crop	60	60	60	40	-	-			
Fourth crop	60	-	-	60	60	60			
Fifth crop	60	-	-	40	-	-			
Sixth crop	-	-	-	40	-	-			

* based on the Technology descriptor of CSR&TI, Mysore, Berhampore and Pampore.

tree plantation is preferred in areas where Rainfed plantations are maintained. Tree plantations are used for fruit production and in such cases only branches are trimmed to give the plant a sturdy framework.

Pests and Diseases*

In India

The most common pests of mulberry are *Maconellicoccus hirsutusi* (Tukra), *Diaphania pulverulentalis* (Thrips), and *Spilarctia oblique* (Bihar hairy caterpillar). Besides, whitefly (caused by *Aleroclava pentatuberculata* and *Dialeeuroporade cempuncta*) is also a major pest in the mulberry growing states of Eastern and NE India. To prevent

---

* See also Chapter 15.

tukra, clipping off the affected apical portion or releasing the predatory lady bird beetle *Cryptalaemus montrouzieri* at rate of 250 adults/acre is found effective. For control of thrips (*Pseudodendrothrips mori*; Thysanoptera: Thripidae), after harvest of leaves, mulberry field should thoroughly cleaned by removing small branches dead leaves and weeds in order to eliminate any developmental stages of thrips left in the field. Spray of 1.5% neem oil (1500 ppm; 1.5 ml/litre of water), 0.1% Dimethoate 30EC (3 ml/litre) or 0.015% Thiamethoxam 25WG (w/v) 15 to 25 days after pruning is also effective in controlling thrips chemically. However, last two pesticides need atleast 14 days of safe-period before silkworm consumption. Release of *Micraspis discolor* to mulberry garden (single female can feed 4462 eggs or 1835 nymphs) also minimizes the attack of thrips. Bihar hairy caterpillar infestation is effectively controlled by the spray of 1% aqueous soap solution or 0.1% Dichlorovos 76 EC (1.5 ml/liter). Sucking pest Whitefly can be controlled by using mechanical, chemical and biological ways. Among these, installation of yellow sticky traps at 150 nos/ha; size: 24" x 12", after 15 days of pruning, Spray of 1.5% neem oil (1500 ppm; 1.5 ml/litre of water), 0.1% Dimethoate 30EC (3.3 ml/litre) or 0.015% Thiamethoxam 25WG (5 g/litre) after 15–25 days of pruning and releasing of *Brumoides sutularis* ca 1500 pairs/ha are most effective.

According to Reed (1976), insects causing most damage to mulberry are: *Phenococcus hirsutus* (sucks sap of stem, leaf or petiole, causing severe curling and crinkling of leaves and swelling and twisting of apical regions), *Pseudodendro thripsornatissima* (thrips), and white grubs and termites. There are also other insects: *Hyphantrya cunea* Drury, *Eulecanium corni* Bouche, *Tetranycusurticrie* Koch.

The major diseases that cause more than 15–20% crop loss are powdery mildew caused by *Phyllactinia corylea*, leaf spot caused by *Cercospora moricola* (in south India) or *Myrothecium roridum* (in eastern and north eastern regions) and leaf rust caused by *Aecidium mori* (Teotia and Sen 1994). Spraying 0.2% Carathian or 0.2% Bavistin on the lower surface of the leaf will check the spread of diseases. A few soil borne diseases were also reported to cause major crop loss. Significant among them are root rot caused by *Fusarium solani* and *Botryodiplodia theobromae* and root knot caused by *Meloidogyne incognita*. The root rot can be effectively controlled by the soil application of Mancozeb 75% WP (10 g/plant). On the other hand, it is reported that soil application of 1000 kg neem oil cake/ha/year in 4 split doses after proper digging and weeding or 0.2% carbofuran followed by application of same concentration of salicylic acid control the root knot (Naik and Sharma 2007). Some other diseases like stem canker, cutting rot, collar rot and dieback also affects the mulberry.

Since continuous use of chemicals pose hazards to the environment, efforts have been made to use biological agents and botanical formulation to control the pests of mulberry in India (Tables 5 and 6). The predators and parasitoids are mass multiplied and released in the field as and when pests are noticed. The botanical formulations in Table 6 are prepared by soaking specified quantities of plant materials in fresh cow urine and mixed with the

Table 5. Important predators and parasitoids used to control mulberry pests (Source: Sakthivel et al. 2014).

Sl. No.	Name of the insect pest	Name of the biocontrol agent	Numbers to be released/ acre/crop
1.	Pink mealybug *Maconellicoccushirsutus*	Predators A) *Cryptolaemusmontrouzieri* B) *Scymnuscoccivora*	250 adults 500 adults
2.	Papaya mealy bug *Paracoccus marginatus*	Parasitoids A) *Acerophaguspapayae* B) *Pseudleptomastixmexicana* C) *Anagyrusloecki*	100–250 adults 50–100 adults 100–200 adults
3.	Thrips *Pseudodendrothripsmori*	Predator *Chrysoperla* spp.	4000–8000 eggs
4.	Spiralling whitefly *Aleurodicusdispersus*	Predators A) *Axinoscymnusputtarudriahi* B) *Scymnuscoccivora*	250 adults 250 adults
5.	Leaf webber *Diaphaniapulverulentalis*	Parasitoids A) *Trichogrammachilonis*- egg B) *Braconbrevicomis*- larval C) *Tetrastichushowardii*- pupal	3 cc of eggs 200 adults 1 lakh adults in 3 splits

Table 6. Important botanical formulation developed for controlling major diseases in mulberry (Source: Sakthivel et al. 2014).

Name of the plants	Parts	Quantity (g)		
		A. dispersus	*P. marginatus*	*P. mori*
Azadirachtaindica (L.)	Leaf	500	500	--
Aloe vera (L.)	Leaf	500	500	--
Cassia auriculata (L.)	Leaf	500	500	--
Zingiberofficinale (Roscoe)	Rhizome	50	50	50
Curcuma longa (L.)	Rhizome	50	50	50
Allium sativum (L)	Bulb	50	50	50
Capsicum annuum (L.)	Green Chilly	50	50	50
Acoruscalamus (L.)	Rhizome	50	50	50
Ocimum sanctum	Leaf	--	500	--
Neriumindicum	Leaf	--	500	--
Calotropisgigantea L.	Leaf	--	--	500
Annona squamosa L.	Leaf	--	--	500
Leucasaspera Willd.	Leaf	--	--	500
Cow urine		4.51	6.51	6.51

paste of *Zingiber officinale, Curcuma longa, Alliumsativum, Capsicum annuum* and *Acorus calamus*. The mixture was allowed to ferment for 13 days, filtered and collected in a separate container. With the botanical formulations it is possible to control spiralling whitefly (*Aleurodicus disperses*), papaya mealy bug (*Paracoccus marginatus*) and mulberry thrips (*Pseudodendrothrips mori*) and other pests, effectively.

In Romania

In European countries especially in Romania, the mulberry bacteriosis is caused by *Pseudomonas moris*. This bacteria has the optimal development conditions between 9–15°C and 95–100% relative humidity. In aerial parts, leaves infected by the bacteria initially appear with small yellow spots, which later develop into large, brown spots, while on the stem, brown spots appear and develop subsequently into deep, cancerous lesions. The control measures include cutting and burning of the affected parts and application of chemical bacteriocides (Pătruică 2007). Anthracnose, the Brown stain of leaves, is another disease of mulberry caused by *Mycosphaerella mori,* which appears on the leaf surface as brown corner spots with a reddish edge. It can be controlled by burning and burial of infected leaves. Mulberry Mildew caused by *Uncinula mori* occurs very rarely in Romania. The pathogen affects the leaves leading to development of large spots. The control measures include cutting and burning of the affected plant parts and application of chemical mixture Bordeaux broth at 2%.

In Romania, the most common pest found in mulberry is Hairy caterpillar *Hyphantriacunea*, with two generations occurring during May–June and July–August, respectively. The pest can be mechanically controlled by destroying the nests of caterpillars followed by pesticide application.

Economic Significance of Mulberry

Mulberry is a multipurpose tree providing good sources of income and almost all parts of mulberry have high economic potential. Mulberry leaf is the only natural feed available for rearing the silk producing insect *Bombyx mori* L. (Vijayan et al. 2018). If proper care is taken, the mulberry establishes well during the first six months as in most of areas the planting season will be either just prior to the rainy season or immediately after the rainy season. The leaf harvesting can be started after one year of establishment, the good plantation yield can be obtained consistently for 15 to 20 years without a much variation in leaf yield. Based on the type of plantation and the variety, leaf yield under irrigation condition varies from 40–60 Mt/ha/yr in India and from 54 to 95 MT/ha/yr in Romania/Europe (Table 7a & 7b).

Besides supporting sericulture industry and other products of economic value the cost of mulberry cultivation including initial plantation and maintenance has been worked out by all major mulberry producing countries including India (Tables 8 & 9). It is pertinent to note that the silk–mulberry ratio is approximately 1:120 kg and net return in mulberry sericulture from one acre is approximately USD 1100 with a cost benefit ratio of 1:1.50. During 2018 the world silk production was approximately 159648 MT and out of which more than 90% is coming from mulberry silkworm, the *Bombyx mori* L. Thus, it is obvious that a large areas of mulberry cultivation is taking place across the globe and mulberry is contributing significantly to the welfare of the man by generating considerable amount of employment and economy to the rural populations across the 60 countries where sericulture is being practiced. For instance, in India alone silk industry provides employment to nearly 8.0 million people and in China nearly 1.0 million (Anonymous 2019). Thus, mulberry is one of the most economically important crops in Asian country.

In addition to the use of leaf for silkworm rearing, it is also used as fodder for cattle like goats (Uribe and Sanchez 2001). Besides leaf, mulberry also produces several other important economic products, prominent among them being the fruit, which is highly delicious and contains number of health promoting compounds such as sugar, carbohydrate, alkaloids, vitamins, fats, minerals, aminoacids, carotenoids, flavonoids, antioxidents, etc. (Table 10; Asano et al. 2001, Ercisli and Orhan 2007, Hassimotto et al. 2007, Yang et al. 2010, Singhal et al. 2009, 2010, Yigit et al. 2010). Different plant parts of mulberry have been used as an herbal medicine due to its antihyperglycemic (Singab et al. 2005), antiallergic (Chai et al. 2005), antineoplastic (Chen et al. 2006), and immunomodulatory (Hou et al. 2011) activities. The genus *Morus* is known to be rich in flavonoids, including quercetin 3-(6-malonylglucoside), maclurin, rutin, isoquercitin (Katsube et al. 2006, Cui et al. 2019) and non-flavonoid polyphenols Stilbenes including resveratrol, oxyresveratrol and mulberroside A (Wang et al. 2017). Root

Table 7a. Popular mulberry varieties being cultivated in India* (Source: Vijayan et al. (2018)).

Sl. No.	Variety	Leaf Yield potential (Mt/ha/yr)	Region and cultural conditions
1	G4	55–60	South India, irrigated
2	C2038	50–55	Eastern and NE India Irrigated
3	Tr23	15–20	Hills of Eastern India, Rainfed
4	Victory-1	55–60	South India, Irrigated
5	Vishala	55–60	Pan India, Irrigated
6	Anantha	50–55	South India, Irrigated
7	DD	50–55	South India, Irrigated
8	S-13	12–15	South India, Rainfed
9	S-34	12–15	South India, Rainfed
10	S-1	35–40	Eastern and NE India, Irrigated
11	S-7999	35–40	Eastern and NE India, Irrigated
12	S-1635	40–45	Eastern and NE India, Irrigated
13	S-36	40–45	South India Irrigated
14	S-146	14–16	Hilly regions of North and North-Western India, Rainfed
15	Tr-10	11–15	Hills of Eastern India, Rainfed
16	BC-$_2$59	11–15	Hills of Eastern India, Rainfed
17	ChakMajra	25–30	Sub-temperate regions of North India, Rainfed
18	China White	25–30	Temperate regions of North & North Western India, Rainfed

* Based on the report of CSB, Bengaluru.

Table 7b. Popular mulberry varieties being cultivated in Romania (Craiciu 1972).

Sl. No.	Variety	Leaf Yield potential (Mt/ha/yr)	Special characteristics
	Land races		
1	Lugoj	65	High Resistance to frost
2	Orşova	64	Resistant to frost and bacteriosis
3	Calafat	62	Resistant to frost and drought
4	Galicea	76	High Resistance to frost, drought and bacteriosis
5	Eforie	68	Frost resistance
6	Basarabi	60	Resistant to frost and drought
7	Comun mulberry		Frost resistance
	Improved Varieties		
8	China 3	58	Resistance to frost
9	Ukraine 107	82	Resistant to frost and bacteriosis
10	Tbilisuri	80	Resistant to frost and diseases
11	Bulgaria 59	65	Low resistance to frost, High resistance to drought
12	Kokuso 21	95	Low frost resistance
13	Ichinose	78	Low frost resistance
14	China Hibrid	54	Resistance to frost

Table 8. The cost of establishment of one acre of mulberry garden in India.*

Sl. No.	Particulars	USD
1	Tractor tilling (4 hr), Harrowing (2 hr)	42
2	Final land preparation (Bullock power)	33
3	Farm yard manure (MT)	167
4	Saplings	125
5	Making trenches with tractor (hr)	28
6	Planting (MD)	55
7	Fertilizer	56
8	Fertilizer application charges (MD)	6
9	Irrigation (MD)	28
10	Hoeing/Weeding - 3 times (MD)	83
11	Miscellaneous expenditure	7
	Total establishment cost	630

* Based on the technology descriptor of CSR&TI, Mysuru Central Silk Board.

bark contains phenolics. Further, mulberry root and bark contains other secondary metabolites like morin, tannins, phlobaphens, phytosterol, ceryl alcohol, calcium malate, fatty acids and phosphoric acid (Wu et al. 2010, Chang et al. 2011). Two useful medicines for diabetics namely deoxynojirimycin (DNJ) and fagomin are reported to be present in the root barks and leaves (Hansawasdi and Kawabata 2006, Hao et al. 2018). Its wood also has good use for manufacturing of sports articles and construction of buildings, which is comparable to teakwood in shock resistance, strength and hardness (Sharma et al. 2013). It is also used as an ideal tree species for landscaping due to

Table 9. Maintenance cost of one acre of irrigated mulberry garden in India.*

Sl. No.	Particulars	USD
1	Farm yard manure	133
2	Fertilizer	185
3	Manure and fertilizer application	69
4	Irrigation water	104
5	Irrigation	55
6	Inter-cultivation	55
7	Inter-cultivation	83
8	Shoot harvest	417
9	Pruning and cleaning of plants	14
10	Plant protection Chemicals, Micro nutrients, Growth promoters, etc.	21
Total cost for leaf production in one acre		1136

* Based on the technology descriptor of CSR&TI, Mysuru Central Silk Board.

Table 10. Chemical composition of mulberry fruit (Source: Singhal et al. 2009).

Sl. No.	Chemical constituents	Quantity
1	Carbohydrates	7.8–9.0%
2	Protein	0.5–1.4%
3	Fatty acids (linoleic, stearic and oleic acids in seeds	0.3–0.5%
4	Free acid (mainly malic acid)	1.1–1.8%
5	Fiber	0.9–1.3%
6	Ash	0.8–1.0%
7	Moisture	85–88%
8	Calcium	0.17–0.39%
9	Potassium	1.00–1.49%
10	Magnesium	0.09–0.10%
11	Sodium	0.01–0.02%
12	Phosphorus	0.18–0.21%
13	Sulphur	0.05–0.06%
14	Iron	0.17–0.19%
15	Carotene	0.16–0.17%
16	Ascorbic acid	11.0–12.5 mg/100 g
17	Nicotinic acid	0.7–0.8 mg/100 g
18	Thiamine	7.0–9.0 µg/100 g
19	Riboflavin	165–179 µg/100 g

its excellent features in tree form (Tipton 1994). Since mulberry has good resistance to drought and flood, it is used for water conservation and prevention of soil erosion by raising plantations along riversides, field edges, slopes, roadsides, public parks and other recreation places (Jian et al. 2012).

Mulberry Genetic Resources and their Conservation*

Mulberry genetic resources are very precious materials and countries across the world have acquired a good amount of mulberry genetic resources. China, being the leading silk producer in the world, maintains more than 1860 germplasm accessions in various provinces like Zhejiang, Jiangsu, Guangdong, Guangxi, Shandong, Sichun, Anhui, Hubei, Hunan, Hebei, Shanxi, Shuanxi and Xinjiang (Pan 2000). Japan has more than 1375 germplasm accessions and India maintains (1346 accessions) (Machii et al. 1999). Likewise, countries like Bulgaria, has 140 mulberry accessions at SES-Vratza (Tzenov 2002) and Romania 45 mulberry tree varieties. Some of the best biological materials for import are: kokuso 21, Ichinose from Japan, Hu San 3 from China and 107 from Ukraine. Most productive local varieties are: Galicea, Calafat, Lugoj, and Eforie due to low-temperature (frost) resistance, Korea maintain 664 accessions. These genetic resources mainly consist of traditional and modern varieties, elite lines, wild relative species, which are collected from the native habitats during extensive exploratory surveys on distribution, pattern of genetic variability, presence of new alleles etc, and also through exchange of materials between research institutes and countries. Efforts have also been made to conserve these genetic resources without being subjected to genetic erosion.

Among the different techniques of plant genetic conservation, seeds or pollen preservation is the most economical and easiest method for plant genetic resources because of the less requirement of space and resources. However, mulberry being a highly heterozygous, anemophilous, cross breeding plant, conservation of the genetic resources through vegetative propagules (stem or buds) is preferred to (sexually propagated) seeds (Vijayan et al. 2011b). Thus, the conservation strategy for mulberry genetic resources adopted are those suitable for clonally propagated (micropropagated) plants, which include conserving the genetic resources either in the original home (*in situ*), or *ex situ* preservation in laboratories under *in vitro* conditions such as tissue culture or cryopreservation or as DNA in DNA banks. Considering the merits and demerits of these techniques (Table 11), the most commonly practiced preservation method is *ex situ* conservation (Tikader et al. 2009). In this process mulberry which is also a vegetatively propagated crop can be preserved besides *in vitro* in the form of whole plants in the "field gene bank" at well protected areas like research institutes, botanical garden, national parks, etc. (Nino 1995). For *ex situ* conservation plant samples have to be collected and brought to the place of conservation. While collecting the samples care has to be taken to note the name of the location, details of the geographic features and environment conditions and important morphological features expressed in the habitat. The collected genetic resources are then brought to the germplasm station and subjected to systematic characterization, evaluation and documentation. Characterization is the initial process of the assessment wherein recording of morphological, anatomical, reproductive, biochemical features along with response to pest and diseases is made following standard methods (Tikader et al. 2014). The genetic resources in *ex situ* gene bank are generally maintained in two ways such as an active collection meant for research and distribution to other Institutes and a base collection for long-term preservation. In general mulberry accessions in germplasm are maintained as a dwarf tree with a spacing of 2.4 × 2.4 m between plants with a crown height of 1.5 m following recommended cultural practices of the region.

Conservation of mulberry through tissue culture and cryopreservation is also being practiced to a certain extent. These methods offer several advantages as in cryopreservation, mulberry germplasm is maintained at an ultra-low temperature where the cell division and metabolic activities remain suspended, thus the materials remain preserved for long periods with greater genetic stability. Another advantage it enjoys is the minimum requirement of space and maintenance cost (Niino 1995). Owing to these obvious advantages, cryopreservation of mulberry genetic resources has been practiced by many countries. In India, 908 accessions of *M. indica, M. alba, M. latfolia, M. cathayana, M. laevigata, M. nigra, M. australis, M. bombycis, M. sinensis, M. multicaulis* and *M. rotundioba* are being maintained through cryopreservation (Ananda Rao et al. 2007). The general procedure for mulberry cryopreservation is to treat the shoot segments at –3°C for 10 days, –5°C for three days, –10°C for 1 day and –20°C for one day and immerse it in liquid nitrogen. The winter buds in this way can also be stored in liquid nitrogen upto 3–5 years without any significant genetic changes.

---

* For more details see Chapter 17.

Table 11. Merits and demerits of different methods used for conserving the genetic resources of mulberry.* Source: Vijayan et al. (2011b).

In situ	*Ex situ*	*In vitro*	DNA banking
Apt for forest species and wild crop relatives	Only option for the asexually reproducing plants	Suitable for both sexually and asexually reproducing plants	Suitable for both sexually and asexually reproducing plants
Field oriented, laborious and expensive	Field oriented, laborious and expensive	Laboratory oriented minimum space and less laborious	Laboratory oriented minimum space and less laborious
Allows evolution to continue	Evolution restricted	No chance of evolution	No chance of evolution
Increases genetic diversity.	Less prone to genetic variability	No genetic variation	No genetic variation
Vulnerable to disease and other natural calamities	Vulnerable to disease and other natural calamities	Well protected against disease and other natural calamities	Well protected against disease and other natural calamities
Strengthens the link between conservationists and local people who traditionally maintain the plant	Minimum interactions	No interactions	No interactions
Exchange of materials is difficult	Exchange of materials possible but needs extra care	Easy exchange of materials	Easy exchange of materials

* Based on the reports of CSGRC, Hosur, Central Silk Board.

Thus, various procedures are available for conserving the precious mulberry genetic resources essential for continual supply of economic parts of mulberry species particularly to silk, pharmaceutical and other industries.

Conclusion

Mulberry is one of the economically important crops being cultivated across the world, particularly in Asian countries like China, India, Japan, and some parts of Europe, which serves the mankind in many different ways. Apart from sericulture, many pharmaceuticals and nutraceuticals have also been prepared from mulberry leaf, fruits, root and barks. Considering the hardy nature and sufficient genetic plasticity of the plant, it can be cultivated in a wide range of climatic conditions. The input-output ratio of mulberry cultivation is such that farmers have less opportunity to doubt on the sustainability of the crop. Understanding this great economic potential, countries across the world have been collecting, as well as conserving huge amount of mulberry genetic resources. These genetic resources need to be utilized to the maximum extent, for which lack of adequate information on the genetic makeup of the species remains as the biggest obstacle. The high heterozygosity, out breeding nature, easy interspecific hybridization pose problems for species identification in mulberry for better utilization in breeding. Thus, it is high time that an amalgamation of both conventional and modern techniques is to be used to elucidate the genetics of this very important crop.

References

Ananda Rao, A., R. Chaudhury, S. Kumar, D. Velu, R.P. Saraswat and C.K. Kamble. 2007. Cryopreservation of mulberry germplasm core collection and assessment of genetic stability through ISSR markers. Intl. J. Indust. Entomol. 15: 23–33.

Anonymous. 2019. Websites of International sericulture Commission, Bangalore, India (https://inserco.org/en/statistics).

Asano, N., T. Yamashita, K. Yasuda, K. Ikeda, H. Kizu, Y. Kameda, A. Kato, R.J. Nash, H.S. Lee and S.R. Kang. 2001. Polyhydroxylated alkaloids isolated from mulberry trees (*Morus alba* L.) and silkworms (*Bombyx mori* L.). J. Agri. Food Chem. 49: 4208–4213.

Azinzan, D. and A. Sonboli. 2001. Chromosome counts for five species of Moraceae. Iranian J. Bot. 9: 103–106.

Barber, E.J.W. 1991. Prehistoric textiles: The development of cloth in the Neolithic and bronze ages with special reference to the Aegean. Princeton University Press, USA.

Basavaiah, R., S.B. Dandin and M.V. Rajan. 1989. Microsporogenesis in hexaploid *Morus serrata* Roxb. Cytologia 54: 747–751.

Berg, C.C. 2001. Moreae, Artocarpeae, and Dorstenia (Moraceae). Flora Neotropica. Monogr 83, New York Botanical Garden, Bronx, New York, USA, pp. 24–32.

Chai, O.H., M.S. Lee, E.H. Han, H.T. Kim and C.H. Song. 2005. Inhibitory effects of *Morus alba* on compound 48/80 induced anaphylactic reaction and anti-chiken gamma globulin IgE-mediated mast cell activation. Biol. Pharma Bull. 28: 1852–1858.

Chang, L.W., L.J. Juang, B.S. Wang, M.Y. Wang, H.M. Tai, W.J. Hung, Y.J. Chen and M.H. Huang. 2011. Antioxidant and antityrosinase activity of mulberry (*Morus alba* L.) twigs and root bark. Food Chem. Toxicol. 49: 785–90.

Chase, M.W. et al. 2009. An update of the Angiosperm Phylogeny Group classification for the orders and families of flowering plants: APG III. Bot. J. Linn. Soc. 161: 105–121.

Chen, P.N., S.C. Chu, H.L. Chiou, W.H. Kuo, C.L. Chiang and Y.S. Hsieh. 2006. Mulberry anthocyanins, cynidin 3-rutinoside and cynidin 3-glucoside exhibited an inhibitory effect on the migration and invasion of a human lung cancer cell line. Cancer Lett. 235: 248–259.

Craiciu, E. 1972, Tehnologiaplantatiilor de dud, Ed. Revistelor Agricole, Academia de Ştiinţe Agricolesi Silvice, Bucuresti, România, 138–156 p.

Cui, H., T. Lu, M. Wang, X. Zou, Y. Zhang, X. Yang and H. Zhou. 2019. Flavonoids from *Morus alba* L. leaves: Optimization of extraction by response surface methodology and comprehensive evaluation of their antioxidant, antimicrobial, and inhibition of α-amylase activities through analytical hierarchy process. Molecules 24(13): 2398.

Das, B.C. 1961. Cytological studies on two *Morus indica* L. and *Morus laevigata* Wall. Caryologia 14: 159–162.

Datta, R.K. 2000. Mulberry cultivation and utilization in India. FAO Electronic conference on mulberry for animal production (*Morus* L.) http://www.fao.org/DOCREP/005/ X9895E/x9895e04.htm#TopOfPage Accessed 15 July 2017.

Dutta, M. 1954. Cytogenetical studies on two species of *Morus*. Cytologia 19: 86–95.

Ercisli, S. and E. Orhan. 2007. Chemical composition of white (*Morus alba*), red (*Morus rubra*) and black (*Morus nigra*) mulberry fruits. Food Chem. 103: 1380–1384.

FAO. 1990. Sericulture Training Manual. Food and Agriculture Organization (FAO) Agricultural Services Bulletin 80, Rome, Italy, 117 p.

Hansawasdi, C. and J. Kawabata. 2006. Alpha-glucosidase inhibitory effect of mulberry (*Morus alba*) leaves on Caco-2. Fitoterapia 6: 77: 568–73.

Hao, J.Y., Y. Wan, X.H. Yao, W.G. Zhao, R.Z. Hu, C. Chen, L. Li, D.Y. Zhang and G.H. Wu. 2018. Effect of different planting areas on the chemical compositions and hypoglycemic and antioxidant activities of mulberry leaf extracts in Southern China. PLoS ONE 13: e0198072 doi.org/10.1371/journal.pone.0198072.

Hassimotto, N.M.A., M.I. Genovese and F.M. Lajolo. 2007. Identification and characterization of anthocyanins from wild mulberry (*Morus nigra* L.) growing in Brazil. Food Sci. Technol. Int. 13: 17–25.

He, N., C. Zhang, X. Qi, S. Zhao, Y. Tao, G. Yang and M. Lu. 2013. Draft genome sequence of the mulberry tree *Morus notabilis*. Nat. Commun. 4: 2445.

Hou, R.H., S.T. Liao, F. Liu, Y.X. Zou and Y.Y. Deng. 2011. Immunomodulatory effect of polysaccharides from mulberry leaves (PML) in mice. Journal of Food Science 32(13): 280–283.

Jian, Q., H. Ningjja, W. Yong and X. Zhonghuai. 2012. Ecological issues of mulberry and sustainable development. Journal of Resources and Ecology 3: 330–339.

Jolly, M.S. 1987. Appropriate sericulture techniques, ICTRETS, Mysore.

Katsube, T., T.N. Imawaka, Y. Kawano, Y. Yamazaki, K. Shiwaku K. and Y. Yanane. 2006. Antioxdant flavonol glycosides in *Morus alba* L. leaves isolated based on LDL antioxidant activity. Food Chem. 97: 25–31.

Krishnaswamy, S. 1986. Mulberry cultivation in South India, Omkar, Offset Printers, Bangalore-560 002. India. pp. 1–20.

Le Houerou, H.N. 1980. The role of browse in management of natural grazing lands. *In*: Le Houerou, H.N. (ed.). Browse in Africa. The current state of knowledge: Addis Ababa, Ethiopia, ILCA. pp. 355.

Machii, H. and A. Koyama and H. Yamanouchi. 1999. Fruit traits of genetic mulberry resources. J. Seric Sci. Japan 68(2): 145–155.

Maode, Y. and X. Zhonghuai, F. Lichun, K. Yifu, Z. Xiaoyong and J. Chengjun. 1996. The discovery and study on a natural haploid *Morus notabilis* Schneid. Acta Sericol. Sin. 22: 67–71.

Naik, N.V. and D.D. Sharma. 2007. Efficacy of Pesticides and growth hormones against root disease complex of mulberry (*Morus alba* L.). Int. J. Indust. Entomol. 15(2): 101–106.

Nepal, M.P. and C.J. Ferguson. 2012. Phylogenetics of Morus (Moraceae) inferred from ITS and trnL-tmF sequence data. Systematic Bot. 37: 442–450.

Niino, T. 1995. Cryopreservation of germplasm of mulberry (*Morus* spp.). pp. 102–113. *In*: Bajaj, Y.P.S. (ed.). Biotechnology in Agriculture and Forestry. Springer-Verlag, Berlin.

Pan, Y.L. 2000. Progress and prospect of germplasm resources and breeding of mulberry. ActaSeric Sin. 26: 1–8.

Pătruică, S. 2007. Tehnologia Cresteriiviermilor de mătase, Curs Universitar, EdituraEurobit, Timisoara, România, ISBN 978-973-620-332-9.

Rathore, M.S. and Y. Srinivasulu, R. Kour, G.M. Darzi, Anil Dhar and M.A. Khan. 2011. Integrated soil nutrient management in mulberry under temperate conditions. European Journal of Biological Sciences 3: 105–111.

Reed, C.F. 1976. Information summaries on 1000 economic plants. Typescripts submitted to the USDA.

Rodríguez, C., R. Arias and J. Quiñones. 1994. Efecto de la frecuencia de poda y el nivel de fertilizaciónnitrogenada, sobre el rendimiento y calidad de la biomasa de morera (Morus spp.) en el trópicoseco de Guatemala. En: Benavides, J.E. Arboles y arbustosforrajerosen América Central. CATIE, Turrilaba, Costa Rica. p. 515–529.

Sakthivel, N., J. Ravikumar, Chikknna, M.V. Kirsur, B.B. Bindroo and V. Sivaprasad. 2014. Organic farming in mulberry: Recent Breakthrough. Regional Sericultural Research Station, Central Silk Board, Ministry of Textiles, Govt. of India, Allikkuttai Post, Salem - 636 003, Tamil Nadu.

Sharma, V., S. Chand and P. Singh. 2013. Mulberry: A most common and multi-therapeutic plant. International Journal of Advanced Research 1(5): 375–378.

Singab, A.N., H.A. El-Beshbishy, M. Yonekawa, T. Nomura and T. Fukai. 2005. Hypogycemic effect of Eygiptian *Morus alba* root bark extract: effect in diabetes and lipid peroxidation of streptozotocin induced diabetic rats. J. Ethnopharmacol. 100: 333–338.

Singhal, B.K. and A. Dhar, M.A. Khan, B.B. Bindroo and R.K. Fotedar. 2009. Potential economic additions by mulberry fruits in sericulture industry. Plant Horti. Tech 9: 47–51.

Singhal, B.K., M.A. Khan, A. Dhar, F.M. Baqual and B.B. Bindroo. 2010. Approaches to industrial exploitation of mulberry (*Morus* spp.) fruits. J. Fruit Ornament Plant Res. 18(1): 83–99.

Teotia, R.S. and S.K. Sen 1994. Mulberry diseases in India and their control. Sericologia 34: 1–18.

Tikader, A., M. Shamsudin, K. Vijayan and T. Pavankumar. 1995. Survival potential in different varieties of mulberry (*Morus* spp.). Indian J. Agricult. Sci. 65: 33–35.

Tikader, A., K. Vijayan, B.N. Roy and T. Pavankumar. 1996. Studies on propagation efficiency of mulberry (*Morus* spp.) at ploidy level. Sericologia 36: 345–349.

Tikader, A. and S.B. Dandin. 2007. Pre-breeding efforts to utilize two wild *Morus* species. Curr. Sci. 92: 1072–1076.

Tikader, A., K. Vijayan and C.K. Kamble. 2009. Conservation and management of mulberry germplasm through biomolecular approaches-a review. Biotechnol. Molec. Biol. Rev. 3: 92–104.

Tikader, A., B. Saratchandra, K. Vijayan and R.N. Singh. 2014. Mulberry Germplasm management and utilization. APH Publishing Corporation. New Delhi, India.

Tipton, J. 1994. Relative drought resistance among selected southwestern landscape plants. J. Arboric. 20: 151–155.

Tutin, G.T. 1996. *Morus* L. *In*: Tutin, G.T., N.A. Burges, A.O. Chater, J.R. Edmondson, V.H. Heywood, D.M. Moore, D.H. Valentine, S.M. Walters, D.A. Webb (eds.). Flora Europa, Psilotaceae to Platanaceae, 2nd ed., vol. 1. Cambridge University Press, Australia.

Tzenov, P.I. 2002. Conservation status of mulberry germplasm resources in Bulgaria. Paper contributed to Expert Consultation on Promotion of Global Exchange of Sericulture Germplasm Satellite Session of 19th th ISC Congress, 21–25 Sept. Bangkok, Thailand: http://www.fao.org/DOCREP/ 005/AD107E/ ad107e01.htm.

Uribe, T.F. and M.D. Sanchez. 2001. Mulberry for animal production: animal production and health paper. Rome: FAO, pp. 199–202.

Vijayan, K., P.P. Srivastava, A.K. Awasthi. 2004. Analysis of phylogenetic relationship among five mulberry (*Morus*) species using molecular markers. Genome 47: 439–448.

Vijayan, K. 2010. The emerging role of genomics tools in mulberry (*Morus*) genetic improvement. Tree Genetics and Genomes 6: 613–625.

Vijayan, K., A. Tikader, Z. Weiguo, C.V. Nair, S. Ercisli and C.H. Tsou. 2011a. Mulberry (*Morus* L.). pp. 75–95. *In*: Cole, C. (ed.). Wild Crop Relatives: Genomic & Breeding Resources, Springer, Heidelberg.

Vijayan, K., B. Saratchandra and J.A.T. da Silva. 2011b. Germplasm conservation in mulberry. Scientia Horticulturae 128: 371–379.

Vijayan, K., G. Ravikumar and A. Tikader. 2018. Breeding for higher fruit production in mulberry. *In*: Al-KHayri, J.M., Mohan Jain and D.V. Johnson (eds.). Advances in Plant Breeding Strategies, Vol 3: Fruits. Springer, Cham. https://doi.org/10.1007/978-3-319-91944-7_3.

Vural, U., H. Dumanoglu and V. Erdogan. 2008. Effect of grafting/budding techniques and time on propagation of black mulberry (*Morus nigra* l.) in cold temperate zones. Propagation of Ornamental Plants 8: 55–58.

Wang, C., S. Zhi, C. Liu, F. Xu, A.C. Zhao, X. Wang, Y. Ren, Z. Li. and M. Yu. 2017. J. Agric. Food Chem. DOI: 10.1021/acs.jasc.6b05212.

Wang, Z.Y. and S.D. Tanksley. 1989. Restriction fragment length polymorphism in *Oryza sativa* L. Genome 32: 1113–111.

Wu, D., X. Zhang, X. Huang, X. He, G. Wang and W. Ye. 2010. Chemical constituents from root barks of *Morus atropurpurea*. Zhongguo Zhong Yao ZaZhi 35: 1978–1982.

Yang, X., L. Yang and H. Zheng. 2010. Hypolipidemic and antioxidant effects of mulberry (*Morus alba* L.) fruit in hyperlipidaemia rats. Food Chem. Toxicol. 48: 2374–2379.

Yigit, D., F. Akar, E. Baydas and M. Buyukyildiz. 2010. Elemental composition of various mulberry species. Asian J. Chem. 22(5): 3554–3560.

Yokoyama, T. 1962. Synthesized science of sericulture. Japan, pp. 39–46.

Zeng, Q., H. Chen, C. Zhang, M. Han, T. Li, X. Qi, Z. Xiang and N. He. 2015. Definition of eight mulberry species in the genus *Morus* by internal transcribed spacer-based phylogeny. PLoS ONE10: e0135411. DOI:10.1371/journal.pone.0135411.

CHAPTER 4

Medicinal Uses of Mulberry

Danica Dimitrijević, Biljana Arsić* and Danijela Kostić

INTRODUCTION

Genus *Morus* belongs to Moraceae family and is globally distributed under varied climatic conditions, ranging from tropical, subtropical to temperate (Yuan et al. 2015, Pel et al. 2017). There are 10–16 species, commonly called mulberry, generally accepted by the majority of botanical authorities (Dimitrijević et al. 2013), but the most common are white mulberry (*Morus alba* L.), black mulberry (*Morus nigra* L.) and red mulberry (*Morus rubra* L.). The white mulberry has the origin in eastern and central China and it became naturalized in Europe centuries ago. The red or American mulberry comes from the United States spreading from Massachusetts to Kansas and then to the Gulf coast. The black mulberry is characteristic for western Asia and then it was transferred to Europe before Roman times (Sass-Kiss et al. 2005, Özgen et al. 2009, Jiang and Nie 2015). All mulberry species are deciduous trees, rapidly growing, small to medium size, or about 10–20 m tall. The bark is gray and thick, with many irregular longitudinal cracks. The taxonomic hierarchy of mulberry in the plant kingdom is given in Table 1.

Table 1. Taxonomic hierarchy of mulberry.

Kingdom	Plantae
Subkingdom	Tracheobionta
Superdivision	Spermatophyta
Division	Magnoliophyta
Class	Magnoliopsida
Subclass	Hamamelididae
Order	Urticales
Family	Moraceae
Genus	*Morus*
Species	*Morus alba* L., *Morus rubra* L., *Morus nigra* L., etc.

University of Niš, Faculty of Sciences and Mathematics, Višegradska 33, 18000 Niš, Serbia.
Emails: danicadimitrijevic7@gmail.com; danijelaaakostic@yahoo.com
* Corresponding author: ba432@ymail.com

In China, Korea, and Japan, mulberry fruit is used for folk medicine for its pharmacological effects, including fever reduction, treatment of a sore throat, liver and kidney protection, eyesight improvement and ability to lower blood pressure (Chen et al. 2016, Zhou et al. 2017). The root of the mulberry tree is one of the constituents of a medicine that is used to treat high blood pressure. Root juice agglutinates blood and is very useful in killing worms from the digestive system. These traditional uses have given impetus for numerous pharmacological studies and assessment of chemical composition in order to confirm its ethnopharmacological use (Miljković et al. 2014, Jiang and Nie 2015, Sanchez-Salcedo et al. 2015).

Because of the high amount of protein in leaves white mulberry is very common and extensively farmed as ideal fodder for the nourishment of silkworms which are involved in industrial silk production. Leaves are also used as a foodstuff for cows and sheep in restricted areas.

Figure 1. Leaves and fruits of *Morus alba* L., *Morus nigra* L. (https://www.phillyorchards.org/2016/07/01/the-plucky-mulberry-morus/), and *Morus rubra* L.

Biochemical Components of Fruit

Morus spp. fruit contains abundant protein, lipid, carbohydrate, fiber, minerals, and vitamins which have low calories and can be a healthy food choice for consumers, e.g., 100 g of the fresh mulberry fruit can produce 1.44 g of protein. This is higher than that of strawberries and raspberries and comparable to that of blackberries (1.39 g/100 g). A total of 18 amino acids, including all nine essential amino acids required by humans, are found in mulberry fruit. The order of abundance of fatty acids in mulberry fruit is polyunsaturated fatty acids (PUFA) > monounsaturated fatty acids (MUFA) > saturated fatty acids (SFA). PUFA were the major fraction of fatty acids, representing at least 76.68%, which was determined to be higher than that of strawberries. Fourteen fatty acids from mulberry fruit with different compositions and contents were found. The fatty acid profile can be affected by variables including environmental conditions and genetic factors. In general, linoleic acid (C18:2), palmitic acid (C16:0), and oleic acid (C18:1) were the major fatty acids in mulberry fruit. The predominant fatty acid was linoleic acid (C18:2), which is necessary for human development, health promotion, and disease prevention, and it must be obtained from exogenous sources because humans cannot synthesize it.

Mulberry fruit also contains fatty acids which are important for cell membrane formation, proper development and functioning of nervous system, production of eicosanoids and many inflammatory responses (Ercisli and Orhan 2007). The sugar content in mulberry fruit increases during ripening. The principal sugars are glucose and fructose.

Mulberry fruit is an outstanding source of minerals, particularly potassium, followed by calcium and phosphorus. The metals like Zn, Mn, Fe, Pb, Ni, Cu and Cd present in soil, are found in leaves and mulberry fruits (*Morus alba* L. and *Morus nigra* L.). A positive correlation was determined between metal concentration in soil, and that of mulberry leaves and fruits (Randjelović et al. 2014). A high amount of ascorbic acid content at 36.4 mg/100 g of fresh weight (FW) is present in mulberry fruit. In addition, mulberry fruit also provides other vitamins, e.g., thiamin, riboflavin, niacin, folate, vitamin A, vitamin B-6, vitamin E, and vitamin K. These reported nutrients in the mulberry fruit are beneficial for human health (Yuan and Zhao 2017).

Mulberry leaves and fruits contain many bioactive components such as alkaloids, anthocyanins, and flavonoids (Vijayan et al. 2004, Hassimotto et al. 2008). Flavonoids commonly occur in plant kingdom and mulberry is found to contain at least four flavonoids including rutin, quercetin, myricetin, and kaempferol (Miljkovic et al. 2015). Natić and coworkers found that the most abundant phenolic compound is rutin, contributing 44.66% of the total phenolics in 11 mulberry samples (Natić et al. 2015). These findings were consistent with previous findings (Zhang et al. 2008, Gundogdu et al. 2011, Chan et al. 2015). Flavonoids have been recognized to possess anti-inflammatory, antioxidant, antiallergic, antithrombotic, hepatoprotective, antiviral and carcinogenic activities in human beings (Daimon et al. 2010, Miljković et al. 2018). Red and black mulberries are a rich source of phenolic compounds, including flavonols, phenolic acids, and anthocyanins. Anthocyanins are a major contributor to flower and fruit color ranging from red, blue to purple. They are also the best source of health benefit compounds which are antioxidant and anti-inflammatory (Nuengchamnong et al. 2007). Anthocyanins have a high inhibitory effect on lipid oxidation. Mulberry anthocyanin extract has the antimetastatic activity to inhibit migration of B16-F1 cells (Ozgen et al. 2009). The principal anthocyanin in mulberry fruit is cyanidin-3-*O*-glucoside (C3G), followed by cyanidin-3-*O*-rutinoside (C3R), and a small amount of pelargonidin-3-*O*-rutinoside (Liu et al. 2008a, Qin et al. 2010, Chang et al. 2013, Chen et al. 2016). White mulberry is rich with phenolic glycosides and prenylated flavonoids. Among flavanols glycosides, catechin, epigallocatechin gallate, epicatechin, procyanidin B1, and procyanidin B2 have been found in mulberry fruit (Natić et al. 2015). Prenylated flavonoids inhibit LDL (Low-Density Lipoprotein) oxidation and hence atherosclerosis (Butt et al. 2008). In mulberry leaves, alkaloid such as 1-deoxynojirimycin (DNJ) is the most potent glycosidase inhibitor that decreases blood-sugar levels (Kimura et al. 2007, Nakagawa et al. 2010). The detailed biochemical components of fruits of black, red and white mulberry are mentioned in Table 2.

Mulberry fruit contains a significant amount of polysaccharides (Liu and Lin 2012). A glycoprotein with a carbohydrate content (28.4%) and protein content (71.6%) isolated from the lyophilized powder of mulberry fruit juice (yield 10.6%) showed better antiapoptotic activity in comparison to strawberry fruit polysaccharides (Liu and Lin 2014). The mulberry fruit polysaccharide has a rhamnogalacturonan type I (RG I) backbone composed of the repeating disaccharide fragments [4-α-D-GalpA-1 → 2-α-L-Rhap-1 →]. The arabinan side chain is composed of (1 → 5)-α-L-Ara attached to the *O*-4 position of α-L-Rhap. The Arabinogalactan type II (AG II) side chain was found to have a (1 → 6)-β-D-galactan core branched at *O*-3 by α-L-Araf. The connection patterns between AG II and RG I remain unclear (Choi et al. 2016). Structure of polysaccharides remains to be further determined to provide more information for elucidating the structure-activity relationship (Yuan and Zhao 2017).

In general, there are inconsistencies in the polyphenolic composition as well as content in mulberry fruit which are due to a large number of factors, such as genotype of cultivars, extraction procedures, analytical methods, genetic differences, the degree of ripening, and environmental conditions like temperature, light (sun exposure), humidity, etc. (Mahmood et al. 2012, Butkhup et al. 2013, Natić et al. 2015, Sanchez-Salcedo et al. 2015).

Mineral Content

The human body requires both metal and non-metal elements within certain permissible limits for growth and good health. Therefore, the determination of elemental compositions in food and related products is essential for understanding their nutritive importance. Mulberry fruit is an outstanding source of minerals, particularly potassium ranging from 834 to 1731 mg/100 g fresh matter (Kostić et al. 2013a), which is used by the body to create energy necessary to power cells, followed by calcium, magnesium, sodium and phosphorus (Ercisli and Orhan 2007, Imran et al. 2010, Micić et al. 2013). Trace elements in some tested extracts of mulberry were in the range from 0.4 to 59.2 mg/100 g fresh matter, Fe and Zn being the most dominant elements and Cu the least dominant element in relation to all detected elements (Kostić et al. 2013a). Randjelović et al. (2014) used atomic absorption spectroscopy for determination of the concentration of Zn, Mn, Fe, Pb, Ni, Cu and Cd in the soil, leaves and mulberry fruits (*Morus alba* L., *Morus nigra* L.) from southeast Serbia. They found that the iron content was several times higher than that of other metals, ranging from 18569.7 to 27987.7 ppm (mg/kg) in the soil, 115.097 to 206.63 ppm in leaves, and 9.57 to 26.89 ppm in fruit. Zn concentrations ranged from 40.02–125.78 ppm in soil, 18.65–55.37 ppm in leaves, and 1.369–7.18 ppm in fruit. Copper ranges from 20.37 to 40.0 ppm in soil, 4.33–16.52 ppm in leaves, and 1.06–2.586 ppm in fruit. Cadmium was 0.502–0.907 ppm in soil, 0.00–0.225 ppm in leaves and in fruit it was below the detection limit. Nickel was detected in soil (6.351–12.54 ppm), but it was detected neither in leaves nor fruit (Randjelović et al. 2014).

According to PUF (plant uptake factors) which measures the efficiency of metal accumulation, Cd, Cu and Zn are more efficiently concentrated in leaves, while Mn and Fe less. Cu, Fe, and Zn accumulate more in fruit than in leaves. Fe is deposited mostly in fruit (Gardiner 1997, Randjelović et al. 2014).

Table 2. The biochemical components of mulberry fruit.

Class	Compound	Reference
Anthocyanins	Cyanidin-3-glucoside	Chang et al. 2013, Miljkovic et al. 2015, Chen et al. 2016
	Cyanidin-3-rutinoside	Chang et al. 2013, Chen et al. 2016
	Pelargonidin-3-glucoside	Huang et al. 2008, Liu et al. 2009
	Pelargonidin-3-rutinoside	Huang et al. 2008, Liu et al. 2009
	Cyanidin 3-O-(6″-O-α-rhamnopyranosyl-β-D-glucopyranoside)	Du et al. 2008
	Cyanidin 3-O-(6″-O-α-rhamnopyranosyl-β-D-galactopyranoside)	Du et al. 2008, Miljkovic et al. 2015
	Cyanidin 3-O-β-D-galactopyranoside	Du et al. 2008, Miljkovic et al. 2015
	Cyanidin 7-O-β-D-glucopyranoside	Du et al. 2008, Miljkovic et al. 2015
	Petunidin 3-O-β-glucopyranoside	Sheng et al. 2014
Flavonols	Rutin	Radojković et al. 2012, Miljkovic et al. 2015, Natić et al. 2015
	Quercetin	Radojković et al. 2012, Butkhup et al. 2013, Miljkovic et al. 2015, Natić et al. 2015
	Quercetin 3-O-rutinoside	Jin et al. 2015
	Quercetin 3-O-glucoside	Jin et al. 2015
	Quercetin 3-O-galactoside	Jin et al. 2015
	Myricetin	Butkhup et al. 2013, Miljkovic et al. 2015, Natić et al. 2015
	Kaempferol	Butkhup et al. 2013, Natić et al. 2015
	Kaempferol 3-O-glucoside	Jin et al. 2015
	Kaempferol 3-O-rutinoside	Sánchez-Salcedo et al. 2015
Flavanols	Catechin	Butkhup et al. 2013, Miljkovic et al. 2015, Natić et al. 2015
	Epigallocatechin Gallate	Natić et al. 2015
	Epicatechin	Butkhup et al. 2013, Natić et al. 2015
	Procyanidin B1	Butkhup et al. 2013, Natić et al. 2015
	Procyanidin B2	Butkhup et al. 2013, Natić et al. 2015
Hydroxy cinnamic acid	Chlorogenic acid	Mahmood et al. 2012, Miljkovic et al. 2015
	Ferulic acid	Radojković et al. 2012, Butkhup et al. 2013, Natić et al. 2015
	p-Coumaric acid	Natić et al. 2015
	o-Coumaric acid	Gundogdu et al. 2011
	Cinnamic acid	Butkhup et al. 2013, Natić et al. 2015
	Caffeic acid	Butkhup et al. 2013, Natić et al. 2015
Benzoic acid	Gallic acid	Radojković et al. 2012, Butkhup et al. 2013, Miljkovic et al. 2015, Natić et al. 2015
	p-Hydroxybenzoic acid	Natić et al. 2015
	Syringic acid	Gundogdu et al. 2011
	Protocatechuic acid	Natić et al. 2015
	Vanillic acid	Gundogdu et al. 2011

Antioxidant Potential

The potentially reactive derivatives of oxygen, ascribed as Reactive Oxygen Species (ROS) like singlet oxygen (O_2^-), superoxide anion radical ($O_2^{-\cdot}$), hydroxyl radical ($OH\cdot$), nitric oxide radical ($NO\cdot$), and alkyl peroxyl ($ROO\cdot$) are produced continuously in the cells either as the accidental by-products of metabolism or deliberately, for example, phagocytosis. Protective mechanisms mediate their harmful effects but sometimes ROS overrides the defense capabilities of the body resulting in oxidative damage to molecules and membranes. In such cases, there is a need to strengthen this mechanism by antioxidant supplementation (Arsic et al. 2016).

According to Andallu, mulberry plants contain many active compounds which act as an antioxidant like polyphenols, carotenoids, and vitamin A, C, E (Andallu et al. 2001). *Morus alba* L. extract contain flavonoids having antioxidant properties that scavenging free radicals protect many organs from oxidative stress (Seifried et al. 2007). White mulberry leaves water extract prepared at high temperature contains four important flavonols: quercetin-3-ß-D-glucose, quercetin-3-*O*-glucose-6-acetate, rutin, and quercetin (Wattanapitayakul et al. 2005, Kim and Jang 2011). It was reported that *Morus alba* L. leaf water extract has the highest antioxidant properties as evaluated using ferric reducing/antioxidant power assay (Wattanapitayakul et al. 2005, Kim and Jang 2011). By employing voltammetry techniques, the volatile substance present in *Morus nigra* L. mulberry tea was tested for antioxidant activity in chromatography assays and results showed that consumption of antioxidant beverages prepared from mulberry may be beneficial to human health (Ahmed and Shakeel 2012, Nam et al. 2012). Quercetin exhibited a significant radical scavenging effect on 1,1-diphenyl-2-picrylhydrazyl (DPPH). Absorption and bioavailability of quercetin in mulberry are most significant for antioxidant potential of the mulberry plant (Kim et al. 1999, Seifried et al. 2007, Butt et al. 2008, Lee et al. 2008). Black, white and red mulberry from the area of southeast Serbia has a high content of natural phenolic compounds. Red and black mulberry have a high content of anthocyanins and showed significant antioxidant activity (Kostic et al. 2013b). Dimitrijević and coworkers (2014) found that fresh white mulberry fruit from southeastern Serbia contained high levels of total phenols and flavonoids, more than 1000 mg/kg fresh fruit, while the extracts did not contain anthocyanins. The highest content of phenols was in the ethanolic extract and the highest content of flavonoids was in the methanolic extract. All investigated extracts showed high antioxidant activity. The highest activity against DPPH was observed in the methanolic extract. Figure 2 shows histograms of antioxidant activities of different types of mulberry extracts using DPPH method. Kostic et al. (2013b) found that the content of total phenols in the ethanol, ethanol/water and water extracts of black mulberry fruit from southeastern Serbia ranged from 90.26 mg GAE/100 g (Gallic Acid Equivalent/100 g) to 118.84 mg GAE/100 g. Water extract displayed the highest content of phenolic compounds. Total flavonoids were in the range 141.70 mg (water extract) to 183.90 mg CE/100 g (ethanol/extract) while anthocyanin content was between 114.83 mg cyanidin 3-glucoside/100 g for water extract and 128.68 mg cyanidin 3-*O*-glucoside/100 g for ethanol-water extract. All the extracts exhibited good scavenging activity (40.55 to 71.41%) against DPPH radicals. Based on ABTS test, the activity ranged from 41.80 to 55.43%. Ethanol extract exhibited the highest level of polymeric color (37.67%) followed by the water and ethanol/water extracts with 28.26 and 22.80%, respectively. Kostic et al. (2013b) found a high degree of correlation between monomeric anthocyanin content and antioxidant activity based on ABTS assay (0.9993), and between the content of polymeric color and antioxidant activity based on DPPH assay (0.9137).

Anti-tumor and Anti-cancer Activities

Over the years, it has been reported that not only crude extracts but also compounds obtained from mulberry fruit exhibited antitumor activity through different pathways. This could be one of the most important biological activities of mulberry fruit (Yuan and Zhao 2017). Chon et al. (2009) determined the antiproliferative activity of methanolic extracts of *Morus alba* L. leaves against human pulmonary carcinoma, human breast adenocarcinoma, and human colon carcinoma using Calu-6, MCF-7, and HCT-116 cell lines, respectively. Anticancer effect of mulberry leaf extracts was evaluated in a study using human hepatoma HepG2 cell line. Different types of leaf extracts of *Morus alba* L. were prepared. Various parameters including cytotoxicity, apoptosis, expression of topoisomerase IIα, and cell cycle progression were investigated. For HepG2 cell inhibition, IC_{50} values were found to be 33.1, 79.4, 35.6, and 204.2 µg/mL for 100% methanol, 50% methanol, butanol, and water extracts, respectively. It was concluded

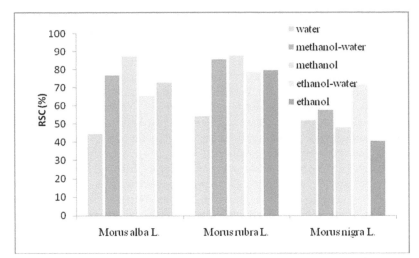

Figure 2. Comparative overview of antioxidant activity of three types of mulberry extracts using DPPH and ABTS method (Dimitrijević 2014).

that phenolic-rich organic extracts of *Morus alba* L. leaves had antiproliferative potential against HepG2 hepatoma cells *via* cell cycle arrest in the G2/M phase, caspases release and topoisomerase IIα inhibition (Naowaratwattana et al. 2010). Three prenylated flavonoids, kuwanon E, cudraflavone B, and -*O*-methylkuwanon E isolated from *Morus alba* L. were tested for cytotoxic activity against THP-1 cells. On the contrary, rests of flavonoids were moderately capable of antiproliferative activity against THP-1 cells (Kollar et al. 2013).

Mulberry extracts inhibit the initiation and development of cancer. Its mechanism of action is omnidirectional and consists in the elimination of the reactive form of oxygen and nitrogen, which reduces the number of negative mutations, reducing inflammation, and promoting apoptosis (Grajek et al. 2015). Anthocyanins and flavonoids are the group of natural phenolic compounds with major effect in lowering the risk of cardiovascular diseases and cancer because of its anti-inflammatory, antioxidant, and chemoprotective properties. They also have an anti-cancer effect. Methanolic extract of mulberry leaves shows efficient cytotoxic behavior against cancer cells. Singh (2008) identified many compounds like kuwanon S, 8-granilapigenin, cyclomulberrin, cyclomorusin, morusin, atalantoflavones, kaempferol with strong action on cell lines HeLa, MCF-7 and Hep3B (Singh 2008). Zhang and coworkers (Zhang et al. 2009a) conducted a short-term study on root bark of *Morus alba* L. and isolated a flavanics, i.e., glycoside 5,2'-dihydroxyflavanone-7,4'-di-*O*-D-glucoside, which prevents cell proliferation of human ovarian cancer cell HO-8910.

Antimicrobial Activity

Investigations on antimicrobial activities of methanol extracts of the plants from the genus *Morus* showed that *M. nigra* L. extract was more effective than extracts of *M. alba* L. and *M. rubra* L. (Miljković et al. 2018). The minimum inhibitory concentration (MIC) of kuwanon G isolated from the ethyl acetate fraction of foot bark methanol extract of *Morus alba* L. was determined and compared with several commercially available agents used for control of caries. The MIC value against *Streptococcus* was 8 µg/ml, that is less in comparison to 32 µg/ml for sanguinarine, 125 µg/ml carvacrol, 500 µg/ml thymol and eucalyptol, which are commercial agents with antibacterial activity. This fact indicates that kuwanon G has stronger antibacterial activity compared to commercial agents. Antibacterial activity of kuwanon G can be compared to antibiotics such as vancomycin and chlorhexidine, which, however, possess harmful side effects like a change in the color of teeth, vomiting, diarrhea, and decline in immunity (Park et al. 2003). Chalcomoracin, a leaf phytoalexin of mulberry tree exhibited considerable antibacterial activity against methicillin-resistant *Staphylococcus aureus* (Fukai et al. 2005). Prenylated flavonoids isolated from *Morus alba* L. showed antibacterial, antifungal and antiviral activities (Sohn et al. 2004). Oxyresveratrol and arylbenzofuran moracin M2 showed bactericidal activity against *Staphylococcus aureus* [minimal bactericidal concentration (MBC) = 125 and 62.5 µg/mL, respectively]; arylbenzofuran moracin M2 also showed bactericidal activity against *Streptococcus faecalis* [MBC = 500 and 250 µg/mL, respectively]). The structure-activity relationship of cyclomorusin and morusin showed that the cyclization of the prenyl unit at C-3 in 3 reduced the activity, whereas the presence of a free prenyl unit at C-3 in 4 enhanced the activity. By comparing the activities of 4 morusin and kuwanon C6, it was observed that, when the prenyl unit at C-3 was cyclized (kuwanon C6), then the C-8 attached prenyl unit should be open to have better activity. Cyclomorusin has both prenyl units cyclized, and kuwanon C5 has both prenyl units open; they have lower activities than the compounds morusin and kuwanon C6, which have one of the prenyl units open. Though it is inconclusive, the results show that the electron donating group (–OH) at positions 7 and/or 2' was vital for increased activity, as observed for morusin and kuwanon C6 (Mazimba et al. 2011).

Dimitrijević et al. (2014) evaluated the antimicrobial activity of different *Morus alba* L. fruit extracts by measuring the zone of inhibition. The fruit was collected in southeastern Serbia. The water extract was active against the bacteria *Salmonella typhimurium* and *Staphylococcus aureus*. The methanol–water extract was active only against the bacteria *S. typhimurium*. The methanolic extract showed the highest antioxidant activity and was active against the bacteria *Salmonella typhimurium*, *Staphylococcus aureus*, *Bacillus subtilis*, and *Escherichia coli*.

Anti-inflammatory Activity

Inflammation is the complex biological response of the organism to the internal and external harmful stimuli such as by pathogens, dead cells, metabolic disorders, physical damage, and irritant substances. This reaction aims to protect the body by removing harmful stimulus and the initiation of a tissue healing process. Sometimes it is necessary to alleviate the symptoms of the above-mentioned immune response and for this purpose anti-inflammatory drugs are applied. Mulberry plants have a long history in traditional medicine as anti-inflammatory agents. According to Chao et al. (2009), anti-inflammatory properties of *Morus alba* L. extract might be a result of the inhibition of pro-inflammatory mediators (NO and PGE). The root epidermis of *Morus alba* L. was shown to have anti-inflammatory effects (Wang et al. 2013a). Butanol extract of *Morus alba* L. significantly reduced LPS lipopolysaccharide (LPS)-induced PGE2 production, tumor necrosis factor alpha (TNF-alpha) and cyclooxygenase 2 (COX-2) expression in RAW264.7 macrophages. Methanolic extract of *Morus alba* L. leaves is capable of suppressing inflammatory mediators (Choi and Hwang 2005). Ethyl acetate fractions of *Morus alba* L. significantly decreased NF-kB (nuclear factor kappa light chain enhancer of activated B cells) luciferase activity and also secretion of NO and PGE in LPS/IFN-gamma stimulated peritoneal macrophages. Cudraflavone B, a prenylated flavonoid found in large amounts in the roots of *Morus alba* L., causes a significant inhibition of inflammatory mediators in selected *in vitro* models. It was a potent inhibitor of TNF-alpha by blocking the translocation of NF-kB from the cytoplasm to the nucleus. The NF-kB activity reduction also resulted in the inhibition of COX-2 gene expression (Hošek et al. 2011). Molecular docking studies have shown that moracin M from *Morus alba* L. can inhibit the activity of a variety of phosphodiesterase-4 (PDE-4) enzymes. Inhibitors of PDE4 enzymes have therapeutic applications as an anti-inflammatory agent because inhibition of PDE-4 will lead to the accumulation of cAMP (cyclic adenosine monophosphate) and thus attenuated inflammatory responses in various cell types (Chen et al. 2012). Flavonoids isolated from *Morus nigra* L. and *Morus alba* L. bark showed potent anti-inflammatory activity. Morin, a flavonoid present in mulberry, exhibited considerable anti-inflammatory activity based on studies of the role of morin on the disposition of CsA (cyclosporin A) in lymphoid and non-lymphoid tissues and immune cells in mice. Decreased level of CsA in tissues was correlated to increased doses of morin (Yu et al. 2007, Ahmad et al. 2013).

Other Activities

Antiatherosclerosis Activity

Atherosclerosis is widely accepted as the main cause of cardiovascular diseases (Libby 2002). Oxidative low-density lipoprotein (ox-LDL) is an important atherogenic factor (Hertog et al. 1993). Consumption of a diet rich in natural antioxidants is associated with slowing of the development of atherosclerosis (Kaliora et al. 2006, Gendron et al. 2010). Mulberry water extract and mulberry anthocyanin extract exhibited antioxidation and atherosclerogenesis abilities *in vitro* and these extracts could decrease macrophage death induced by ox-LDL as well as inhibit the formation of foam cells (Liu et al. 2008b). Anthocyanin components in mulberry extracts could prevent atherosclerosis (Liu et al. 2008b). Additionally, the anti-atherogenic effect of mulberry leaf extracts was investigated and it was found that polyphenolic extracts contained quercetin (11.70%), naringenin (9.012%) and gallocatechin gallate (11.02%). Both the extracts inhibited oxidation and lipid peroxidation of LDL but the polyphenolic extract was more potent. Hence, mulberry leaf polyphenolic extract potentially could be developed as an anti-atherogenic agent (Yang et al. 2011).

Activity against Alzheimer's Disease

Alzheimer's disease produces general mental enfeeblement. Evidence suggests that amyloid beta-peptide (1–42) plays an important role in the etiology of the disease, forming plaques and fibrils disturbing the neuron network in the brain. The performed investigation suggests that mulberry extract provides viable treatment for Alzheimer's disease through inhibition of amyloid beta-peptide fibril formation and attenuation of neurotoxicity induced by amyloid beta-peptide (Niidome et al. 2007).

Anti-obesity Activity

There are studies that show how obesity plays a major role in contributing to dyslipidemia. Liu et al. (2008b) treated high-fat-diet-induced obese mice with a combination of mulberry leaf and mulberry fruit extracts at low to high doses and observed a significant decrease in body weight gain. Also, male hamsters fed with mulberry fruit water extract for 12 wk had a lower high-fat-diet-induced body weight, lower visceral fat, accompanied by a decrease in cholesterol, serum triacylglycerol, LDL/HDL ratio, and free fatty acids. The authors suggested that mulberry fruit water extracts regulate lipogenesis and lipolysis, which can be used to reduce body weight (Peng et al. 2011).

Hypolipidemic Effects

Hyperlipidemia is characterized by excess cholesterol and fatty substances in the blood (Chobanian 1991). Since a majority of the hypolipidemic drugs can potentially cause side effects, besides being expensive, an increasing focus on research was to determine the effectiveness of natural alternative medicine in reducing blood lipid levels (Arabshahi-Delouee and Urooj 2007). The effect of mulberry fruit consumption on lipid profiles in hypercholesterolemic subjects (aged 30–60 yr) was studied. The level of LDL cholesterol significantly decreased as compared to the control group, thereby indicating that mulberry fruit could improve lipid profiles in hypercholesterolemic patients (Sirikanchanarod et al. 2016). Liu and coworkers (Liu et al. 2002) conducted the study on rats and found that mulberry leaf extract rich in flavonoids, could work as the scavenger of blood lipid radicals in sugar metabolism and antioxidation in rats. According to them, mulberry extract showed the hypolipidemic effects, which elevate LDLR (Low-density lipoprotein receptor) gene expression, the clearance proficiency of Low-density lipoprotein, and a decline in the lipid biosynthesis. Moracin present in mulberry leaves is capable of inhibiting lipid peroxidation that strongly indicates their role as a scavenger (Sharma et al. 2001). In addition, polysaccharides could also be investigated to elucidate the hypolipidemic mechanism of mulberry fruit because many polysaccharides from natural products show obvious hypolipidemic effects (Liu et al. 2002, Zhao et al. 2012).

Anti-hyperglycemic Activity

Diabetes mellitus has become nowadays the third most life-threatening disease worldwide (Wang et al. 2013b). In traditional Chinese medicine, mulberry leaves have been known to prevent and treat diabetes (Miyahara et al.

2004) because of the presence of chemical compounds that can suppress high blood sugar levels (hyperglycemia). Mulberry contains 1-deoxynojirimycin (DNJ) and some of its derivatives like alpha-glucosidase inhibitors that have been used as medicines to treat diabetes mellitus (Mudra et al. 2007, Asai et al. 2011). Mulberry therapies that were conducted on type-2 streptozoticin induced diabetic rats showed improvement in decreasing blood glucose levels. Quercetin, the quantitatively major flavonoids glycoside in mulberry leaves, effectively suppressed the blood glucose levels (Katsube et al. 2010). In Thailand, there is a belief that beverages which contain mulberry leaf (*Morus alba* L.) can be beneficial against diabetes. It was based on study of the effect of long-term administration of ethanolic extracts of mulberry leaf on blood glucose. Daily administration of 1 g/kg of *Morus alba* L. for 6 wk decreased blood glucose by 22% which was comparable to the effect of 4 v/kg insulin. Findings indicated that long-term supplement of *Morus alba* L. has anti-hyperglycemic effects in chronic diabetic rats (Naowaboot et al. 2009). Wang and coworkers (Wang et al. 2013b) investigated the antidiabetic effect of mulberry fruit extract *in vitro* and *in vivo*. Mulberry fruit extract rich in phenolics and flavonoids appeared to be a potent inhibitor of α-glucosidase, which has been confirmed recently (Wang et al. 2013b, Xiao et al. 2017).

Activity against Skin Diseases

Melanin present in the skin protects from UV induced hyperpigmentation, wrinkling, melasma, and cancer. Tyrosinase is an important enzyme in melanin production. *Morus alba* L. leaf extract exhibited potent inhibitory effects on mushroom tyrosinase, mammalian tyrosinase, and melanin synthesis in Melan-a cells (Lee et al. 2002). TMBC, a chalcone from the stem of *Morus nigra* L. modulated melanogenesis by inhibiting tyrosinase. It inhibited the L-dopa oxidase activity of mushroom tyrosinase which was more potent than kojic acid, a well-known tyrosinase inhibitor (Zhang et al. 2009b). Topical applications of mulberroside A, oxyresveratrol, and oxyresveratrol-3-*O*-glucoside clearly caused depigmentation, reduced melanin indices, inhibited tyrosinase activity, and decreased melanin content in UV induced hyperpigmentation in guinea pig skin. Oxyresveratrol and oxyresveratrol-3-*O*-glucoside more potently inhibited melanogenesis than mulberroside A. This treatment decreased the expression of MATF (Melanogenesis Associated Transcription Factor) gene, that is regulating the transcription of proteins involved in melanocyte pigmentation (Park et al. 2011).

Conclusion

For the good health, longevity, remedy in pain and discomfort, fragrance, flavor, and food, all over the world, mankind depends on plants. Medicinal plants still play important role as there is growing public attention given to diverse natural source-based compounds for the treatment of various diseases and in the improvement of the quality of life. Mulberry like the traditional herb is used in medicine for centuries and represents one of the most studied plants. Due to its pharmacological properties, all parts of the plant are used as medicine. Mulberry fruits possess an outstanding nutritional value due to their high levels of phenolic compounds as well as their high antioxidant activity and they could be successfully used in obtaining functional foods. There are many opportunities for the food and healthcare industry to explore the health benefits of mulberry fruits. Here, we have presented results that endorse ethnopharmacological usage of mulberry as anti-oxidative, anti-inflammatory, anti-microbial, anti-tumor, anti-cancer, anti-atherosclerosis, anti-Alzheimer's disease activity, anti-obesity, anti-hyperlipidemic and anti-hyperglycaemic. There are other useful effects such as protecting the liver, improving eyesight, facilitating discharge of urine, immune-modulation and chemoprotective properties, etc. It is important to continue the research on unidentified biocompounds to ascertain their biological values and mechanisms by which they exhibit activity. It seems that more research now will be focused on the natural source-based compounds to treat various diseases and mulberry species are definitely one of those to be explored more extensively.

References

Ahmad, A., G. Gupta, M. Afzal, I. Kazmi and F. Anwar. 2013. Antiulcer and antioxidant activities of a new steroid from *Morus alba*. Life Sci. 92(3): 202–210.

Ahmed, S. and F. Shakeel. 2012. Voltammetric determination of antioxidant character in *Berberis lycium* Royel, *Zanthoxylum armatum* and *Morus nigra* Linn plants. Pak. J. Pharm. Sci. 25(3): 501–507.

Andallu, B., V. Suryakantham, B.L. Srikanthi and G.S. Reddy. 2001. Effect of mulberry (*Morus indica* L.) therapy on plasma and erythrocyte membrane lipids in patients with type 2 diabetes. Clin. Chim. Acta 314(1-2): 47–53.

Arabshahi-Delouee, S. and A. Urooj. 2007. Antioxidant properties of various solvent extracts of mulberry (*Morus indica* L.) leaves. Food Chem. 102(4): 1233–1240.

Arsic, B., D. Dimitrijevic and D. Kostic. 2016. Mineral and vitamin fortification. pp. 1–40. *In*: Grumazescu, A. (ed.). Nutraceuticals: Nanotechnology in the Agri-food Industry. Elsevier, USA.

Asai, A., K. Nakagawa, O. Higuchi, T. Kimura, Y. Kojima, J. Kariya, T. Miyazawa and S. Oikawa. 2011. Effect of mulberry leaf extract with enriched 1-deoxynojirimycin content on postprandial glycemic control in subjects with impaired glucose metabolism. J. Diabetes Investig 2(4): 318–323.

Butkhup, L., W. Samappito and S. Samappito. 2013. Phenolic composition and antioxidant activity of white mulberry (*Morus alba* L.) fruits. Int. J. Food Sci. Technol. 48(5): 934–940.

Butt, M.S., A. Nazir, M.T. Sultan and K. Schroen. 2008. *Morus alba* L. nature's functional tonic. Trends Food Sci. Technol. 19(10): 505–512.

Chan, K.-C., H.-P. Huang, H.-H. Ho, C.-N. Huang, M.-C. Lin and C.-J. Wang. 2015. Mulberry polyphenols induce cell cycle arrest of vascular smooth muscle cells by inducing NO production and activating AMPK and p53. J. Funct. Foods 15: 604–613.

Chang, J.-J., M.-J. Hsu, H.-P. Huang, D.-J. Chung, Y.-C. Chang and C.-J. Wang. 2013. Mulberry anthocyanins inhibit oleic acid induced lipid accumulation by reduction of lipogenesis and promotion of hepatic lipid clearance. J. Agric. Food Chem. 61(25): 6069–6076.

Chao, W.-W., Y.-H. Kuo, W.-C. Li and B.-F. Lin. 2009. The production of nitric oxide and prostaglandin E2 in peritoneal macrophages is inhibited by *Andrographis paniculata*, *Angelica sinensis* and *Morus alba* ethyl acetate fractions. J. Etnopharmacol. 122(1): 68–75.

Chen, S.-K., P. Zhao, Y.-X. Shao, Z. Li, C. Zhang, P. Liu, X. He, H.-B. Luo and X. Hu. 2012. Moracin M from *Morus alba* L. is a natural phosphodiesterase-4 inhibitor. Bioorg. Med. Chem. Lett. 22(9): 3261–3264.

Chen, Y., W. Zhang, T. Zhao, F. Li, M. Zhang, J. Li, Y. Zou, W. Wang, S.J. Cobbina, X. Wu and L. Yang. 2016. Adsorption properties of macroporous adsorbent resins for separation of anthocyanins from mulberry. Food Chem. 194: 712–722.

Chobanian, A.V. 1991. Single risk factor intervention may be inadequate to inhibit atherosclerosis progression when hypertension and hypercholesterolemia coexist. Hypertension 18(2): 130–131.

Choi, E.-M. and J.-K. Hwang. 2005. Effects of *Morus alba* leaf extract on the production of nitric oxide, prostaglandin E$_2$ and cytokines in RAW264.7 macrophages. Fitoterapia 76(7-8): 608–613.

Choi, J.W., A. Synytsya, P. Capek, R. Bleha, R. Pohl and Y.I. Park. 2016. Structural analysis and anti-obesity effect of a pectic polysaccharide isolated from Korean mulberry fruit Oddi (*Morus alba* L.). Carbohydr. Polym. 146: 187–196.

Chon, S.-U., Y.-M. Kim, Y.-J. Park, B.-G. Heo, Y.-S. Park and S. Gorinstein. 2009. Antioxidant and antiproliferative effects of methanol extracts from raw and fermented parts of mulberry plant (*Morus alba* L.). Eur. Food Res. Technol. 230(2): 231–237.

Daimon, T., C. Hirayama, M. Kanai, Y. Ruike, Y. Meng, E. Kosegawa, M. Nakamura, G. Tsujimoto, S. Katsuma and T. Shimada. 2010. The silkworm *Green b* locus encodes a quercetin 5-*O*-glucosyltransferase that produces green cocoons with UV-shielding properties. Proc. Natl. Acad. Sci. USA 107(25): 11471–11476.

Dimitrijević, D.S., D.A. Kostić, G.S. Stojanović, S.S. Mitić, M.N. Mitić and R. Micić. 2013. Polyphenol contents and antioxidant activity of five fresh fruit *Morus* spp. (Moraceae) extracts. Agro Food Ind. Hi. Tech. 24(5): 34–37.

Dimitrijević, D.S. 2014. Analiza hemijskog sastava i antioksidativne aktivnosti ekstrakata duda (*Morus* spp., Moraceae). Ph.D. Thesis. University of Niš (in Serbian).

Dimitrijević, D.S, D.A. Kostić, G.S. Stojanović, S.S. Mitić, M.N. Mitić and A.S. Đorđević. 2014. Phenolic composition, antioxidant activity, mineral content and antimicrobial activity of fresh fruit extracts of *Morus alba* L. J. Food Nutr. Res. 53: 22–30.

Du, Q., J. Zheng and Y. Xu. 2008. Composition of anthocyanins in mulberry and their antioxidant activity. J. Food Compost. Anal. 21(5): 390–395.

Ercisli, S. and E. Orhan. 2007. Chemical composition of white (*Morus alba*), red (*Morus rubra*) and black (*Morus nigra*) mulberry fruits. Food Chem. 103(4): 1380–1384.

Fukai, T., K. Kaitou and S. Terada. 2005. Antimicrobial activity of 2-arylbenzofurans from *Morus* species against methicillin-resistant *Staphylococcus aureus*. Fitoterapia 76(7-8): 708–711.

Gardiner, W.P. 1997. Statistical Analysis Methods for Chemists: A Software Based Approach. Royal Society of Chemistry, London. UK.

Gendron, M.-E., J.-F. Theoret, A.M. Mamarbachi, A. Drouin, A. Nguyen, V. Bolduc, N. Thorin-Trescases, Y. Merhi and E. Thorin. 2010. Late chronic catechin antioxidant treatment is deleterious to the endothelial function in aging mice with established atherosclerosis. Am. J. Physiol. Heart Circ. Physiol. 298(6): H2062–H2070.

Grajek, K., A. Wawro and D. Pieprzyk-Kokocha. 2015. Bioactivity of *Morus alba* L. extracts–an overview. Int. J. Pharm. Sci. Res. 6(8): 3110–3122.

Gundogdu, M., F. Muradoglu, R.I. Gazioglu Sensoy and H. Yilamz. 2011. Determination of fruit chemical properties of *Morus nigra* L., *Morus alba* L. and *Morus rubra* L. by HPLC. Sci. Hortic. 132: 37–41.

Hassimotto, N.M.A., M.I. Genovese and F.M. Lajolo. 2008. Absorption and metabolism of cyanidin-3-glucoside and cyanidin-3-rutinoside extracted from wild mulberry (*Morus nigra* L.) in rats. Nutr. Res. 28(3): 198–207.

Hertog, M.G.L., E.J.M. Feskens, D. Kromhout, M.G.L. Hertog, P.C.H. Hollman and M.B. Katan. 1993. Dietary antioxidant flavonoids and risk of coronary heart disease: the Zutphen Elderly Study. Lancet 342(8878): 1007–1011.

Hošek, J., M. Bartos, S. Chudik, S. Dall'Acqua, G. Innocenti, M. Kartal, L. Kokoška, P. Kollar, Z. Kutil, P. Landa, R. Marek, V. Zavalova, M. Žemlička and K. Šmejkal. 2011. Natural compound cudraflavone B shows promising anti-inflammatory properties *in vitro*. J. Nat. Prod. 74(4): 614–619.

https://www.phillyorchards.org/2016/07/01/the-plucky-mulberry-morus/.

Huang, H.-P., Y.-W. Shih, Y.-C. Chang, C.-N. Hung and C.-J. Wang. 2008. Chemoinhibitory effect of mulberry anthocyanins on melanoma metastasis involved in the Ras/PI3K pathway. J. Agric. Food Chem. 56(19): 9286–9293.

Imran, M., H. Khan, M. Shah, R. Khan and F. Khan. 2010. Chemical composition and antioxidant activity of certain *Morus* species. J. Zhejiang Univ. Sci. B 11(12): 973–980.

Jiang, Y. and W.-J. Nie. 2015. Chemical properties in fruits of mulberry species from the Xinjiang province of China. Food Chem. 174: 460–466.

Jin, Q., J. Yang, L. Ma, J. Cai and J. Li. 2015. Comparison of polyphenol profile and inhibitory activities against oxidation and α-glucosidase in mulberry (genus *Morus*) cultivars from China. J. Food Sci. 80(11): C2440–C2451.

Kaliora, A.C., G.V.Z. Dedoussis and H. Schmidt. 2006. Dietary antioxidants in preventing atherogenesis. Atherosclerosis 187(1): 1–17.

Katsube, T., M. Yamasaki, K. Shiwaku, T. Ishijima, I. Matsumoto, K. Abe and Y. Yamasaki. 2010. Effect of flavonol glycoside in mulberry (*Morus alba* L.) leaf on glucose metabolism and oxidative stress in liver in diet-induced obese mice. J. Sci. Food Agric. 90(14): 2386–2392.

Kim, G.-N. and H.-D. Jang. 2011. Flavonol content in the water extract of the mulberry (*Morus alba* L.) leaf and their antioxidant capacities. J. Food Sci. 76(6): C869–C873.

Kim, S.Y., J.J. Gao, W.-C. Lee, K.S. Ryu, K.R. Lee and Y.C. Kim. 1999. Antioxidative flavonoids from the leaves of *Morus alba*. Arch. Pharm. Res. 22(1): 81–85.

Kimura, T., K. Nakagawa, H. Kubota, Y. Kojima, Y. Goto, K. Yamagishi, S. Oita, S. Oikawa and T. Miyazawa. 2007. Food-grade mulberry powder enriched with 1-deoxynojirimycin suppresses the elevation of postprandial blood glucose in humans. J. Agric. Food Chem. 55(14): 5869–5874.

Kollar, P., T. Barta, J. Hošek, K. Souček, V. Muller Zavalova, S. Artinian, R. Talhouk, K. Šmejkal, P. Suchy Jr. and A. Hampl. 2013. Prenylated flavonoids from *Morus alba* L. cause inhibition of G1/S transition in THP-1 human leukemia cells and prevent the lipopolysaccharide-induced inflammatory response. Evid. Based Complement Alternat. Med. 2013: 1–13.

Kostić, D.A., D.S. Dimitrijević, S.S. Mitić, M.N. Mitić, G.S. Stojanović and A.V. Živanović. 2013a. Phenolic content and antioxidant activities of fruit extracts of *Morus nigra* L. (*Moraceae*) from Southeast Serbia. Trop. J. Pharm. Res. 12(1): 105–110.

Kostic, D.A., D.S. Dimitrijevic, G.S. Stojanovic, S.S. Mitic and M.N. Mitic. 2013b. Phenolic composition and antioxidant activity of fresh fruit extracts of mulberries from Serbia. Oxid. Commun. 36(1): 4–14.

Lee, C.Y., S.M. Sim and H.M. Cheng. 2008. Phenylacetic acids were detected in the plasma and urine of rats administered with low-dose mulberry leaf extract. Nutr. Res. 28(8): 555–563.

Lee, S.H., S.Y. Choi, H. Kim, J.S. Hwang, B.G. Lee, J.J. Gao and S.Z. Kim. 2002. Mulberroside F isolated from the leaves of *Morus alba* inhibits melanin biosynthesis. Biol. Pharm. Bull. 25(8): 1045–1048.

Libby, P. 2002. Inflammation in atherosclerosis. Nature 420: 868–874.

Liu, C.-J. and J.-Y. Lin. 2012. Anti-inflammatory and anti-apoptotic effects of strawberry and mulberry fruit polysaccharides on lipopolysaccharide-stimulated macrophages through modulating pro-/anti-inflammatory cytokines secretion and Bcl-2/Bak protein ratio. Food Chem. Toxicol. 50(9): 3032–3039.

Liu, C.-J. and J.-Y. Lin. 2014. Protective effects of strawberry and mulberry fruit polysaccharides on inflammation and apoptosis in murine primary splenocytes. J. Food Drug Anal. 22(2): 210–219.

Liu, H.-H., W.-C. Ko and M.-L. Hu. 2002. Hypolipidemic effect of glycosaminoglycans from the sea cucumber *Metriatyla scabra* in rats fed a cholesterol-supplemented diet. J. Agric. Food Chem. 50(12): 3602–3606.

Liu, H., N. Qiu, H. Ding and R. Zao. 2008a. Polyphenols contents and antioxidant capacity of 68 Chinese herbals suitable for medical or food uses. Food Res. Int. 41(4): 363–370.

Liu, L.-K., H.-J. Lee, Y.-W. Shih, C.-C. Chyau and C.-J. Wang. 2008b. Mulberry anthocyanin extracts inhibit LDL oxidation and macrophage-derived foam cell formation induced by oxidative LDL. J. Food Sci. 73(6): H113–H121.

Liu, L.-K., F.-P. Chou, Y.-C. Chen, C.-C. Chyau, H.-H. Ho and C.-J. Wang. 2009. Effects of mulberry (*Morus alba* L.) extracts on lipid homeostasis *in vitro* and *in vivo*. J. Agric. Food Chem. 57(16): 7605–7611.

Mahmood, T., F. Anwar, M. Abbas, M.C. Boyce and N. Saari. 2012. Compositional variation in sugars and organic acids at different maturity stages in selected small fruits from Pakistan. Int. J. Mol. Sci. 13: 1380–1392.

Mazimba, O., R.R.T. Majinda and D. Motlhanka. 2011. Antioxidant and antibacterial constituents from *Morus nigra*. Afr. J. Pharm. Pharmacol. 5(6): 751–754.

Micić, R.J., D.S. Dimitrijević, D.A. Kostić, G.S. Stojanović, S.S. Mitić, M.N. Mitić, A.N. Pavlović and S.S. Ranđelović. 2013. Content of heavy metals in mulberry fruits and their extracts-correlation analysis. Am. J. Analyt. Chem. 4: 674–682.

Miljković, V.M., G.S. Nikolić, Lj.B. Nikolić and B.B. Arsić. 2014. *Morus* species through centuries in pharmacy and as food. Adv. Technol. 3(2): 111–115.

Miljkovic, V., Lj. Nikolic, N. Radulovic, B. Arsic, G. Nikolic, D. Kostic, Z. Bojanic and J. Zvezdanovic. 2015. Flavonoids in mulberry fruit: Identification of nonanthocyanin phenolics in some mulberry fruit species (*Morus alba* L., *Morus rubra* L. and *Morus nigra* L.). Agro Food Ind. Hi Tech. 26(3): 38–42.

Miljković, V., G. Nikolić, T.M. Mihajlov-Krstev and B. Arsić. 2018. Antibacterial activities of fruits extracts of three mulberry species (*Morus alba* L., *Morus rubra* L. and *Morus nigra* L.) and bilberry (*Vaccinium myrtillus* L.). Acta Medica Medianae 57(3): 5–12.

Miyahara, C., M. Miyazawa, S. Satoh, A. Sakai and S. Mizusaki. 2004. Inhibitory effects of mulberry leaf extract on postprandial hyperglycemia in normal rats. J. Nutr. Sci. Vitaminol. 50(3): 161–164.

Mudra, M., N. Ercan-Fang, L. Zhong, J. Furne and M. Levitt. 2007. Influence of mulberry leaf extract on the blood glucose and breath hydrogen response to ingestion of 75 g sucrose by type 2 diabetic and control subjects. Diabetes Care 30(5): 1272–1274.

Nakagawa, K., K. Ogawa, O. Higuchi, T. Kimura, T. Miyazawa and M. Hori. 2010. Determination of iminosugars in mulberry leaves and silkworms using hydrophilic interaction chromatography-tandem mass spectrometry. Anal. Biochem. 404(2): 217–222.

Nam, S., H.W. Jang and T. Shibamoto. 2012. Antioxidant activities of extracts from teas prepared from medicinal Plants, *Morus alba* L., *Camellia sinensis* L., and *Cudrania tricuspidata*, and their volatile components. J. Agric. Food Chem. 60(36): 9097–9105.

Naowaboot, J., P. Pannangpetch, V. Kukongviriyapan, B. Kongyingyoes and U. Kukongviriyapan. 2009. Antihyperglycemic, antioxidant and antiglycation activities of mulberry leaf extract in streptozotocin-induced chronic diabetic rats. Plant Foods Hum. Nutr. 64(2): 116–121.

Naowaratwattana, W., W. De-Eknamkul and E.G. De Mejia. 2010. Phenolic-containing organic extracts of mulberry (*Morus alba* L.) leaves inhibit HepG2 hepatoma cells through G2/M phase arrest, induction of apoptosis, and inhibition of topoisomerase IIα activity. J. Med. Food 13(5): 1045–1056.

Natić, M.M., D.Č. Dabić, A. Papetti, M.M. Fotirić Akšić, V. Ognjanov, M. Ljubojević and Ž. Lj. Tešić. 2015. Analysis and characterisation of phytochemicals in mulberry (*Morus alba* L.) fruits grown in Vojvodina, North Serbia. Food Chem. 171: 128–136.

Niidome, T., K. Takahashi, Y. Goto, S. Goh, N. Tanaka, K. Kamei, M. Ichida, S. Hara, A. Akaike, T. Kihara and H. Sugimoto. 2007. Mulberry leaf extract prevents amyloid beta-peptide fibril formation and neurotoxicity. Neuroreport 18(8): 813–816.

Nuengchamnong, N., K. Ingkaninan, W. Kaewruang, S. Wongareonwanakij and B. Hongthongdaeng. 2007. Quantitative determination of 1-deoxynojirimycin in mulberry leaves using liquid chromatography-tandem mass spectrometry. J. Pharm. Biomed. Anal. 44(4): 853–858.

Özgen, M., S. Serçe and C. Kaya. 2009. Phytochemical and antioxidant properties of anthocyanin-rich *Morus nigra* and *Morus rubra* fruits. Sci. Hortic. 119(3): 275–279.

Park, K.M., J.S. You, H.Y. Lee, N.I. Baek and J.K. Hwang. 2003. Kuwanon G: an antibacterial agent from the root bark of *Morus alba* against oral pathogens. J. Ethnophamacol. 84(2-3): 181–185.

Park, K.-T., J.-K. Kim, D. Hwang, Y. Yoo and Y.-H. Lim. 2011. Inhibitory effect of mulberroside A and its derivatives on melanogenesis induced by ultraviolet B irradiation. Food Chem. Toxicol. 49(12): 3038–3045.

Pel, P., H.-S. Chae, P. Nhoek, Y.-M. Kim and Y.-W. Chin. 2017. Chemical constituents with proprotein convertase subtilisin/kexin type 9 mRNA expression inhibitory activity from dried immature *Morus alba* fruits. J. Agric. Food Chem. 65(26): 5316–5321.

Peng, C.-H., L.-K. Liu, C.-M. Chuang, C.-C. Chyau, C.-N. Huang and C.-J. Wang. 2011. Mulberry water extracts possess an anti-obesity effect and ability to inhibit hepatic lipogenesis and promote lipolysis. J. Agric. Food Chem. 59(6): 2663–2671.

Qin, C., Y. Li, W. Niu, Y. Ding, R. Zhang and X. Shang. 2010. Analysis and characterisation of anthocyanins in mulberry fruit. Czech J. Food Sci. 28(2): 117–126.

Radojković, M.M., Z.P. Zeković, S.S. Vidović, D.D. Kočar and P.Z. Mašković. 2012. Free radical scavenging activity and total phenolic and flavonoid contents of mulberry (*Morus* spp. L., Moraceae) extracts. Hem. Ind. 66(4): 547–552.

Randjelović, S.S., D.A. Kostić, B.B. Arsić and G. Stojanović. 2014. Bioaccumulation of metals in different species of mulberry. Adv. Technol. 3(2): 105–110.

Sánchez-Salcedo, E.M., P. Mena, C. Garcia-Viguera, J.J. Martinez and F. Hernandez. 2015. Phytochemical evaluation of white (*Morus alba* L.) and black (*Morus nigra* L.) mulberry fruits, a starting point for the assessment of their beneficial properties. J. Funct. Foods 12: 399–408.

Sass-Kiss, A., J. Kiss, P. Milotay, M.M. Kerek and M. Toth-Markus. 2005. Differences in anthocyanin and carotenoid content of fruits and vegetables. Food Res. Int. 38(8-9): 1023–1029.

Seifried, H.E., D.E. Anderson, E.I. Fisher and J.A. Milner. 2007. A review of the interaction among dietary antioxidants and reactive oxygen species. J. Nutr. Biochem. 18(9): 567–579.

Sharma, R., A. Sharma, T. Shono, M. Takasugi, A. Shirata, T. Fujimura and H. Machii. 2001. Mulberry moracins: scavengers of UV stress-generated free radicals. Biosci. Biotechnol. Biochem. 65(6): 1402–1405.

Sheng, F., Y. Wang, X. Zhao, N. Tian, H. Hu and P. Li. 2014. Separation and identification of anthocyanin extracted from mulberry fruit and the pigment binding properties toward human serum albumin. J. Agric. Food Chem. 62(28): 6813–6819.

Singh, A. 2008. A note on variation of active principles in Indian medicinal plants and TIM formulations. Ethnobotanical Leaflets 12: 603–606.

Sirikanchanarod, A., A. Bumrungpert, W. Kaewruang, T. Senawong and P. Pavadhgul. 2016. The effect of mulberry fruits consumption on lipid profiles in hypercholesterolemic subjects: a randomized controlled trial. J. Pharm. Nutr. Sci. 6: 7–14.

Sohn, H.-Y., K.H. Son, C.-S. Kwon, G.-S. Kwon and S.S. Kang. 2004. Antimicrobial and cytotoxic activity of 18 prenylated flavonoids isolated from medicinal plants: *Morus alba* L., *Morus mongolica* Schneider, *Broussnetia papyrifera* (L.) Vent, *Sophora flavescens* Ait and *Echinosophora koreensis* Nakai. Phytomedicine 11(7-8): 666–672.

Vijayan, K., P.P. Srivastava and A.K. Awasthi. 2004. Analysis of phylogenetic relationship among five mulberry (*Morus*) species using molecular markers. Genome 47(3): 439–448.

Wang, J., W. Jin, W. Zhang, Y. Hou, H. Zhang and Q. Zhang. 2013b. Hypoglycemic property of acidic polysaccharide extracted from *Saccharina japonica* and its potential mechanism. Carbohydr. Polym. 95(1): 143–147.

Wang, Y., L. Xiang, C. Wang, C. Tang and X. He. 2013a. Antidiabetic and antioxidant effects and phytochemicals of mulberry fruit (*Morus alba* L.) polyphenol enhanced extract. PloS ONE 8(7): e71144.

Wattanapitayakul, S.K., L. Chularojmontri, A. Herunsalee, S. Charuchongkolwongse, S. Niumsakul and J.A. Bauer. 2005. Screening of antioxidants from medicinal plants for cardioprotective effect against doxorubicin toxicity. Basic Clin. Pharmacol. Toxicol. 96(1): 80–87.

Xiao, T., Z. Guo, B. Sun and Y. Zhao. 2017. Identification of anthocyanins from four kinds of berries and their inhibition activity to α-glycosidase and protein tyrosine phosphatase 1B by HPLC–FT-ICR MS/MS. J. Agric. Food Chem. 65(30): 6211–6221.

Yang, M.-Y., C.-N. Huang, K.-C. Chan, Y.-S. Yang, C.-H. Peng and C.-J. Wang. 2011. Mulberry leaf polyphenols possess antiatherogenesis effect *via* inhibiting LDL oxidation and foam cell formation. J. Agric. Food Chem. 59(5): 1985–1995.

Yu, Z., W.P. Fong and C.H.K. Cheng. 2007. Morin (3,5,7,2',4'-pentahydroxyflavone) exhibits potent inhibitory actions on urate transport by the human urate anion transporter (hURAT1) expressed in human embryonic kidney cells. Drug Metab. Dispos. 35(6): 981–986.

Yuan, Q., Y. Xie, W. Wang, Y. Yan, H. Ye, S. Jabbar and X. Zeng. 2015. Extraction optimization, characterization and antioxidant activity in vitro of polysaccharides from mulberry (*Morus alba* L.) leaves. Carbohydr. Polym. 128: 52–62.

Yuan, Q. and L. Zhao. 2017. The mulberry (*Morus alba* L.) fruit—a review of characteristic components and health benefits. J. Agric. Food Chem. 65(48): 10383–10394.

Zhang, M., R.-R. Wang, M. Chen, H.-Q. Zhang, S. Sun and L.-Y. Zhang. 2009a. A new flavanone glycoside with anti-proliferation activity from the root bark of *Morus alba*. Chin. J. Nat. Med. 7(2): 105–107.

Zhang, W., F. Han, J. He and C. Duan. 2008. HPLC-DAD-ESI-MS/MS analysis and antioxidant activities of nonanthocyanin phenolics in mulberry (*Morus alba* L.). J. Food Sci. 73(6): C512–C518.

Zhang, X., X. Hu, A. Hou and H. Wang. 2009b. Inhibitory effect of 2,4,2',4'-tetrahydroxy-3-(3-methyl-2-butenyl)-chalcone on tyrosinase activity and melanin biosynthesis. Biol. Pharm. Bull. 32(1): 86–90.

Zhao, L.-Y., W. Huang, Q.-X. Yuan, J. Cheng, Z.-C. Huang, L.-J. Ouyang and F.-H. Zeng. 2012. Hypolipidaemic effects and mechanisms of the main component of *Opuntia dillenii* Haw. polysaccharides in high-fat emulsion-induced hyperlipidaemic rats. Food Chem. 134(2): 964–971.

Zhou, M., Q. Chen, J. Bi, Y. Wang and X. Wu. 2017. Degradation kinetics of cyanidin 3-*O*-glucoside and cyanidin 3-*O*-rutinoside during hot air and vacuum drying in mulberry (*Morus alba* L.) fruit: a comparative study based on solid food system. Food Chem. 229: 574–579.

CHAPTER 5

Pharmacological Uses of Mulberry Products

Halil Koyu

INTRODUCTION

Among various mulberry species, *Morus alba* (white mulberry) and *Morus nigra* (black mulberry) have been standing out with many studies focused on pharmacological activities. Besides, there are reports on pharmacological activity potential of *M. rubra*, *M. indica*, *M. mongolica* and *M. australis*. The use of the leaves, branches, stem and root parts, especially the fruits, of *Morus* species for various purposes have also been recorded. Most widely known has been the use of syrup that was prepared from its fruits, against mouth and throat wounds, especially in children. Ethnobotanical usage of different mulberry parts with therapeutic aims include the use of (a) fruits against urinary tract problems and kidney diseases, diabetes, depression, anemia, constipation, liver and asthma diseases; (b) leaves against diabetes, asthma, hypertension, kidney diseases and for wound healing; (c) stem and root barks against antihelmintic, laxative, purgative and antipyretic purposes (Abbasi et al. 2014, Younus et al. 2016, Thaipitakwong et al. 2018).

In this chapter, the pharmacological properties of preparations, extracts and active compounds obtained from mulberry species are presented under the sections *ex vivo* and *in vivo* studies, clinical trials, pharmacokinetics and toxicity data.

Critical points regarding methodology and outcome of the researches like scientific name of mulberry species, parts used, extraction method and solvent composition, method of investigative activity and results of activity studies (with statistical data) are summarized with mention of articles for providing up to date information on those aspects in order to represent the wide spectrum of pharmacological use of mulberry products obtained from different parts of mulberry tree species.

Ex Vivo Studies

Antioxidant Activity

Antioxidant activity of methanol: chloroform (1:1) and water extracts of *M. nigra* stem and leaves were determined as 28.7 and 37.2 mg ascorbic acid equivalent (AAE)/g dry weight with the reducing power method and total antioxidant capacities were found as 36.5 and 15.6 mg AAE/g dry weight, respectively (Akhtar et al. 2015).

Isolated compounds from methanol extract of *M. alba* root bark as moracin M, P, O and S were reported to possess EC_{50} (half maximal effective concentration) values of 6, 40, 43 and 7 μM for ABTS$^+$ (2,2'-Azinobis-(3-

Department of Pharmaceutical Botany, Faculty of Pharmacy, Izmir Katip Celebi University, 35620 Cigli-Izmir, Turkey.
Email: halilkoyu@yahoo.com, halil.koyu@ikcu.edu.tr

ethylbenzothiazoline-6-sulfonic acid) diammonium salt) radical scavenging activity, respectively, while Trolox's was 10 µM (Seong et al. 2018).

Methanol (80%) and acetone (80%) extracts of *M. alba* and *M. nigra* fruits were investigated by DPPH (2,2-diphenyl-1-picrylhydrazyl radical) and TEAC (Trolox equivalent antioxidant activity) methods. For *M. alba* and *M. nigra*, radical scavenging activity EC_{50} values were 79 and 48 µg/ml; total antioxidant capacity values were 0.75 and 1.25 mmole Trolox equivalent (TE)/g extract for methanol extracts while 66 and 58 µg/ml for radical scavenging activity; 0.78 and 1.19 mmole TE/g extract for total antioxidant capacity for acetone extracts, respectively (Arfan et al. 2012).

The antioxidant capacities of the extracts obtained with different solvent mixtures from *M. nigra* fruits were compared with FRAP, DPPH and ORAC methods where ethanol:water:acetic acid (50:49.5:0.5) extract was found as having the highest values as 1491 mmole Fe^{+2}/kg, 395 mmole TE/kg and 1128 mmole TE/kg dry weight, respectively (Boeing et al. 2014).

Antioxidant activity of methanol (80%) extracts of *M. alba*, *M. nigra* and *M. rubra* leaves were determined with DPPH, ABTS and FRAP methods and results were found to be ranging between 1.89–2.12, 6.12–9.89 and 0.56–0.97 mM TE/g dry leaves, respectively (Iqbal et al. 2012).

Determined antioxidant activity values for methanol extracts of *M. nigra* leaves with DPPH and FRAP methods were 131.95 µg DPPH/ml and 0.15 mmole Fe^{+2}/g, respectively (Žugić et al. 2014).

Antibacterial Activity

Ethanol extract of *M. alba* leaves were determined with antibacterial activity against *Staphylococcus aureus* and *Pseudomonas aeruginosa* with minimum inhibitory concentration (MIC) of 512 and 1,021 µg/ml, respectively (de Oliveira et al. 2015).

Antibacterial activity of methanol:chloroform (1:1) and water extracts from *M. nigra* stem and leaf parts was investigated by disc diffusion method. Inhibition zone diameter of methanol:chloroform extract at 100 µg/disc was 6 mm against *S. aureus* (e.g., cefixime 20 mm) and *Escherichia coli* (e.g., roxithromycin 31 mm), while water extract's was 8 mm against *E. coli*, 11 mm against *Enterobacter aerogenes* (e.g., cefixime 7.5 mm) and 9 mm against *Bordetella bronchiseptica* (e.g., cefixime 15.4 mm) (Akhtar et al. 2015).

Among the minimum bactericidal concentrations of ethanol (48%) extracts obtained from fruits of *M. alba*, *M. nigra* and *M. mongolica*, *M. nigra* extracts had the best values against *E. coli*, *P. aeruginosa* and *S. aureus* as 2, 2 and 1.8 mg/ml, respectively (Chen et al. 2018).

MIC of ethanol extract of *M. nigra* leaves 0.195 mg/ml was found to be antibacteria particularly *Bacillus cereus*, *Enterococcus faecalis*, *E. coli* and 0.39 mg/ml against *Klebsiella pneumoniae* (Souza et al. 2018).

Antituberculosis Activity

IC_{50} values of kuwanon G and kuwanon H compounds isolated from acetone extracts of *M. nigra* roots for the inhibition of *Mycobacterium tuberculosis* protein tyrosine phosphatase A and B, were found as 5.89 and 0.83 µM; 1.49 and 0.36 µM, respectively. MIC of both compounds for *M. tuberculosis* were found to be 32 µg/ml (Mascarello et al. 2018).

Antifungal Activity

Ethanol extract of *M. alba* leaves were determined with antifungal activity against *Candida albicans*, *C. tropicalis*, *C. krusei* and *Aspergillus flavus* with MIC of 1,024, 512, 512 and 512 µg/ml, respectively (de Oliveira et al. 2015).

Inhibition zone diameters of water extract of *M. nigra* stem and leaf parts against *A. flavus* and *Mucor* sp. were found as 6.2 mm (reference terbinafine 25.3 mm) and 18.4 mm (terbinafine 30.6 mm) at 100 µg/disc concentration by disc diffusion method (Akhtar et al. 2015).

Inhibition zone diameter of methanol extract of *M. indica* leaves were 30 and 29 mm against *Aspergillus* sp. and *Penicillum* sp. whereas reference drug fluconazole's were 11 and 10 mm, respectively (Niratker and Preeti 2015).

Antiinflammatory Activity

Isolated compounds from chloroform extracts of *M. alba* and *M. nigra* root bark which were named as morusinol ($P < 0.01$) and sanggenon E ($P < 0.05$), were reported as significantly inhibiting the secretion of TNF-α (Tumor necrosis factor alpha) at 1 μM in LPS (lipopolysaccharide)-stimulated macrophages (Zelova et al. 2014).

Ethanol (95%) extract of *M. australis* roots was reported as significantly inhibiting nitrite and prostaglandin E2 production with the dosage of 10 μg/ml ($P < 0.05$) to 40 μg/ml ($P < 0.01$) in LPS-stimulated RAW264.7 cells (Tseng et al. 2018).

Antityrosinase Activity

Optimized condition extracts of *M. nigra* fruits obtained by subcritical water and microwave assisted extraction systems provided IC_{50} (half maximal inhibitory concentration) of 1.71 mg/ml and 1.60 mg/ml, respectively whereas anthocyanin rich fraction isolated from microwave extracts had an IC_{50} of 0.16 mg/ml for tyrosinase inhibitory activity (Koyu et al. 2017, Koyu et al. 2018).

Ethanol (95%) extract of *M. nigra* leaves had an IC_{50} of 5 μg/ml which was close to the tyrosinase inhibitory activity potency of the standard drug, kojic acid (IC_{50}: 3 μg/ml) (de Freitas et al. 2016).

An isolated compound from *M. nigra* roots, named as morachalcone A, was reported as having remarkable tyrosinase inhibitory activity with IC_{50} of 0.14 μM when compared to kojic acid (IC_{50}: 47 μM) (Zheng et al. 2010).

A chalcone derivative isolated from ethanol (95%) extract of *M. nigra* stem was shown to possess stronger tyrosinase inhibitory activity (IC_{50}: 0.95 μM) than kojic acid (IC_{50}: 25 μM) (Zhang et al. 2009).

Cytotoxic Activity

Supercritical carbondioxide and ethanol (70%) extracts of *M. nigra* leaves had IC_{50} of 26, 27, 30 μg/ml and 9, 16, 13 μg/ml against Hep2c (Human cervical carcinoma), RD (human rhabdomyosarcoma), L2OB (murine fibroblast) cell lines while IC_{50} of *M. alba* leaf extracts were 56, 350, 50 μg/ml and 29, 42, 25 μg/ml, respectively (Radojković et al. 2016).

DMSO extract of *M. nigra* fruits presented cytotoxic activity against PC-3 (human prostate carcinoma) cell line with IC_{50} of 370 μg/ml while cisplatin's was 0.6 μg/ml (Turan et al. 2017).

Isolated compounds from chloroform extract of *M. alba* and *M. nigra* root bark, namely as morusinol and sanggenon E, was shown to possess cytotoxic activity against THP-1 (human monocytic leukemia) cell line with IC_{50} of 4.3 and 4.0 μM, respectively (Zelova et al. 2014).

Antidiabetic Activity

Chalcone derivative compounds isolated from methanol extract of *M. alba* root bark were determined as showing strong α-glucosidase inhibitory activity with IC_{50} values ranging between 2.90–3.64 μM through competitive type inhibition while IC_{50} of acarbose was 203.97 μM (Ha et al. 2018).

Methanol (70%) extracts obtained from fruits, leaves, branches and barks of *M. alba* were reported as having α-glucosidase inhibitory activity with IC_{50} of 0.63, 1.03, 0.40 and 0.08 mg/ml, respectively, as utmost values among 24 different samples (Chen et al. 2018).

Isolated compounds, namely a 4H-chromen-4-one derivative and sanggenol H, from ethanol (90%) extract of *M. nigra* twigs were found to provide strong α-glucosidase inhibitory activity with IC_{50} of 1.63 and 1.43 μM, respectively, whereas acarbose's IC_{50} was 987.90 μM (Xu et al. 2018).

In Vivo Studies

Antioxidant Activity

Freeze-dried *M. alba* fruits were given to high fat diet fed male Wistar rats for 4 weeks. The content of lipid peroxidation product, thiobarbituric acid related substances, was determined as significantly decreased in the serum and liver, while superoxide dismutase and blood glutathione peroxidase activities were found increased ($P < 0.05$) in liver (Yang et al. 2010).

Alcohol (95%) extract of *M. nigra* fruits was given to mice at 100-400 mg/kg doses for 30 days. When compared to the control group, the decrease of malonaldehyde levels in serum and liver ($P < 0.01$) and increase of superoxide dismutase (400 mg/kg group ($P < 0.01$) to 100 and 200 mg/kg group $P < 0.05$)), and catalase, glutathione peroxidase (200 and 400 mg/kg group ($P < 0.01$) to 100 mg/kg group ($P < 0.05$)) activity was noticed, while potent antioxidant activity was reported in the intervention group (Feng et al. 2015).

Antiinfective Activity

A product prepared from *M. nigra* leaves provided a significant reduction ($P < 0.05$) in parasitemia in male Swiss mice which were infected with *Trypanosoma cruzi* (Chagas disease). Stimulation of antioxidant defence and reduction of inflammatory tissue damage were identified as the action mechanisms against Chagas disease progression (Montenote et al. 2017).

Antiinflammatory Activity

Ethanol (70%) extract of *M. alba* leaves was administered orally at 150 mg/kg dose to Swiss mice with carrageenan induced air pouch. Inhibitory activity on leukocyte migration was determined as 63% which was equivalent to the results obtained by the intraperitoneal administration of standard drug indomethacin at 10 mg/kg dose. Antiinflammatory action of mulberry extract was attributed to the presence of chlorogenic acid and flavonoids in the composition of extracts (Oliveira et al. 2016).

Methanol extract of *M. alba* root bark was given at 300 mg/kg dose for 3 weeks to male albino Wistar rats with ethanol and cerulein induced pancreatitis. Biochemical investigations on serum caspase-1, IL-1beta, and IL-18 levels indicate significant antiinflammatory activity of mulberry extract which was also supported with histopathological observations ($P = 0.000$). Activity of the extract was associated with the presence of flavonoid compounds (Kavitha and Geetha 2018).

Dichloromethane fraction of ethanol (50%) extract of *M. nigra* leaves exerted significant anti-inflammatory activity (50% and 53% inhibition, $P < 0.05$) and inhibited granulomatous tissue formation (32% and 49% inhibition, $P < 0.01$) as a result of oral administration of 100 and 300 mg/kg doses in male rats with carrageenan induced paw edema. In the extracts, betulinic acid, β-sitosterol and germanicol were pointed out as the main compounds responsible for the antiinflammatory activity (Padilha et al. 2010).

Antinociceptive Activity

Flavonoid rich extracts obtained from fruits of *M. alba* (48 seconds) and *M. nigra* (60 seconds) were found as exerting significant antinociceptive activity ($P < 0.05$) when compared to control group (122 seconds) by reducing the duration of inflammatory pain phase, while *M. mongolica* extracts did not show such activity (Chen et al. 2018).

Dichloromethane extract of *M. nigra* leaves was shown to possess significant antinociceptive activity in visceral and central pain models when administered orally at 100 and 300 mg/kg to male Swiss mice ($P < 0.05$). The extract had a stronger effect at 300 mg/kg dose than indomethacin (5 mg/kg, p.o. (per oral)) and morphine (10 mg/kg, p.o. (per oral); de Mesquita Padilha et al. 2009).

Wound Healing Activity

Ethanol (70%) extracts of *M. alba* leaves and fruits were combined with 2:1 ratio and given orally 5 times a week at 500 mg/kg/day dose for 12 weeks to male C57BL/6 mice with high fat diet induced obesity. Supplementation of combined mulberry extracts were denoted as having significant wound healing activity ($P < 0.05$) in obese mice compared to nontreated group through regulation of NLRP3 inflammasome which stimulates secretion of IL-1 beta (Eo and Lim 2016).

Antidiabetic Activity

Polysaccharide rich fractions obtained from hot water (80°C) extract of *M. alba* fruits were intragastrically given at 400 mg/kg for 7 weeks to male Wistar rats with streptozocin and high-fat diet induced diabetes. In the intervention group, significant decrease in fasting blood glucose, serum insulin and glycated serum protein levels were determined ($P < 0.01$; Jiao et al. 2017).

A preparation containing ethanol (50%) extract of *M. alba* leaves was orally given at 480 mg/kg dose once a day during 21 days to male Sprague-Dawley rats with streptozocin induced diabetes. Significant blood glucose lowering effect of mulberry extract at 30 min (20%) and 60 min (29%) after starch administration was reported in the intervention group compared to the control ($P < 0.05$; Hwang et al. 2016).

Ethanol (60%) extract of *M. alba* branches were given at 1,000 mg/kg dose for 22 days to male ICR mice with streptozocin induced diabetes. Significant effect of mulberry extract on the decrease of fasting blood and plasma glucose levels were determined when compared to the control group ($P < 0.05$; Ahn et al. 2017).

Ethanol (80%) extract of *M. nigra* leaves was given orally at doses of 10–50 mg/kg body weight for 1 week to male albino mice with nicotinamide-streptozocin induced diabetes. Fasting and 2 hours postprandial blood glucose levels were 189.16 ± 50.95 mg/dl and 255.00 ± 12.78 mg/dl; 107.50 ± 32.58 mg/dl and 100.00 ± 28.11 mg/dl for 10 mg/kg and 50 mg/kg doses, respectively. Extracts were defined as having significant antidiabetic activity ($P < 0.05$; AbouZid et al. 2014).

Aqueous extract of *M. rubra* leaves was administered at 400 mg/kg for 21 days to male Wistar rats with streptozocin induced diabetes using orogastric cannula. Significant reduction in glycosylated haemoglobin with increase in plasma insulin and C-peptide levels which were supported by the histopathological findings showing the increase in number of pancreatic islets and β-cells, indicated the antidiabetic activity of mulberry extracts in the treatment group ($P < 0.01$; Sharma et al. 2010).

M. nigra leaf extract (70% ethanol:water, 1:1) was administered at a dose of 500 mg/kg for 30 days to female albino Fischer rats with alloxan induced diabetes. The leaf extract was found as decreasing the superoxide dismutase/catalase ratio, carbonylated protein levels, oxidative damage, matrix metalloproteinase activity and increasing insulinemia. It was evaluated that leaf extract might serve as a protective and therapeutic agent as it was effective against oxidative damage and complications related to diabetes (Araujo et al. 2015).

Oral administration of ethanol (90%) extract of *M. nigra* leaves at 400 mg/kg dose for 14 days to male Wistar rats with alloxan induced diabetes was shown to provide decrease in fasting and postprandial glucose levels, decrease of lipolysis and proteolysis, and improved oral glucose tolerance. It was determined that blood cholesterol, triglyceride, VLDL levels were decreased; HDL and decreased glutathione levels were increased in the intervention group compared to the control group ($P < 0.05$). The acute use of the extract was suggested as preventing hyperglycemia and diabetic comorbidities (Junior et al. 2017).

Antidepressant Activity

Ethanol (95%) extract of *M. alba* leaves was orally given at 100 mg/kg dose for 8 weeks to female Wistar rats with high cholesterol diet induced obesity for the investigation of mood, cognitive and motor activity deficits. Motor deficit, declined memory, depression and anxiety like behaviors linked to visceral adipose tissue were found as attenuated with the administration of mulberry extract in the intervention group compared to nontreated hypercholesterolemic group ($P < 0.05$). Activity was attributed to the presence of polyphenolic and flavonoid compounds in the extracts (Metwally et al. 2019).

Hot water extracts of *M. nigra* leaves when administered orally at 3–100 mg/kg dose to male Swiss mice for 7 days demonstrated anti-depressant like activity in open-field tests as compared to fluoxetine. Syringic acid was pointed out as the main compound that was responsible for the antidepressant activity (Dalmagro et al. 2017).

Antihyperlipidemic Activity

Freeze-dried *M. alba* fruits were added to diet of male Wistar rats which were on high fat diet during 4 weeks. Serum and liver triglyceride, total cholesterol, low-density lipoprotein cholesterol levels and atherogenic index had significantly decreased while serum high-density lipoprotein cholesterol levels increased in the intervention group compared to the non-treated group ($P < 0.05$; Yang et al. 2010).

A tea preparation obtained with *M. alba* leaves was given for 4 weeks to male Sprague-Dawley rats with streptozocin induced diabetes. In the treatment group, serum total cholesterol levels were determined as significantly decreased and HDL-cholesterol levels as increased compared to the diabetic control group ($P < 0.05$; Wilson and Islam 2015).

Methanol extract of *M. alba* root bark was administered at 160 mg/kg dose for 4 weeks to male Wistar rats with high fat emulsion induced hyperlipidemia. Levels of serum total cholesterol, triglyceride ($P < 0.05$), LDL-cholesterol ($P < 0.01$) and activity of ALT, AST ($P < 0.05$) were reported as significantly lowered while HDL cholesterol levels had increased ($P < 0.01$; Qi et al. 2016).

Hot water extract of *M. nigra* leaves was given by gavage at 100–400 mg/kg dose twice a day to male Wistar rats with Triton WR-1339 induced hyperlipidemia for three consecutive days. Significant decrease of serum total cholesterol, LDL, triglyceride levels with an increase of HDL levels were determined compared to the control group ($P < 0.05$; Zeni et al. 2017).

Gastrointestinal Motility Regulating Activity

Aqueous methanol (70%) extract of *M. nigra* fruits was investigated for the effect on gastrointestinal motility. When the fruit extract was orally given to the mice at 30–70 mg/kg dose (34% activity), similar laxative activity to carbachol (43% activity) was observed ($P < 0.01$). At higher doses as 100–500 mg/kg, fruit extract showed significant protective activity against castor oil-induced diarrhoea ($P < 0.01$). The protective efficacy of extracts at 500 mg/kg dose (82% activity) against diarrhoea was equivalent to loperamide (80%) reference drug (Akhlaq et al. 2016).

Hepatoprotective Activity

Hepatoprotective activity of lectin compound isolated from *M. nigra* was investigated with lipopolysaccharide induced oxidative stress model in rats showing decreased glutathione and antioxidant enzyme levels, increased serum glucose and bilirubin levels, increased APL and transaminase activity, increased liver weight and decreased body weight conditions. It was observed that lipid peroxidation levels and serum glucose, bilirubin, AST, ALT levels and APL activity decreased and liver GSH levels increased with lectin administration. The antioxidant enzymes glutathione-peroxidase and glutathione-S-transferase returned to normal levels. The findings of liver histological examination also support the improvement of biochemical parameters (Ahlem and Youcef 2016).

Methanol (70%) extract of *M. nigra* leaves was orally administered at 250 and 500 mg/kg doses for 7 days to albino Swiss mice with paracetamol induced hepatotoxicity. As a result, significant decrease in ALT, AST, ALP and total bilirubin levels were reported which was equivalent to silymarin ($P < 0.001$). Histopathological findings also indicated the recovery from hepatocyte necrosis, inflammation and sinusoidal constriction (Mallhi et al. 2014).

Ethanol (80%) extract of *M. nigra* fruits was given at 100 mg/kg dose by gavage to male rats fed with high-fat diet. Protective effect of mulberry extract against high-fat diet induced obesity, hepatic steatosis and insulin resistance, increase of fatty acid oxidation and decrease of cholesterol synthesis ($P < 0.05$) was determined (Song et al. 2016).

Ethanol (50%) extract of *M. nigra* leaves was administered at 500 mg/kg dose for 14 days to male albino rats with methotrexate induced hepatoxicity. Significant reduction in AST, ALT, ALP and LDH levels occurred in the intervention group ($P < 0.05$). Moderate protection on liver cells against methotrexate induced injury was supported with histopathological findings (Tag 2015).

Radioprotective Activity

Aqueous ethanol extracts from *M. alba and M. nigra* leaves were shown to possess antimutagenic activity in Wistar rats exposed to gamma rays (3 Gy) by decreasing cellular chromosomes aberrations 58% to 64%. Levels of chromosome mutations induced by sodium fluoride (20 mg/100 g animal weight) were effectively reduced by administration of the extracts as 53% and 66%, respectively (Agabeyli 2012).

The water extract of *M. nigra* fruits was administered intraperitoneally at a dose of 200 mg/kg for 6 days (3 days before and after irradiation) to male Wistar rats that were exposed to gamma rays (3 and 6 Gy). Significant reduction in the frequency of micronucleated polychromatic erythrocytes and micronucleated normchromatic erythrocytes were determined ($P < 0.05$). Decrease of malonaldehyde and superoxide dismutase levels with an increase of liver total thiol content and catalase activity were reported in the intervention group compared to nontreated irradiated group (Ghasemnezhad Targhi et al. 2017).

Nephroprotective Activity

Powdered *M. alba* leaves were added to diet of male Wistar rats with streptozocin induced diabetes. Supplementation of mulberry leaves to diet was reported (Abignan Gurukar et al. 2012) providing significant reduction in the synthesis of extracellular matrix components like laminin and fibronectin in kidneys, which ameliorate increased albumin excretion and glomerular filtration rate as compared to nontreated diabetic group ($P < 0.001$).

Hydroalcoholic extract of *M. nigra* fruits provide a significant decrease in urea levels when given at 800 mg/kg for 8 weeks to rats having alloxan induced diabetes compared to the control group ($P < 0.05$). Histopathological findings also showed mulberry extract's effectiveness against kidney tissue damage (Rahimi-Madiseh et al. 2017).

Gastroprotective Activity

Methanol extract of *M. nigra* fruits given at 300 mg/kg dose to female Swiss mice with acidified ethanol induced acute gastric ulcer. The reduction in ulcer area to 64% followed the reduction of lipoperoxides and glutathione amount in the gastric mucosa (Nesello et al. 2017) demonstrated as the indicators of gastroprotectice activity ($P < 0.01$).

Neuroprotective Activity

Acetone (70%) extract of *M. alba* leaves were administered intraperitoneally at 100 µg/ml/kg dose for 2 weeks to female Wistar rats with glyphosate (100 mg/kg) induced toxicity. Study results indicated the neuroprotective activity of mulberry extract in the brain by alleviating oxidative stress, LDH levels, protein carbonylation ($P < 0.01$) and increasing superoxide dismutase activity ($P < 0.05$) when compared to non-treated group (Rebai et al. 2017).

Clinical Trials

In a randomized, double blind clinical trial, standardized preparations containing ethanol (50%) extract of *M. alba* leaves were given to 46 female patients of 50–70 years with impaired glucose tolerance. Significant decrease in blood glucose levels 30 min after the meal was determined ($P < 0.05$). Lowering effect on postprandial glucose was attributed to the presence of deoxynojirimycin (5.20 mg/g extract) in the mulberry extracts (Hwang et al. 2016).

A randomized, double-blind, placebo-controlled clinical trial was conducted on 45 menopausal participants (45–60 years) to investigate the effect of consuming a *M. alba* leaf extract preparation (1500 mg/day) for improving bone turnover markers. At the end of 8 weeks trial, serum osteocalcin ($P < 0.01$) and alkaline phosphatase ($P < 0.05$)

levels were detected as significantly increased while serum beta CTX levels decreased ($P < 0.01$) compared to the placebo group. Study results indicate enhanced bone formation and decreased bone resorption, which stated that consumption of the mulberry leaf extract could decrease osteoporosis risk (Wattanathorn et al. 2018).

In an 8 weeks, randomized clinical trial, a total of 72 volunteers (54 male, 18 female, aged 25–65 years) with hyperlipidemia were divided into two groups. About 300 ml *M. nigra* fruit juice with pulp was given to intervention group per day while control group followed their regular diet. At the end of the study, apolipoprotein A1 ($P = 0.015$) and HDL cholesterol ($P = 0.001$) levels in the intervention group increased significantly compared to initial levels of apolipoprotein B ($P = 0.044$), LDL cholesterol ($P = 0.04$), while systolic blood pressure ($P = 0.005$) got significantly decreased. In *M. nigra* juice consuming group when compared to the control group, significant differences were observed for apolipoprotein A1 ($P = 0.005$), HDL ($P = 0.014$), hs-CRP ($P = 0.01$) levels and apolipoprotein B/apolipoprotein A1 ratio ($P = 0.009$) (Aghababaee et al. 2015).

Pharmacokinetics

Cyanidin-3-glycoside (79%) and cyanidin-3-rutinoside (19%) rich extracts obtained from *M. nigra* fruits were orally administered to Wistar rats. Anthocyanin concentrations in plasma, kidney and gastrointestinal tract were determined. Following 15 min of oral administration, total cyanidin level reached maximum concentration in plasma (5.7 ± 1.4 µg total cyanidin/ml) and kidneys (10.6 ± 2.5 µg total cyanidin/g) and at the end of 8 hours anthocyanins were not detected. Further after ingestion, it was found that the cyanidin glycosides reached the maximum concentration in 15 min in the stomach and small intestine, but 3 hours in the colon. At the end of 8 hours, 0.11% of the initial amount of anthocyanin-rich extract was absorbed from the gastrointestinal tract (Hassimotto et al. 2008).

Toxicity

Intragastric administration of polysaccharide rich fractions obtained from hot water (80°C) extract of *M. alba* fruits at 1,000 mg/kg dose to Wistar rats for 1 week did not cause significant changes in behaviors that was observed in emaciation, posture and respiratory distress conditions. Absence of mortality during 1st week of the study was observed as mulberry extract did not cause acute toxicity (Jiao et al. 2017).

Acute oral toxicity of the ethanol (50%) extract of *M. alba* leaves was determined with the administration of oral doses up to 2,000 mg/kg for 14 days to ICR mice. No abnormal symptoms like decrease in weight or death was reported since LD_{50} was found higher than 2,000 mg/kg (Hwang et al. 2016).

Water:ethanol (80%) extract of *M. nigra* leaves did not cause any toxic symptoms or death to mice after 48 hours with the oral dose of 2,000 mg/kg body weight (AbouZid et al. 2014).

M. nigra tincture was given orally to pregnant rats at doses of 0.4, 0.8 and 1.0 mg/kg/day from the first day to the twentieth day of pregnancy. No effect on weight gain, plasma urea, creatinine, AST, ALT levels and fetal viability during pregnancy was observed. At given doses, it did not cause morphological anomalies, maternal toxicity and teratogenicity (Almeida et al. 2014).

Flavonoid rich extracts obtained with ethanol:water (48%) from fruits of *M. alba*, *M. nigra* and *M. mongolica* did not cause significant weight change in mice at a dose of 5 g/kg while dexamethasone caused weight loss and side effects at 1.5 mg/kg (Chen et al. 2018).

Water extract of *M. nigra* leaves was given orally two times a day at 1 mg/kg dose to streptozocin-induced diabetic mice for 42 days. Both in the control and intervention groups, there was significant increase in serum creatinine level, SGPT and SGOT activities. The results further interpreted that the use of *M. nigra* leaf extracts in diabetes should also be monitored for liver and kidney functions (Hemmati et al. 2010).

Dichloromethane extract of *M. nigra* leaves was reported as not causing any effect on the behavior or death to Swiss male mice at 0.5–5 g/kg following seven days of oral administration. The oral LD_{50} value was found higher than 5 g/kg (Padilha et al. 2010).

Ethanol (70%) extract of *M. nigra* leaves did not cause toxic effects on embryo development and female reproductive system at 15th day of gestation when given orally to female Wistar rats in the range of 25–700 mg/kg (de Queiroz et al. 2012).

Conclusion

In this chapter, pharmacological properties of preparations, extracts and active compounds obtained from various *Morus* species were presented under the sections as *ex vivo*, *in vivo* studies and clinical trials. An assuring evaluation regarding the results of *ex vivo* and *in vivo* activity studies could only be maintained by the researches conducted by separate groups in distinct laboratories. Therefore, only consistent results could be the major indicator of reliability.

Variations within the same species resulting from climatic and regional differences have remarkable impact on the results of present studies. Standardization of cultivation conditions hold a high level of importance, so about standardization of extracts and preparations while avoiding contamination and adulteration risks.

Results of clinical trials have great importance as they bring out the highest level of evidence. Most outstanding activities noted were antidiabetic and antihyperlipidemic activities that could be confirmed with clinical trials following *in vivo* studies. Also, antiinflammatory and hepatoprotective activities were prominent. Active compounds in extracts from different parts of mulberry such as phenolics, flavonoids, anthocyanins and alkaloids could be pointed out as the key factor for the high potency of mulberry products against a range of diseases related to oxidative and inflammatory mechanisms. Nephroprotective, gastroprotective and wound healing activities that were included in folk medicine practices, had also been confirmed *in vivo*. While evaluating and comparing the pharmacological potential of different preparations, various parameters such as the used plant parts, extraction conditions and solvent composition should be carefully considered. Results of clinical studies confirming the activity potentials determined by *ex vivo* and *in vivo* studies including toxicity data, can lead to the development of standardized pharmaceutical products for the treatment of diseases within the aspect of evidence based therapy.

References

Abbasi, A.M., M.A. Khan, M. Ahmad et al. 2014. Ethnobotanical and taxonomic screening of genus *Morus* for wild edible fruits used by the inhabitants of Lesser Himalayas-Pakistan. J. Med. Plant. Res. 8: 889–898.

Abignan Gurukar, M.S., C.D. Nandini, S. Mahadevamma and P.V. Salimath. 2012. Ocimum sanctum and *Morus alba* leaves and Punica granatum seeds in diet ameliorate diabetes-induced changes in kidney. J. Pharm. Res. 5: 4729–4733.

AbouZid, S.F., O.M. Ahmed, R.R. Ahmed, A. Mahmoud, E. Abdella and M.B. Ashour. 2014. Antihyperglycemic effect of crude extracts of some Egyptian plants and algae. J. Med. Food 17: 400–6.

Agabeyli, R.A. 2012. Antimutagenic activities extracts from leaves of the *Morus alba, Morus nigra* and their mixtures. Int. J. Biol. 4: 166–172.

Aghababaee, S.K., M. Vafa, F. Shidfar et al. 2015. Effects of blackberry (*Morus nigra* L.) consumption on serum concentration of lipoproteins, apo A-I, apo B, and high-sensitivity-C-reactive protein and blood pressure in dyslipidemic patients. J. Res. Med. Sci. 20: 684–91.

Ahlem, B. and N. Youcef. 2016. Hepatoprotective and anti-inflammatory effects of new lectin purified from *Morus nigra* against lipopolysaccharide induced oxidative stress in rats. Toxicol. Lett. 258: S105.

Ahn, Eunyeong, Jimin Lee, Young-Hee Jeon, Sang-Won Choi and Eunjung Kim. 2017. Anti-diabetic effects of mulberry (*Morus alba* L.) branches and oxyresveratrol in streptozotocin-induced diabetic mice. Food Sci. Biotechnol. 26: 1693–1702.

Akhlaq, A., M.H. Mehmood, A. Rehman et al. 2016. The prokinetic, laxative, and antidiarrheal effects of morus nigra: possible muscarinic, Ca(2+) channel blocking, and antimuscarinic mechanisms. Phytother. Res. 30: 1362–76.

Akhtar, Nosheen, Haq Ihsan ul and Bushra Mirza. 2015. Phytochemical analysis and comprehensive evaluation of antimicrobial and antioxidant properties of 61 medicinal plant species. Arab. J. Chem.

Almeida, Thais F., Daniele O. Benedicto, Giovana Sabadim, Thaisa B. Pickler, Denise Grotto and Marli Gerenutti. 2014. Reproductive performance of pregnant rats: Effects of Morus nigra tincture. Reprod. Toxicol. 48: 24.

Araujo, C.M., P. Lucio Kde, M.E. Silva et al. 2015. Morus nigra leaf extract improves glycemic response and redox profile in the liver of diabetic rats. Food Funct. 6: 3490–9.

Arfan, M., R. Khan, A. Rybarczyk and R. Amarowicz. 2012. Antioxidant activity of mulberry fruit extracts. Int. J. Mol. Sci. 13: 2472–80.

Boeing, J.S., E.E.O. Barizao, B.C.E. Silva, P.F. Montanher, V.D. Almeida and J.V. Visentainer. 2014. Evaluation of solvent effect on the extraction of phenolic compounds and antioxidant capacities from the berries: application of principal component analysis. Chem. Cent. J. 8: 48.

Chen, H., W. Yu, G. Chen, S. Meng, Z. Xiang and N. He. 2018. Antinociceptive and antibacterial properties of anthocyanins and flavonols from fruits of black and non-black mulberries. Molecules 23.

Chen, Z., X. Du, Y. Yang, X. Cui, Z. Zhang and Y. Li. 2018. Comparative study of chemical composition and active components against alpha-glucosidase of various medicinal parts of *Morus alba* L. Biomed. Chromatogr. 32: e4328.

Dalmagro, A.P., A. Camargo and A.L.B. Zeni. 2017. *Morus nigra* and its major phenolic, syringic acid, have antidepressant-like and neuroprotective effects in mice. Metab. Brain Dis. 32: 1963–1973.

de Freitas, M.M., P.R. Fontes, P.M. Souza et al. 2016. Extracts of *Morus nigra* L. leaves standardized in chlorogenic acid, rutin and isoquercitrin: Tyrosinase inhibition and cytotoxicity. PLoS One 11:e0163130.

de Mesquita Padilha, M., F.C. Vilela, M.J. da Silva, M.H. dos Santos, G. Alves-da-Silva and A. Giusti-Paiva. 2009. Antinociceptive effect of the extract of *Morus nigra* leaves in mice. J. Med. Food 12: 1381–1385.

de Oliveira, Alisson Mac, Matheus da Silva Mesquita, Gabriela Cavalcante da Silva et al. 2015. Evaluation of toxicity and antimicrobial activity of an ethanolic extract from leaves of *Morus alba* L. (Moraceae). Evid. Based Complement. Alternat. Med. 2015: 7.

de Queiroz, G.T., T.R. Santos, R. Macedo et al. 2012. Efficacy of *Morus nigra* L. on reproduction in female Wistar rats. Food Chem. Toxicol. 50: 816–822.

Eo, H. and Y. Lim. 2016. Combined mulberry leaf and fruit extract improved early stage of cutaneous wound healing in high-fat diet-induced obese mice. J. Med. Food 19: 161–169.

Feng, Rui-Zhang, Qin Wang, Wen-Zhi Tong et al. 2015. Extraction and antioxidant activity of flavonoids of Morus nigra. Int. J. Clin. Exp. Med. 8: 22328–22336.

Ghasemnezhad Targhi, Reza, Mansour Homayoun, Somaieh Mansouri, Mohammad Soukhtanloo, Shokouhozaman Soleymanifard and Masoumeh Seghatoleslam. 2017. Radio protective effect of black mulberry extract on radiation-induced damage in bone marrow cells and liver in the rat. Radiat. Phys. Chem. 130: 297–302.

Ha, M.T., S.H. Seong, T.D. Nguyen et al. 2018. Chalcone derivatives from the root bark of *Morus alba* L. act as inhibitors of PTP1B and alpha-glucosidase. Phytochemistry 155: 114–125.

Hassimotto, N.M., M.I. Genovese and F.M. Lajolo. 2008. Absorption and metabolism of cyanidin-3-glucoside and cyanidin-3-rutinoside extracted from wild mulberry (*Morus nigra* L.) in rats. Nutr. Res. 28: 198–207.

Hemmati, M.R., A. Hemmati, M.J. Hemmati, I.R. Hemmati and H.T. Hemmati. 2010. Effects of aqueous extract of black mulberry (*Morus nigra*) on liver and kidney of diabetic mice. Toxicol. Lett. 196: S213–S214.

Hwang, Seung Hwan, Hong Mei Li, Soon Sung Lim, Zhiqiang Wang, Jae-Seung Hong and Bo Huang. 2016. Evaluation of a standardized extract from *Morus alba* against α-glucosidase inhibitory effect and postprandial antihyperglycemic in patients with impaired glucose tolerance: A randomized double-blind clinical trial. Evid. Based Complement. Alternat. Med. 2016: 8983232.

Iqbal, S., U. Younas, Sirajuddin, K.W. Chan, R.A. Sarfraz and K. Uddin. 2012. Proximate composition and antioxidant potential of leaves from three varieties of Mulberry (*Morus* sp.): a comparative study. Int. J. Mol. Sci. 13: 6651–6664.

Jiao, Yukun, Xueqian Wang, Xiang Jiang, Fansheng Kong, Shumei Wang and Chunyan Yan. 2017. Antidiabetic effects of Morus alba fruit polysaccharides on high-fat diet- and streptozotocin-induced type 2 diabetes in rats. J. Ethnopharmacol. 199: 119–127.

Junior, Iids, H.M. Barbosa, D.C.R. Carvalho et al. 2017. Brazilian *Morus nigra* attenuated hyperglycemia, dyslipidemia, and prooxidant status in alloxan-induced diabetic rats. ScientificWorld Journal 2017: 5275813.

Kavitha, Y. and A. Geetha. 2018. Anti-inflammatory and preventive activity of white mulberry root bark extract in an experimental model of pancreatitis. J. Tradit. Complement. Med. 8: 497–505.

Koyu, Halil, Aslihan Kazan, Taylan Kurtulus Ozturk, Ozlem Yesil-Celiktas and Mehmet Zeki Haznedaroglu. 2017. Optimizing subcritical water extraction of *Morus nigra* L. fruits for maximization of tyrosinase inhibitory activity. J. Supercrit. Fluids 127: 15–22.

Koyu, Halil, Aslihan Kazan, Serdar Demir, Mehmet Zeki Haznedaroglu and Ozlem Yesil-Celiktas. 2018. Optimization of microwave assisted extraction of *Morus nigra* L. fruits maximizing tyrosinase inhibitory activity with isolation of bioactive constituents. Food Chem. 248: 183–191.

Mallhi, Tauqeer Hussain, M. Imran Qadir, Yusra Habib Khan and Muhammad Ali. 2014. Hepatoprotective activity of aqueous methanolic extract of *Morus nigra* against paracetamol-induced hepatotoxicity in mice. Bangladesh J. Pharmacol. 9: 60–66.

Mascarello, A., A.C. Orbem Menegatti, A. Calcaterra et al. 2018. Naturally occurring Diels-Alder-type adducts from *Morus nigra* as potent inhibitors of Mycobacterium tuberculosis protein tyrosine phosphatase B. Eur. J. Med. Chem. 144: 277–288.

Metwally, Fateheya Mohamed, Hend Rashad and Asmaa Ahmed Mahmoud. 2019. *Morus alba* L. Diminishes visceral adiposity, insulin resistance, behavioral alterations via regulation of gene expression of leptin, resistin and adiponectin in rats fed a high-cholesterol diet. Physiol. Behav. 201: 1–11.

Montenote, M.C., V.Z. Wajsman, Y.T. Konno et al. 2017. Antioxidant effect of *Morus nigra* on Chagas disease progression. Rev. Inst. Med. Trop. Sao Paulo 59: e73.

Nesello, L.A.N., Mlml Beleza, M. Mariot et al. 2017. Gastroprotective value of berries: Evidences from methanolic extracts of *Morus nigra* and Rubus niveus fruits. Gastroenterol. Res. Pract. 2017: 7089697.

Niratker, C.R. and M.S. Preeti. 2015. Antimicrobial activity of leaf extract of *Morus indica* (Mulberry) from Chhattisgarh. Asian J. Plant Sci. Res. 5: 28–31.

Oliveira, Alisson Macário de, Matheus Ferreira do Nascimento, Magda Rhayanny Assunção Ferreira et al. 2016. Evaluation of acute toxicity, genotoxicity and inhibitory effect on acute inflammation of an ethanol extract of *Morus alba* L. (Moraceae) in mice. J. Ethnopharmacol. 194: 162–168.

Padilha, M.M., F.C. Vilela, C.Q. Rocha et al. 2010. Antiinflammatory properties of *Morus nigra* leaves. Phytother. Res. 24: 1496–1500.

Qi, Shi-Zhou, Na Li, Zheng-Dong Tuo et al. 2016. Effects of *Morus* root bark extract and active constituents on blood lipids in hyperlipidemia rats. J. Ethnopharmacol. 180: 54–59.

Radojković, Marija, Zoran Zeković, Pavle Mašković et al. 2016. Biological activities and chemical composition of *Morus* leaves extracts obtained by maceration and supercritical fluid extraction. J. Supercrit. Fluids 117: 50–58.

Rahimi-Madiseh, M., A. Naimi, E. Heydarian and M. Rafieian-Kopaei. 2017. Renal biochemical and histopathological alterations of diabetic rats under treatment with hydro alcoholic *Morus nigra* extrac. J. Renal Inj. Prev. 6: 56–60.

Rebai, O., M. Belkhir, A. Boujelben, S. Fattouch and M. Amri. 2017. *Morus alba* leaf extract mediates neuroprotection against glyphosate-induced toxicity and biochemical alterations in the brain. Environ. Sci. Pollut. Res. Int. 24: 9605–9613.

Seong, S.H., M.T. Ha, B.S. Min, H.A. Jung and J.S. Choi. 2018. Moracin derivatives from *Morus* Radix as dual BACE1 and cholinesterase inhibitors with antioxidant and anti-glycation capacities. Life Sci. 210: 20–28.

Sharma, S.B., S. Gupta, R. Ac, U.R. Singh, R. Rajpoot and S.K. Shukla. 2010. Antidiabetogenic action of *Morus rubra* L. leaf extract in streptozotocin-induced diabetic rats. J. Pharm. Pharmacol. 62: 247–255.

Song, H., J. Lai, Q. Tang and X. Zheng. 2016. Mulberry ethanol extract attenuates hepatic steatosis and insulin resistance in high-fat diet-fed mice. Nutr. Res. 36: 710–718.

Souza, G.R., R.G. Oliveira-Junior, T.C. Diniz et al. 2018. Assessment of the antibacterial, cytotoxic and antioxidant activities of *Morus nigra* L. (Moraceae). Braz. J. Biol. 78: 248–254.

Tag, H.M. 2015. Hepatoprotective effect of mulberry (*Morus nigra*) leaves extract against methotrexate induced hepatotoxicity in male albino rat. BMC Complement. Altern. Med. 15: 252.

Thaipitakwong, T., Surawej Numhom and Pornanong Aramwit. 2018. Mulberry leaves and their potential effects against cardiometabolic risks: a review of chemical compositions, biological properties and clinical efficacy. Pharm. Biol. 56: 109–118.

Tseng, T.H., W.L. Lin, C.K. Chang, K.C. Lee, S.Y. Tung and H.C. Kuo. 2018. Protective effects of *Morus* root extract (mre) against lipopolysaccharide-activated RAW264.7 cells and CCl4 induced mouse hepatic damage. Cell. Physiol. Biochem. 51: 1376–1388.

Turan, I., S. Demir, K. Kilinc et al. 2017. Antiproliferative and apoptotic effect of *Morus nigra* extract on human prostate cancer cells. Saudi Pharm. J. 25: 241–248.

Wattanathorn, J., W. Somboonporn, S. Sungkamanee, W. Thukummee and S. Muchimapura. 2018. A double-blind placebo-controlled randomized trial evaluating the effect of polyphenol-rich herbal congee on bone turnover markers of the perimenopausal and menopausal women. Oxid. Med. Cell. Longev. 2018: 11.

Wilson, R.D. and M.S. Islam. 2015. Effects of white mulberry (*Morus alba*) leaf tea investigated in a type 2 diabetes model of rats. Acta Pol. Pharm. 72: 153–160.

Xu, L., M. Yu, L. Niu et al. 2018. Phenolic compounds isolated from *Morus nigra* and their alpha-glucosidase inhibitory activities. Nat. Prod. Res. 1–8.

Yang, Xiaolan, Lei Yang and Haiying Zheng. 2010. Hypolipidemic and antioxidant effects of mulberry (*Morus alba* L.) fruit in hyperlipidaemia rats. Food Chem. Toxicol. 48: 2374–2379.

Younus, I., A. Fatima, S.M. Ali et al. 2016. A review of ethnobotany, phytochemistry, antiviral and cytotoxic/anticancer potential of Morus alba Linn. Int. J. Adv. Res. Rev. 1: 84–96.

Zelova, H., Z. Hanakova, Z. Cermakova et al. 2014. Evaluation of anti-inflammatory activity of prenylated substances isolated from Morus alba and *Morus nigra*. J. Nat. Prod. 77: 1297–303.

Zeni, A.L.B., T.D. Moreira, A.P. Dalmagro et al. 2017. Evaluation of phenolic compounds and lipid-lowering effect of *Morus nigra* leaves extract. An. Acad. Bras. Cienc. 89: 2805–2815.

Zhang, X., X. Hu, A. Hou and H. Wang. 2009. Inhibitory effect of 2,4,2',4'-tetrahydroxy-3-(3-methyl-2-butenyl)-chalcone on tyrosinase activity and melanin biosynthesis. Biol. Pharm. Bull. 32: 86–90.

Zheng, Z.P., K.W. Cheng, Q. Zhu, X.C. Wang, Z.X. Lin and M. Wang. 2010. Tyrosinase inhibitory constituents from the roots of Morus nigra: a structure-activity relationship study. J. Agric. Food Chem. 58: 5368–5373.

Žugić, Ana, Sofija Đorđević, Ivana Arsić et al. 2014. Antioxidant activity and phenolic compounds in 10 selected herbs from Vrujci Spa, Serbia. Ind. Crop. Prod. 52: 519–527.

Section B

Genetic Improvement of Mulberry using Conventional and Non-Conventional Methods

Sexual Hybridisation and Other Breeding Procedures in Mulberry (*Morus* spp.) to Improve Quality and Productivity of Leaf as well as Other Traits

Kunjupillai Vijayan

INTRODUCTION

Mulberry (*Morus* L.) is a deciduous, perennial tree growing luxuriously in most of the Asian Countries and it has assumed great economic importance because of the use of its leaf for rearing the highly domesticated silkworm *Bombyx mori* L. in order to produce silk fibers (Vijayan et al. 1997a). The rearing silkworm for producing silk is called sericulture which is one of the major avocations, particularly in the villages of India, China, Thailand, and Vietnam, etc. Today, sericulture is a billion dollar business and the annual global silk production stands as 192692 tones (Table 1). Because of the high economic returns, continuous efforts have been made to improve the silk productivity. As a result tremendous growth has been recorded in silk production among major silk producing countries like China and India. Though four types of silks viz., mulberry, eri, tasar and muga are produced in India at commercial scale, 70% of the silk is coming from mulberry silkworm only (Figure 1). Thus, mulberry holds a very important role in the development of silk industry in the country. Mulberry leaf is also being used as animal fodder as it is highly palatable and nutritious (Uribe and Sanchez 2001). The fruit of mulberry is highly delicious and contains a number of health-promoting compounds (Singhal et al. 2009, 2010, Yang et al. 2010). The bark of mulberry root and stem has purgative, anthelmintic, and astringent properties, and is also rich in phenolic compounds such as maclurin, rutin, isoquercitrin, resveratrol and morin (Chang et al. 2011, Wu et al. 2010). Mulberry wood is also used for manufacturing sports articles, turnery items, house construction, agricultural implements, furniture, spokes, poles, shafts and bent parts of carriage and carts (Sharma et al. 2013). Mulberry is also used for landscaping and water conservation (Qin et al. 2012). An elaborated description of these aspects is given in other chapters of this book. Thus, mulberry plays a very important role in the welfare of human being.

Origin and Distribution

It is believed that mulberry originated in the lower slopes of Himalaya and later spread into tropics of the southern hemisphere (Vijayan et al. 2018). Today, mulberry is seen in all regions between 50 °N Lat. and 10 °S Lat. (Vijayan

Research Coordination Section, Central Silk Board, BTM Layout, Madiwala, Bangalore 560068, Karnataka.
Emails: kvijayan01@yahoo.com, nrsinghvi@rediffmail.com

Table 1. World raw silk production during recent years (Source: ISC website: https://www.inserco.org/en/statistics).

#	Country	2010	2011	2012	2013	2014	2015	2016
1	China	115000.0	104000.0	126000.0	130000.0	146000.0	170000.0	158400.0
2	India	21005	23060	23679	26480	28708	28523	30348
3	Uzbekistan	940	940	940	980	1100	1200	1256
4	Thailand	655	655	655	680	692	698	712
5	Brazil	770	558	614	550	560	600	650
6	Vietnam	550	500	450	475	420	450	523
7	North Korea	-	300.0	300.0	300.0	320.0	350.0	365.0
8	Philippines	1.0	10	0.89	1.0	1.1	1.2	182
9	Iran	75	120	123	123	110	120	125
10	Bangladesh	40.0	38.0	42.5	43.0	44.5	44	44
11	Japan	54.0	42.0	30.0	30.0	30.0	30.0	32.0
12	Turkey	18.0	22.0	22.0	25.0	32.0	30.0	32.0
13	Bulgaria	9.4	6.0	8.5	8.5	8.0	8.0	9.0
14	Madagascar	16.0	16.0	18.0	18.0	15.0	5.0	6.0
15	Indonesia	20.0	20.0	20.0	16.0	10.0	8.0	4.0
16	Tunisia	0.12	3.0	3.95	4.0	4.0	3.0	2.0
17	Egypt	0.30	0.70	0.70	0.70	0.82	0.83	1.20
18	South Korea	3.0	3.0	1.50	1.60	1.20	1.0	1.0
19	Syria	0.60	0.50	0.50	0.70	0.50	0.3	0.25
20	Colombia	0.60	0.60	0.60	0.60	0.50	0.5	
Total		139158.0	130285.0	152910.0	159737.0	178057.0	202072.0	192692.0

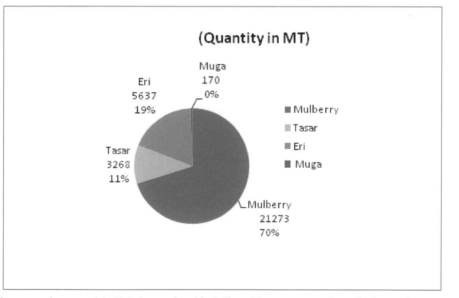

Figure 1. The four type of commercial silk being produced in India and its percentage of contribution to the total silk production (Source: http://csb.gov.in/statistics/silk-production/).

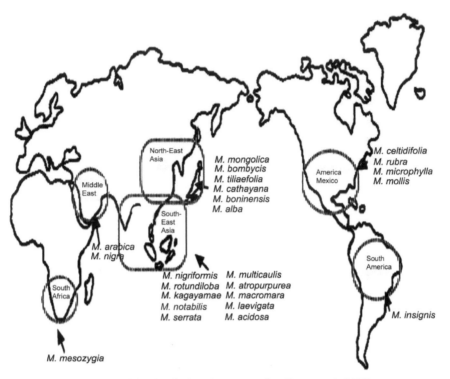

Figure 2. Global distribution of *Morus* species. Sharma et al. (2000).

et al. 2018), from sea level to elevations of 4000 m (Machii et al. 1999), which include Asia, Europe, North and South America, and Africa (Figure 2). Also read Chapter 2.

Taxonomy and Cytology

Although numerous attempts have been made to classify the genus *Morus* based on morphological and phenological characters (Koidzumi 1917, Hotta 1954, Katsumata 1972), none of them could resolve its taxonomy comprehensively to get wider acceptance. One of the major reasons for this ambiguity in the classification is the presence of large number of natural hybrids (Tikader and Dandin 2007) due to natural inter-specific hybridization which is very common in mulberry. Nonetheless, Koidzumi (1917) forwarded a classification dividing the genus *Morus* into two sections: *Dolichostylae* (long style) and *Macromorus* (short style), and further erected two groups in each section, Papillosae and Pubescentae. Based on the nature of stigmatic hairs, morphological characters of leaf, inflorescence and sorosis, Koidzumi (1917) divided the genus into 24 species and one subspecies (Table 2). Later, Hotta (1954) divided the genus into two sections, namely *Dolychocystolithiae* and *Brachycystolithiae*, according to the shape and position of cystolith cells in the leaf, and thus recognized 35 species. These classifications got some acceptance among the taxonomists as well as breeders. Further discussion on taxonomy of mulberry read Chapters 1 and 2.

Cytologically, mulberry exists in different ploidy levels varying from haploid (n = x = 14) to decasoploidy (2n = 22x = 308). Most of the plants from species *M. alba*, *M. indica*, *M. latifolia*, *M. multicaulis*, *M. australis* are diploids while plants from *M. laevigata*, *M. cathayana*, and *M. boninensis* are tetraploids. Hexaploidy is very common in *M. serrata* and *M. tiliaefolia*. A natural haploid *M. notabilis* with 14 chromosomes was also reported in mulberry (Maode et al. 1996).

Table 2. Mulberry species recognized by Koidzumi (1917).

Species	Species
M. bombycis Koidz.	*M. latifolia* Poir.
M. alba L.	*M. acidosa* Griff.
M. indica L.	*M. rotunbiloba* Koidz.
M. kagayamae Koidz.	*M. notabilis* C. K. Schn.
M. boninensis Koidz.	*M. nigriformis* Koidz.
M. atropurpurea Roxb.	*M. serrata* Roxb.
M. laevigata Wall.	*M. nigra* L.
M. formosensis Hotta	*M. rubra* L.
M. mesozygia Stapf.	*M. celtidifolia* Kunth
M. cathayana Hemsl.	*M. tiliaefolia* Makino
M. microphylla Bickl.	*M. macroura* Miq.
M. rabica Koidz.	*M. multicaulis* Perr.

Morphological Characteristics

Mulberry basically a tree is trimmed into bush for the sake of collection of leaves. The color of the shoot varies in different shades like black-brown, yellow-brown, reddish brown, bluish brown, dark gray, and green gray depending on the species. The architecture of the branches also varies from erect to horizontal depending on the species. Mulberry leaf is simple, alternate, stipulate, petiolate, entire or lobed. Though the sex expression in mulberry is greatly influenced by the physiological and chemical processes of the plant, besides being modulated by environmental factors, the dioeciousness is common while monoeciousness is not very rare (Jaiswal and Kumar 1981, Ogure et al. 1980, Sikdar et al. 1988, Tikader et al. 1995). The inflorescence is a catkin with pendent or drooping peduncle bearing unisexual flowers. Male catkins are usually longer than the female catkins and the florets are arranged loosely whereas the female catkin is short and florets are arranged compactly. The male flowers have four perianths and four stamens, while the female flower with four perianths has single celled ovary with bifid stigma and a pendulus ovule. Pollination is anemophilous. Fruit is a sorosis composed of a number of individual achenes. The seed is oval shaped with a nearly flat surface at the micropylar region.

Crop Improvement through Traditional Breeding

Developing mulberry varieties that perform equally well throughout the year irrespective of the seasonal changes is much needed to improve the silk productivity in regions with distinct seasonal variations as the growth and development of mulberry are influenced considerably by temperature and soil moisture. Mulberry growth reduces significantly by low temperature as when the ambient temperature falls below 13°C the growth stops completely and the plant enters into a period of dormancy. Similarly, when the soil moisture falls below 60% of the field capacity the growth of mulberry becomes sluggish and below 30% the growth stops altogether (Ohyama 1966).

Need of Germplasm Resources

Development of varieties with higher productivity requires utilization of well characterized genetic resources. Among selected parents for mulberry hybridization, screening of hybrids, selection of the desired hybrids, and propagation of these hybrids through vegetative means appears as the major mode of crop improvement. Thus, choice of parents plays a significant role in achieving the desired objective. Therefore, large number of germplasm accessions have been maintained in sericulturally important countries. For instance, China has more than 1860 germplasm accessions in Zhejiang, Jiangsu, Guangdong, Guangxi, Shandong, Sichun, Anhui, Hubei, Hunan,

Hebei, Shanxi, Shuanxi and Xinjiang (Pan 2003), Japan has 1375 germplasm accessions (Machii et al. 1999), Bulgaria has 140 accessions in SES-Vratza (Tzenov 2002), Korea has 614 accessions. In India, Central Sericultural Germplasm Resources Centre, Hosur, Tamil Nadu maintains 1191 accessions belonging to 13 *Morus* species viz., *Morus indica, M. alba, M. laevigata, M. serrata, M. rubra, M. cathayana, M. nigra, M. australis, M. bombycis, M. sinensis, M multicaulis, M. rotundiloba* and *M. tiliaefolia* from 30 countries (Table 3a,b). These genetic resources are generally conserved in *ex situ* gene banks. Before employing the genetic resources in breeding program, it has to be characterized and evaluated for desired traits for their proper conservation and utilization. A detailed account of germplasm conservation of *Mulberry* spp. to sustain sericulture is given in Chapter 17 of this book.

Characterization of Germplasm Accessions

Characterization is an essential step of proper germplasm management and utilization of germplasm, wherein systematic information of the accessions are recorded for identification of different accessions. It also helps to

Table 3a. List of mulberry genetic resources available in the *ex situ* germplasm of Japan, China, India and Korea. Vijayan et al. (2018).

Species	Available accessions of mulberry			
	Japan	China	India	Korea
M. bombycis Koidz.	583	22	15	97
M. latifolia Poir.	349	750	19	128
M. alba L.	259	762	93	105
M. acidosa Griff.	44	-	-	1
M. wittorium Hand-Mazz.	-	8	-	-
M. indica L.	30	-	350	5
M. mizuho Hotta	-	17		-
M. rotundiloba Koidz.	24	4	2	--
M. kagayamae Koidz.	23	-	-	1
M. australis Poir.	-	37	2	-
M. notabilis C.K. Schn.	14	-	-	--
M. mongolica Schneider		55	-	-
M. boninensis Koidz.	11	--	--	--
M. nigriformis Koidz.	3	--	-	--
M. atropurpurea Roxb.	3	120	-	--
M. serrata Roxb.	3	--	18	--
M. laevigata wall.	3	19	32	1
M. nigra L.	2	1	2	3
M. formosensis Hotta.	2	--	-	--
M. rubra L.	1	--	1	--
M. mesozygia Stapf.	1	--	-	--
M. celtifolia Kunth.	1	--	-	--
M. cathayana Hemsl.	1	65	1	--
M. tiliaefolia Makino	1	--	1	14
M. microphylla Bickl.	1	--	-	--
M. macroura Miq.	1	--	-	--
M. multicaulis s Perr.	-	-	15	-
Morus spp. (Unknown)	15	-	720	259

Table 3b. Total mulberry germplasm accessions available in different countries. Vijayan et al. (2018).

Country	Mulberry accession
Japan	1,375
China	2,600
Brazil	---
South Korea	208
India	1,271
Bulgaria	140
Italy	50
France	70
Indonesia	5
Taiwan	5
Argentina	2
Colombia	4
Mexico	5
Peru	2
USA	23

identify redundant accessions which are discarded to reduce the maintenance cost. In the traditional method of characterization, a list of morphological characters are used for describing and distinguishing an accession (Banerjee et al. 2007). However, most of the morphological characters vary greatly depending on the developmental stage and environmental conditions. Thus other parameters which are more stable, easy to define, unbiased, and present in large numbers are tried. Of these molecular markers have been used for genetic characterization of germplasm resources (Vijayan et al. 2004a). Once the genetic resources are characterized and evaluated, suitable parental materials are selected for breeding.

Selection of Parents

In mulberry, leaf is the primary product as far as sericulture is concerned. Higher leaf yield needs faster growth and more biomass production. Thus, mulberry breeding adopts an indirect method of selection through consideration of traits that contribute considerably to the leaf yield. Correlation and path-coefficient studies have identified a number of traits that have significant influence on leaf yield. Important among them are plant height, number of branches, leaf weight, leaf retention capacity, and nodal length (Vijayan et al. 1997b, 2010, Gandhi et al. 2011). The other important traits that have been used by breeders to select parents are adaptability, resistance to pests and diseases, tolerance to abiotic stresses like drought, salinity and cold, higher vegetative propagation ability, better leaf quality, and better coppicing ability.

Controlled Pollinated Hybridization

Once suitable parents are selected, the next step is to effect the hybridization. Hybridization is generally done during the regular flowering seasons. Branches of mulberry containing immature flower buds are selected and bagged with butter paper to ward off any contamination from pollen grains of undesirable parents (Figure 3A). Prior to bagging the tip of the branches are removed to prevent growing stem tips breaking the bags. In each bag 20–25 catkins of more or less same size and age are maintained to ensure uniformity in the maturity of the flower for hybridization. Male catkins are also bagged to ward off any possible contamination with pollens from other plants. Mature male catkins (Figure 3B) are collected in sterilized covered petri-plates at 7.00 am to 7.30 am and pollen grains are

Figure 3. Breeding scheme in mulberry. A–Bagging of catkins, B–Male catkins, C–Female Catkins, D & E–fruits of mulberry, F–Seeds of mulberry, G–seedling raised in pot, H–A single mulberry seed.

extracted by keeping them in sunlight for one hour. Viability of pollen used for pollination is generally assessed with the help of 0.5% acetocarmine staining method as well as germination in 2.0% sucrose solution. In general, pollen viability varies from 70–85% in diploid plants. Pollination is generally done at 11.00–11.45 am with the help of a 'pollen gun' where the nozzle of the pollen gun is inserted into the bag and pollen grains are injected with air pressure (Mukherjee 1965, Vijayan et al. 1997a) during the period when female catkins (Figure 3C) are receptive for the pollen grains. The receptivity of the female flower is indicated by the white color and feathery nature of the stigma. Pollination may be repeated for three alternative days to ensure that all flowers in all catkins are pollinated. After the pollination the female flowers are observed carefully to ensure that pollination has really materialized. If pollination has materialized the color of the stigma changes into brown and the perianth of the flower starts to become succulent and thicker. The pollinated female catkins are kept covered until the fruit ripens to avoid loss due to attack of birds and insects. Fruits (Figures 3D & E) ripen within 22–23 days of pollination, which is evident from the color changes, seeds (Figure F) are collected from the fruits by crushing it and removing the perianth.

Raising and Screening of Seedling

The seeds are sown in the nurseries in specially prepared beds of 4.0 m × 1.5 m size having proper drainage channel. The seed beds are maintained with regular watering and with plant protection measures for three months. After three months of growth, the seedlings (Figures 3G & H) are subjected to visual assessment and those seedlings which have higher growth and vigor are transferred to field and planted in Progeny Row Trial (PRT) at 60 cm × 60 cm spacing or paired row system in a spacing of (90 + 150) cm × 60 cm (depending on the climatic conditions and recommended cultural practice of the area; Figure 4) with suitable fertilizer applications, weeding, tillering and other cultural operations. Data on total shoot length, total weight of all branches and weight of 100 leaves, incidence of pest and diseases, and total leaf yield are recorded from each single plant. Based on the performance, the best ones are selected and multiplied for raising saplings to test them in detail in Primary Yield Trail (PYT). In PYT, large number of hybrids selected from the PRT are tested with limited number of plants per replication. Data on growth, clonal propagation efficiency, resistance to disease and pests, leaf yield, and biochemical components are recorded. Based on PYT performance, 7–8 promising plants with higher leaf yield and other characters are selected and tested under Final Yield Trial (FYT) along with control varieties. In FYT, a minimum of 81 plants per replication is planted at randomized block design with a minimum of three replications. All recommended cultural operations and fertilizer applications, on leaf yield, growth, biochemical and anatomical traits as well as performance of silkworm rearing are recorded. The varieties performing better than the controls are

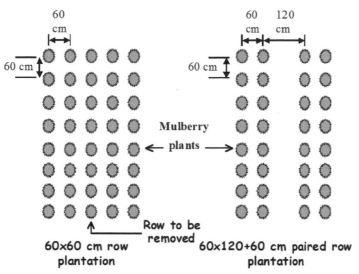

Figure 4. Schematic representation of the Progeny Row Trials for assessing the superiority of mulberry hybrids. Source: Verma and Dandin (2006).

selected and tested in All India Coordinated Experiment on Mulberry (AICEM) at different test centres spanning across the country under different agro-climatic conditions. Varieties that perform the best over the control is selected and recommended for commercial use (Figure 5).

Popular Varieties Developed through Controlled Crosses in India

Development of high yielding mulberry varieties has a significant contribution in improving the production of quality raw silk. The increasing demand of silk in the National vis-à-vis International markets can be accomplished by introduction of high yielding mulberry varieties that are important for sericulture industry in the field considering vertical increment of yield potential in a unit area of land. The improvement envisaged is being achieved through the directional breeding towards leaf yield and quality, tolerance to stresses, and adaptability to new farming areas. Since 1960, a number of mulberry varieties have been developed in India (Table 4), China (Table 5) and other sericulturally important countries using conventional breeding methods and these varieties have improved the leaf productivity considerably. In India, the mulberry leaf productivity has increased from 8–10 MT (Metric Tonnes)/ha/yr in 1960 to 60–65 MT/ha/year in 2018. Nonetheless, other modern biotechnological methods have also been employed to develop mulberry varieties with specific traits/potentials or to expedite the selection process during the variety development.

Other Breeding Methods Used in Mulberry

Polyploidy Breeding: Since mulberry can be multiplied clonally and triploids are known to have several advantages over diploids such as wider adaptability, resistance to environmental stress, higher yield potential, leaf palatability, digestibility and assimilability (Vijayan and Chakraborti 1998), triploids are developed through crossing between diploids and tetraploids. Since majority of the natural triploids are unsuitable for silkworm rearing due to the coarse nature of the leaf, and low propagation efficiency, triploids with desirable traits are engineered through induction of tetraploids from desirable diploids and then crossing the induced tetraploids with diploids. There have been a number of reports on induction of tetraploids in mulberry through colchicine treatment of germinating seeds, seedlings and sprouting vegetative buds (Dwivedi et al. 1986, Eswar Rao 1996, Chakraborti et al. 1998). Utilizing autotetraploids of hybrids of *M. indica* x *M. alba*, Si, K-2, S-30, S-36, S-41, S-54, RFS-135, RFS-175, S-13, S-34 and V-1 for crossing with related diploid parent, five several promising triploids were developed. Triploids like

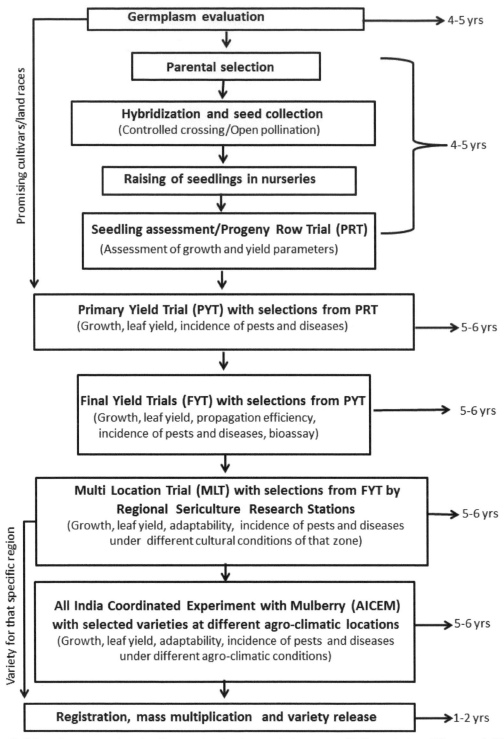

Figure 5. Schematic representation of mulberry variety development in the context of Indian sericulture. Vijayan et al. (2011a).

Table 4. List of mulberry varieties being commercially exploited in India. Vijayan et al. (2018).

Sl. No.	Variety	Breeding method	Area of cultivation
1	G4	Hybrid from *M. multicaulis* x S-13	South India Irrigated
2	G2	*M. multicaulis* x S34	Young age silkworm rearing in irrigated conditions
3	C2038	Hybrid from a cross between CF1 x C763	Eastern and NE India Irrigated
4	Tr23	Hybrid from a cross between T20 x S-162	Hills of Eastern India
5	Victory-1	Hybrid from a cross between S30 x C776	South India Irrigated
6	Vishala	Clonal selection	All India Irrigated condition
7	Anantha	Clonal selection	South India Irrigated
8	DD	Clonal selection	South India Irrigated
9	S-13	Selection from polycross (mixed pollen) progeny	South India Rainfed
10	S-34	Selection from polycross (mixed pollen) progeny	South India Rainfed
11	S-1	Introduction from (Mandalaya, Myanmar)	Eastern and NE India Irrigated
12	S-7999	Selection from open pollinated hybrids	Eastern and NE India Irrigated
13	S-1635	Triploid selection	Eastern and NE India Irrigated
14	S-36	Mutation breeding through EMS treatment of Berhampore Local	South India Irrigated
15	S-54	Mutation breeding through EMS treatment of Berhampore Local	South India Irrigated
16	Sahana	Kanva-2 x Kosen	Coconut shades of South India
17	RC-1	Punjab local x Kosen	Resource Constraint areas of South India
18	RC-2	Punjab local x Kosen	Resource Constraint areas of South India
19	AR-12	S41 (4x) x Ber. C776	Alkaline affected areas of Southern India
20	C776	Hybrid from a cross between English black and *C. multiculis*	Saline soils
21	C-1730	Hybrid from a cross between T-25 × S-162	Rain fed areas
22	C-2028	Hybrid from a cross between China White × S-1532	Water logged areas
23	S-146	Selection from open pollinated hybrids	N. India and Hills of J and K Irrigated
24	Tr-10	Triploid developed from "S1"	Hills of Eastern India
25	BC$_2$59	Back crossing of hybrid of Matigare local x Kosen with Kosen twice	Hills of Eastern India
26	Chak Majra	Selection from natural variability	Sub-temperate
27	China White	Clonal selection	Temperate

Tr8, Tr-10, Tr-23 have been found performing better in hilly areas. Similarly, AR-12 with higher tolerance to soil alkalinity was recommended for cultivation in alkaline soils (Mogili et al. 2017).

Mutation breeding: Mutation is the process by which genes are permanently altered either spontaneously or induced at the chromosome or gene level. Owing to mutation, a character may be lost or enhanced, hence, breeders use the mutation technique in plant breeding to develop new varieties with specific traits which includes quality and productivity of leaf. Mutations can be induced through chemical or physical agents. The most commonly used mutation agents are physical mutagens such as ultra-violet (UV) light, X-rays and neutrons (Koorneef 2002). Chemical mutagens such as nitrous acid, ethyl methane sulfonate (EMS), methyl methane sulfonate (MMS) and diethyl sulfate (DES), have also been used to induce variations in plants. Compared to radiological methods,

Table 5. List of high yielding mulberry varieties developed in China. Source: Vijayan et al. (2011a).

No.	Variety	Selection and breeding method	Suitable zone
1	Xiansang 305	Mutation breeding	The Huanghe River valley
2	Beisangyihao	Selected from local seedling mulberry	The Changjiang River valley, The middle and lower reaches of the Huanghe River
3	Nongsang 8	Hybridization breeding	The Changjiang River valley
4	Huangluxuan	Selection from local variety	The Huanghe River valley
5	Jihu 4	Hybridization breeding	Northeast zone
6	Dazhonghua	Polyploidy breeding	The Changjiang River valley
7	Xinyiyuan	Mutation breeding	The Changjiang River valley, The middle and lower reaches of the Huanghe River
8	Nongsang 14	Hybridization breeding	The Changjiang River valley
9	Yu 237	Hybridization breeding	The Changjiang River valley
10	Xuanqiu 1	Selected from local seedling mulberry	Northeast zone
11	7307	Selected from local seedling mulberry	The Changjiang River valley
12	Husang 32	Selection from local variety	The Changjiang River valley, The middle and lower reaches of the Huanghe River
13	Xiang 7920	Hybridization breeding	The Changjiang River valley
14	Canzhuan 4	Selected from local seedling mulberry	The Changjiang River valley
15	Huamingsang	Selected from local seedling mulberry	ChizhouXuanzhouAnqing in Anhui province and Linyi in Shandong province
16	7946	Hybridization breeding	The Huanghe River valley
17	Yu 2	Hybridization breeding	The Changjiang River valley
18	Shigu 11-6	Mutation breeding	The Changjiang River valley, The middle and lower reaches of the Huanghe River
19	Xuan 792	Selection breeding	The Huanghe River valley
20	Yu 711	Hybridization breeding	The Changjiang River valley, The middle and lower reaches of the Huanghe River
21	Yu 151	Hybridization breeding	The Changjiang River valley
22	Hongxin 5	Hybridization breeding	The Changjiang River valley, The middle and lower reaches of the Huanghe River
23	Lunjiao 40	Selection from local variety	The Zhujiang River valley
24	Wan 7707	Selected from local seedling mulberry	Chizhou, Xuanzhou, Anqing in Anhui province and Linyi in Shandong province
25	Huangsang 14	Selected from local seedling mulberry	The Changjiang River valley
26	Lunjiao 40	Selection from local variety	The Zhujiang River valley
27	Shi 11	Selection from local variety	The Zhujiang River valley
28	Xinyizhilan	Introduced variety	The Changjiang River valley; The Huanghe River valley
29	Jialing 16	Polyploidy breeding	The Changjiang River valley
30	Tang10×Lun 109	Hybrid mulberry seed	The Zhujiang River valley
31	Nongsang 12	Hybridization breeding	The Changjiang River valley

chemical mutagens tend to cause single base-pair (bp) changes, or single-nucleotide polymorphisms (SNPs) rather than deletions and translocations. Of these chemical mutagens, EMS is today the most widely used as it selectively alkylates guanine bases causing the DNA-polymerase to favor placing a thymine residue over a cytosine residue opposite the O-6-ethyl guanine during DNA replication, which results in a random point mutation. A majority of the changes (70–99%) in EMS-mutated populations are GC to AT base pair transitions (Till et al. 2004, 2007). Since, mulberry is propagated vegetatively, mutation breeding is recognized as a useful tool for eliminating certain undesirable characteristics or introducing certain beneficial traits into the plant through altered gene expression. Using mutation breeding, four mulberry varieties viz., s-30, S-36, S-41 and S-54, have been developed and used for cultivation. See for more details in Chapter 7 on polyploidy and mutation breeding in mulberry.

Marker-assisted Breeding in Mulberry

In the past two decades, significant progress has been made in mulberry crop improvement using molecular markers. Molecular markers have been used for characterization of germplasm, identification of genetically diversified accessions, elimination of duplicates from the germplasm to reduce cost of maintenance, and expediting breeding programs through marker-assisted selection. Though initial efforts were to use restriction fragment length polymorphic (RLFP) markers in mulberry, PCR based markers such as random amplified polymorphic DNA (RAPD), inter simple sequence repeats (ISSR), amplified fragment polymorphism (AFLP), simple sequence repeats (SSR), expressed sequence tag (EST) and single nucleotide polymorphism (SNP) have also been utilized in mulberry breeding programmes. RAPD was the first DNA based marker used in mulberry. Using RAPD the genetic diversity among a few cultivars was initially tested (Xiang et al. 1995, Lichuan et al. 1996, Lou et al. 1998, Zhang et al. 1998). Later, Bhattacharya and Ranade (2001), Srivastava et al. (2004), Chatterjee et al. (2004) and Zhao and Pan (2004) used RAPD markers for assessing the interspecific and the intraspecific relationships in mulberry. Considering the inconsistency in the results from RAPD markers, even in the same laboratory, attempts had to be made to explore other DNA markers for mulberry. Consequently, Vijayan and Chatterjee (2003) tested the suitability of ISSR primers in mulberry and found that (AG), (TG), (AC), (ACC), (ATG), (AGC), (GAA), (GATA), (CCCT), (GGAGA) and (GGGGT) produced excellent results. Subsequently, ISSR markers were used to estimate the genetic diversity among indigenous cultivars of *M. alba* and *M. indica* (Awasthi et al. 2004, Vijayan et al. 2004a-d, Vijayan et al. 2005, Vijayan et al. 2006a) and among different ecotypes present in China (Zhao et al. 2006, 2007). Relationships between temperate and tropical mulberry species (Vijayan 2004) and interspecific variability (Vijayan et al. 2004c) were also investigated in detail using ISSR and RAPD markers. Sharma et al. (2000) used AFLP for the first time to elucidate the interrelationships of different mulberry species. In the mean time, efforts were made to develop microsatellite primers for mulberry. Accordingly, Agarwal and Udhaykumar (2004) and Zhao et al. (2005) developed several SSR primers for mulberry. Using these SSR primers Zhao et al. (2005a) were able to discriminate the wild genetic resources from the domesticated ones as wild species such as *M. laevigata*, *M. cathayana*, *M. nigra*, *M. mongolica* and *M. wittiorum* grouped together into a cluster separated from the cluster formed by *M. alba* and other cultivated varieties. In order to utilize the wild mulberry genetic resources in India, Vijayan et al. (2004d) worked out the genetic diversity present in different populations of the wild mulberry *M. serrata*. In a similar study, Chatterjee et al. (2004) investigated the intraspecific diversity among populations of *M. laevigata* collected from different locations in India. This study revealed stronger genetic relationships between *M. laevigata* populations in Andaman Islands with those in Himalayan foot hills of West Bengal, India. Additionally, the genetic relationships between cultivated and wild mulberry species was also investigated using molecular markers (Vijayan et al. 2006b) and found that *M. laevigata* is closer to *M. indica* and *M. alba* than *M. serrata*. In a related investigation, the internal transcribed spacers of nuclear ribosomal RNA (nr*ITS*) and the cpDNA gene (*trn*L-F) were used to elucidate the phylogenetic relationships among different species of mulberry (Zhao et al. 2005b). This study, further, confirmed the close relationship of *M. laevigata* with *M. alba*. These studies generated significant information to help plan breeding strategies to utilize the wild genetic resources of mulberry for the crop improvement in mulberry. Similarly, Vijayan and Chatterjee (2003) identified two RAPD molecular markers viz., 825_{1400} and 835_{750} associated with leaf yield. Further efforts are being made to develop

Table 6. Molecular linkage map developed in mulberry.

Sl. no.	Molecular map	Pedigree of mapping population	Markers	Hereditary nature of marker	Agronomic trait targeted	References
1.	Genetic linkage	S36 x V1	RAPD, ISSR, and SSR	Dominant, co-dominant	No	Venkateswarlu et al. (2006)
2.	QTL map	V1 x Mysore Local	RAPD and ISSR	Dominant	Yield traits	Naik et al. (2014)
3.	QTL map	Himachal Local x MS3	RAPD and ISSR	Dominant	Water use efficiency	Mishra (2014)
4.	QTL map	Dudia White x UP	RAPD and ISSR	Dominant	Root traits	Mishra (2014)
5.	Genetic linkage	Dudia White 9 UP105	SSR	Co-dominant	No	Mathithumilan et al. (2016)

(Source: Sarkar et al. 2017)

genetic linkage maps to identify QTLs for important traits and a few successes have been made in this direction as detailed in Table 6.

Genetic Engineering in Mulberry

Since the development of mulberry varieties exhibiting specific desired traits take a long time because of longer juvenile period, lack of inbreds and dioecious nature of the plant (Vijayan 2010), it becomes essential to follow alternative approach like genetic engineering to develop varieties with specific traits. The two different methods viz., particle bombardment and the *Agrobacterium tumefaciens*-mediated gene delivery systems, have widely been used to insert foreign genes into the mulberry genome. Machii et al. (1996) first used particle bombardment and later Sugimura et al. (1999) reported delivery of the reporter gene GUS into intact leaf tissues and suspension-cultured cells. Bhatnagar et al. (2002) also successfully incorporated the GUS reporter gene into *Morus indica* cv. K2 using hypocotyl, cotyledon, leaf and leaf callus explants. However, due to low transformation efficiency of particle bombardment (Travella et al. 2005), *A. tumefaciens* mediated transformation has been adopted extensively in mulberry (Checker et al. 2012). Using this method, several transgenic mulberry plants having traits such as tolerance to drought and salinity, high temperature, and moisture retention have been developed as detailed in Table 7 (Saeed et al. 2015, Sanjeevan et al. 2017). For applications of genetic engineering in mulberry see Chapter 11 of this book.

Table 7. Transgenesis in mulberry for abiotic stress tolerance.

Gene	Expression profile	Reference
*WAP21**	Cold tolerance	Ukaji et al. (1999)
COR	Cold tolerance	Ukaji et al. (2001)
AlaBlb	Salinity tolerance	Wang et al. (2003)
OC	Insect resistance	Wang et al. (2003)
SHN 1	Drought tolerance	Sanjeevan et al. (2017)
HVA1	Drought and salinity stress	Lal et al. (2008)
bch1	Drought and salinity stress	Saeed et al. (2015)
Osmotin	Drought and salinity stress	Das et al. (2011)

* *AlaBlb*-Soybean glycine gene; *bch-1* inhibitor 2-aminobicyclo-(2, 2, 1)-heptane-2-carboxylic acid; *COR*-cold on regulation; *HVA1*-*Hevea braziliensis* abiotic stress gene; *OC*-osteocalcin *Osmotin*-osmotic stress induced gene; *SHN 1*-schnurri from *Drosophila melanogaster*; *WAP21*-water allocation plan (Source: Vijayan et al. 2011b).

Mulberry Improvement: Present Requirement and Future Strategies

Traditional hybridization has contributed much to the development of high yielding mulberry varieties, particularly under the changing agro-climatic conditions created by global warming and other factors. Considering the escalating cost of inputs including cost of manpower, it is imperative to develop varieties that can grow well and yield quality leaf for rearing highly productive silkworm for the sustainability of sericulture. Mulberry varieties with improved nutrient uptake efficiency, higher biomass production, less susceptibility to micro climatic conditions, and resistance to pest and diseases are essential for sustenance in the future. The genetics of mulberry is yet to be fully resolved owing to lack of inbreds, high heterozygosity, long juvenile periods, and sexual segregation in progeny from monoecious plants. It is therefore essential to employ modern biotechnological tools to develop homozygous plants like doubled-haploidy, genetic engineering, etc. to develop plants with special traits. Further, improved breeding stocks need to be developed through pre-breeding so that conventional breeding produces better varieties to serve in the future.

References

Agarwal, R. and D. Udaykumar. 2004. Isolation and characterization of six novel microsatellite markers for mulberry (*Morus indica*). Mol. Ecol. Notes 4: 477–479.

Awasthi, A.K., G.M. Nagaraja, G.V. Naik, S. Kanginakudru, K. Thangavelu and J. Nagaraju. 2004. Genetic diversity in mulberry (genus *Morus*) as revealed by RAPD and ISSR marker assays. BMC Genet 5: 1: http://www.biomedcentral.com /1471-2156/5/1.

Banerjee, R., S. Roychawdhuri, H. Sahu, B.K. Das, P. Ghosh and B. Saratchandra. 2007. Genetic diversity and interrelationship among mulberry genotypes. Journal of Genetics and Genomics 34(8): 691–7.

Bhattacharya, E. and S.A. Ranade. 2001. Molecular distinction among varieties of Mulberry using RAPD and DAMD profiles. BMC Plant Biol 3: 1: http:// www.biomedcentral.com /1471-2229/1/3.

Bhatnagar, S., A. Kapur and P. Khurana. 2002. Evaluation of parameters for high efficiency gene transfer via particle bombardment in Indian mulberry. Indian J. Exp. Biol. 40: 1387–1393.

Chakraborti, S.P., K. Vijayan, B.N. Roy and S.M.H. Quadri. 1998. *In vitro* induction in tetraploidy in mulberry (*Morus alba* L). Plant Cell Rep. 17: 794–803.

Chang, L.W., L.J. Juang, B.S. Wang et al. 2011. Antioxidant and antityrosinase activity of mulberry (*Morus alba* L.) twigs and root bark. Food Chem. Toxicol. 49: 785–790.

Chatterjee, S.N., G.M. Nagaraja, P.P. Srivastava and G. Naik. 2004. Morphological and molecular variation of *Morus laevigata* in India. Genetica 39: 1612–1624.

Checker, V.G., B. Saeed and P. Khurana. 2012. Analysis of expressed sequence tags from mulberry (*Morus indica*) roots and implications for comparative transcriptomics and marker identification. Tree Genet Genomes 8: 1437–1450.

Das, M., H. Chauhan, A. Chhibbar, Q.M.R Haq and P. Khurana. 2011. High-efficiency transformation and selective tolerance against biotic and abiotic stress in mulberry, *Morusindica* cv. K-2, by constitutive and inducible expression of tobacco osmotin. Transgenic Res. 20(2): 231–246.

Dwivedi, N.K., A.K. Sikdar, S.B. Dandin, C.R. Sastry and M.S. Jolly. 1986. Induced tetraploidy in mulberry I. Morphological, anatomical and cytological investigations in cultivars RFS-135. Cytologia 51: 393–401.

Eswar Rao, M.S. 1996. Improvement of mulberry through polyploidy breeding. Ph.D. Thesis, Bangalore University, Karnataka, India.

Gandhi Doss, S., S.P. Chakraborti, K. Vijayan and P.D. Ghosh. 2011. Character association in improved mulberry genotypes exhibiting delayed senescence. Journal of Ornamental and Horticultural Plants 1(2): 85–95.

Hotta T. 1954. Fundamentals of *Morus* plants classification. Kinugasa Sanpo 390: 13–21 (in Japanese).

Jaiswal, V.S. and A. Kumar. 1981 Modification of sex expression and fruit formation on male plants of *Morus nigra* L. by chloroflurenol. Proc. Indian Acad. Sci. (Plant Sci.) 90: 395–400.

Katsumata, F. 1972. Relationship between the length of styles and the shape of idioplasts in mulberry leaves, with special reference to the classification of mulberry trees. J. Seric. Sci. Jap. 41: 387–395.

Koidzumi, G. 1917. Taxonomy and phytogeography of the genus *Morus*. Bull. SericExpSta Tokyo (Japan) 3: 1–62.

Koornneef, M. 2002. Classical mutagenesis in higher plants. pp. 1–11. *In*: Gilmartin, P.M. and C. Bowler (eds.). Molecular Plant Biology. Oxford University Press, Oxford, UK.

Lal, S., V. Gulyani and P. Khurana. 2008. Overexpression of hva1 gene from barley generates tolerance to salinity and water stress in transgenic mulberry (*Morus indica*). Transgenic Res. 17: 651–663.

Lichuan, F., Y. Guangwei, Y. Maode, K. Yifu, J. Chenjun and Y. Zhonghuai. 1996. Studies on the genetic identities and relationships of mulberry cultivated species (*Morus* L.) via a random amplified polymorphic DNA assay. Canye Kexue 22: 139.

Lou, C.F., Y.Z. Zhang and J.M. Zhou. 1998. Polymorphisms of genomic DNA in parents and their resulting hybrids in mulberry *Morus*. Sericologia 38: 437–445.

Machii, H., Koyama, A. and H. Yamanouchi. 1999. A list of genetic mulberry resources maintained at National Institute of Sericultural and Entomological Science. MiscPubl. Natl. SericEntomol. Sci. 26: 1–77 (in Japanese).

Machii, M., G.B. Sung, H. Yamanuchi and A. Koyama. 1996. Transient expression of GUS gene introduced into mulberry plant by particle bombardment. J. Sericult. Sci. Japan 65: 503–506.

Maode, Y., X. Zhonghuai, F. Lichun, K. Yifu, Z. Xiaoyong and J. Chengjun. 1996. The discovery and study on a natural haploid *Morusnotabilis* Schneid. Acta Sericol. Sin. 22: 67–71.

Mathithumilan, B., R.S. Sajeevan, J. Biradar, T. Madhuri, K.N. Nataraja and S.M. Sreeman. 2016. Development and characterization of genic SSR markers from Indian mulberry transcriptome and their transferability to related species of Moraceae. PLOS One. DOI:10.1371/journal.pone.0162909.

Mishra, S. 2014. Genetic analysis of traits controlling water use efficiency and rooting in mulberry (*Morus* spp.) by molecular markers. Ph.D. Thesis, University of Mysore, Mysuru, India.

Mogili, T., V. Sivaprasad and M. Udaykumar. 2017. Impact of improved varieties and cultivation technology on maximization of leaf yield and sustainability. *In*: National Conference Tree Improvement Research in India: Current Trends and Future Prospects Held at Institute of Wood Science and Technology (ICFRE), Bangalore. p–13.

Mukherjee, S.K. 1965. Breeding of mulberry in India by the use of a pollinator—"the pollen-gun" (A new Japanese device). Science and Culture 31: 101–104.

Naik, V.G., B. Mathithumilan, A. Sarkar, S.B. Dandin, M.V. Pinto and V. Sivaprasad. 2014. Development of genetic linkage map of mulberry using molecular markers and identification of QTLs linked to yield and yield contributing traits. Sericologia 54(4): 221–229.

Ogure, M., N. Harashima, K. Naganuma and M. Matsushima. 1980. Effect of etheral and gibberellin on sex expression in mulberry *Morus* spp. J. Seric Sci. Jap. 49: 335–341.

Ohyama, K. 1966. Effect of soil moisture on growth of mulberry tree. Bull Seric. Expt. Sci. Jap. 20: 333–359.

Pan, Y.L. 2003. Popularization of good mulberry varieties and sericultural development. Acta Seric Sin. 1: 1–6.

Qin, J., N. He, Y. Wang and Z. Xiang. 2012. Ecological issues of mulberry and sustainable development. J. Resour. Ecol. 3(4): 330–339.

Saeed, B., M. Das, Q.M.R. Haq and P. Khurana. 2015. Over expression of beta carotene hydroxylase-1 (bch1) in mulberry, *Morus indica* cv. K2, confers tolerance against high-temperature and high-irradiance stress induced damage. Plant Cell Tissue & Organ Culture 120(3): 1003–1015.

Sajeevan, R.S., K.N. Nataraja, K.S. Shivashankara, N. Pallavi, D.S. Gurumurthy and M.B. Shivanna. 2017. Expression of Arabidopsis SHN1 in Indian mulberry (*Morus indica* L.) increases leaf surface wax content and reduces post-harvest water loss. Front Plant Sci. 8: 418.

Sarkar, T., T. Mogili and V. Sivaprasad. 2017. Improvement of abiotic adaptive traits in mulberry (*Morus* spp.): an update on biotechnological interventions. 3 Biotech 7: 214.

Sharma, A.C., R. Sharma and H. Machii. 2000. Assessment of genetic diversity in a *Morus* germplasm collection using fluorescence-based AFLP markers. Theor. Appl. Genet. 101: 1049–1055.

Sharma, V., S. Chand and P. Singh. 2013. Mulberry: a most common and multi-therapeutic plant. Int. J. Adv. Res. 1(5): 375–378.

Sikdar, A.K., M.S. Jolly and N.K. Dwivedi. 1988. Polyploidization and modification of sex in mulberry by colchicine. Curr. Sci. 57: 736–737.

Singhal, B.K., A. Dhar, M.A. Khan et al. 2009. Potential economic additions by mulberry fruits in sericulture industry. Plant Horti. Tech. 9: 47–51.

Singhal, B.K., A. Dhar, M.A. Khan et al. 2010. Approaches to industrial exploitation of mulberry (*Morus* spp.) fruits. J. Fruit Ornament Plant Res. 18(1): 83–99.

Srivastava, P.P., K. Vijayan, A.K. Awasthi and B. Saratchandra. 2004. Genetic analysis of *Morus alba* through RAPD and ISSR markers. Indian J. Biotechnol. 3: 527–532.

Sugimura, Y., J. Miyazaki, K.Yonebayashi et al. 1999. Gene transfer by electroporation into protoplasts isolated from mulberry call. J Sericult Sci. Japan 68: 49–53.

Tikader, A., K. Vijayan, M.K. Raghunath, S.P. Chakraborti, B.N. Roy and T. Pavankumar. 1995. Studies on sexual variations in mulberry (*Morus* spp.). Euphytica 84: 115–120.

Tikader, A. and S.B. Dandin. 2007. Pre-breeding efforts to utilize two wild *Morus* species. Curr. Sci. 92: 1072–1076.

Till, B.J., S.H. Reynolds, C. Weil, N. Springer, C. Burtner, K. Young, E. Bowers, C.A. Codomo, L.C. Enns, A.R. Odden, E. A. Greene, L. Comai and S. Henikoff. 2004. Discovery of Induced Point Mutations in Maize Genes by TILLING, BMC Plant Biology, vol. 4, article 12.

Till, B.J., J. Cooper, T.H. Tai, E.A. Greene, S. Henikoff and L. Comai. 2007. Discovery of chemically induced mutations in rice by TILLING, BMC Plant Biology, vol. 7, article 19.

Travella, S., S.M. Ross, J. Harden, C. Everett, J.W. Snape and W.A. Harwood. 2005. A comparison of transgenic barley lines produced by particle bombardment and *Agrobacterium*-mediated techniques. Plant Cell Rep. 23: 780–789.

Tzenov, P.I. 2002. Conservation status of mulberry germplasm resources in Bulgaria. Paper contributed to Expert Consultation on Promotion of Global Exchange of Sericulture Germplasm Satellite Session of 19th th ISC Congress, 21–25 Sept.· Bangkok, Thailand: http://www.fao.org/DOCREP/005/AD107E/ad107e01.htm.

Ukaji, N., C. Kuwabara, D. Takezawa, K. Arakawa, S. Yoshida and S. Fujikawa. 1999. Accumulation of small heat shock protein in the endoplasmic reticulum of cortical parenchyma cells in mulberry in association with seasonal cold acclimation. Plant Physiol. 120: 481–489.

Ukaji, N., C. Kuwabara, D. Takezawa, K. Arakawa and S. Fujikawa. 2001. Cold acclimation-induced WAP27 localized in endoplasmic reticulum in cortical parenchyma cells of mulberry tree was homologous to group 3 late embryogenesis abundant proteins. Plant Physio. 126: 1588–1597.

Uribe, T.F. and M.D. Sanchez. 2001. Mulberry for animal production: animal production and health paper. Rome: FAO, pp. 199–202.

Venkateswarlu, M., S. Raje Urs, B.S. Nath, H.E. Shashidhar, M. Maheswaran, T.M. Veeraiah and M.G. Sabitha. 2006. A first genetic linkage map of mulberry (*Morus* spp.) using RAPD, ISSR, and SSR markers and pseudotestcross mapping strategy. Tree Genet. Genom. 3: 15–24.

Verma, S. and S.B. Dandin. 2006. Mechanization in sericulture, Published by the Director, CSR&TI., Central Silk Board, Mysore. Available at http://www.csrtimys.res.in/sites/default/files/ebooks/2006-2.pdfebooks/2006-2.pdf.

Vijayan, K., S. Chauhan, N.K. Das et al. 1997a. Leaf yield component combining abilities in mulberry (*Morus* spp). Euphytica 98: 47–52.

Vijayan, K., A. Tikader, K.K. Das, S.P. Chakraborti and B.N. Roy. 1997b Correlation studies in Mulberry (*Morus* spp.). Indian J. Genetics and Plant Breeding 57: 455–460.

Vijayan K. and S.P. Chakraborti. 1998. Breeding for triploid mulberry varieties in India. (*Review*) Indian J. Seric. 37: 1–7.

Vijayan, K. and S.N. Chatterjee. 2003. ISSR profiling of Indian cultivars of mulberry (*Morus* spp.) and its relevance to breeding programs. Euphytica1. 31: 53–63.

Vijayan, K. 2004. Genetic relationships of Japanese and Indian mulberry (*Morus* spp.) revealed by DNA fingerprinting. Plant Syst. Evol. 243: 221–232.

Vijayan, K., P.P. Srivastava and A.K. Awasthi. 2004a. Analysis of phylogenetic relationship among five mulberry (*Morus*) species using molecular markers. Genome 47: 439–448.

Vijayan, K., P.K. Kar, A. Tikader, P.P. Srivastava, A.K. Awasthi, K. Thangavelu and B. Saratchandra. 2004b. Molecular evaluation of genetic variability in wild populations of mulberry (*Morus serrata* Roxb.). Plant Breed 123: 568–572.

Vijayan, K., A.K. Awasthi, P.P. Srivastava and B. Saratchandra. 2004c. Genetic analysis of Indian mulberry varieties through molecular markers. Hereditas 141: 8–14.

Vijayan, K., S.P. Chakraborti and P.D. Ghosh. 2004d. Screening of mulberry (*Morus* spp.) for salinity tolerance through *in vitro* seed germination. Indian J. Biotechnol. 3: 47–51.

Vijayan, K., C.V. Nair and S.N. Chatterjee. 2005. Molecular characterization of mulberry genetic resources indigenous to India. Genet Resour CropEvol. 52: 77–86.

Vijayan, K., P.P. Srivastava, C.V. Nair, A. Tikader, A.K. Awasthi and S. Raje Urs. 2006a. Molecular characterization and identification of markers associated with leaf yield traits in mulberry using ISSR markers. Plant Breed 125: 298–301.

Vijayan, K., A. Tikader, P.K. Kar, P.P. Srivastava, A.K. Awasthi, K. Thangavelu and B. Saratchandra. 2006b. Assessment of genetic relationships between wild and cultivated mulberry (*Morus*) species using PCR based markers. Genet. Resour. Crop Evol. 53: 873–882.

Vijayan, K. 2010. The emerging role of genomic tools in mulberry (*Morus*) genetic improvement. Tree Genet. Genomes 6: 613–625.

Vijayan, K., S.G. Doss, S.P. Chackraborti and P.D. Ghosh. 2010. Character association in mulberry under different magnitude of salinity stress. Emirates Journal of Food and Agriculture 22: 318–325.

Vijayan, K., B. Saratchandra and J.A.T. da Silva. 2011a. Germplasm conservation in mulberry. Scientia Horticulturae 128: 371–379.

Vijayan, K., AP.P. Srivastava, M.K. Raghunath and B. Saratchandra. 2011b. Enhancement of stress tolerance in mulberry. Sci. Hortic. 129: 511–519.

Vijayan, K., G. Ravikumar and A. Tikader. 2018. Breeding for higher fruit production in mulberry. *In*: Al-Khayri, J., S. Jain and D. Johnson (eds.). Advances in Plant Breeding Strategies: Fruits. Springer, Cham. https://doi.org/10.1007/978-3-319-91944-7_3.

Wang, H., C. Lou, Y. Zhang, J. Tan and F. Jiao. 2003. Preliminary report on *Oryza cystatin*gene transferring into mulberry and production of transgenic plants. Acta Sericol. Sinica 29: 291–294.

Wu, D., X. Zhang, X. Huang et al. 2010. Chemical constituents from root barks of *Morus atropurpurea*. Zhongguo Zhong Yao Za Zhi. 35: 1978–1982.

Xiang, Z., Z. Zhang and M. Yu. 1995. A preliminary report on the application of RAPD in systematics of *Morus alba*. Acta Sericol Sin. 21: 203–207.

Yang, X., L.Yang and H. Zheng. 2010. Hypolipidemic and antioxidant effects of mulberry (*Morus alba* L.) fruit in hyperlipidaemia rats. Food Chem. Toxicol. 48: 2374–2379.

Zhang, Y., L. Chengfu, Z. Jinmei, Z. Hongzi and X. Xiaoming. 1998. Polymorphism studies on genomic DNA of diploids and polyploids in mulberry. J. Zhejiang Agri. Univ. 24: 79–81.

Zhao, W. and Y. Pan. 2004. Genetic diversity of genus *Morus* revealed by RAPD markers in China. Int. J. Agri. Biol. 6: 950–954.

Zhao, W., X. Miao, S. Jia, Y. Pan and Y. Huang. 2005a. Isolation and characterization of microsatellite loci from the mulberry, *Morus* L. Plant Sci. 16: 519–525.

Zhao, W., Y. Pan, Z.J.S. Zhang, X. Miao and Y. Huang. 2005b. Phylogeny of the genus *Morus* (Urticales: Moraceae) inferred from ITS and *trn*L-F sequences. Afr. J. Biotechnol. 4: 563–569.

Zhao, W.G., Z.H. Zhou, X. Miao, S. Wang, L. Zhang, Y. Pan and Y. Huang. 2006. Genetic relatedness among cultivated and wild mulberry as revealed by inter-simple sequence repeat (ISSR) analysis in China. Can. J. Plant Sci. 86: 251–257.

Zhao, Z.Z., M. Xuexia, Z. Yong, W. Sibao, H. Jianhua, X. Hui, P. Yile and H.A. Yongping. 2007. A comparison of genetic variation among wild and cultivated *Morus* species (Moraceae: *Morus*) as revealed ISSR and SSR markers. Biodiv. Conserv. 16: 275–290.

CHAPTER 7

Improvement of Mulberry by Induction of Polyploidy and Mutagenesis

VN Yogananda Murthy[1],* and *HL Ramesh*[2]

INTRODUCTION

Polyploidy occurs in several taxa, it is predominantly prevalent in flowering plants and forms an integral feature of the chromosome evolution of higher plants. The important angiosperm species have evolved through polyploidy as evident from their evolutionary history (Hieter and Griffiths 1999, Shaked et al. 2001, Xing et al. 2010). Compared to their related diploid counterparts, polyploidy in plants frequently express enhanced characteristics such as larger cell and body size (Sugiyama 2005). Polyploidization is a very significant and treasured technique of plant breeding particularly in crop, horticulture and medicinal plants in order to improve various essential morphological and physiological features. This generates new valuable germplasm that can be applied in plant breeding for plant genetic improvement (Tang et al. 2010, Dhooghe et al. 2011). Induced polyploidy has enhanced quality traits in agricultural plants, medicinals and horticultural species, and improved plant resistibility to environmental stresses and diseases (Ahloowalia 1967). Besides, polyploidy induction can lead to content change of secondary metabolites (Majdi et al. 2010). Colchicine is a chromosome doubling agent that has been found to have a significant effect on polyploid induction and is widely used for inducing polyploidy in plants (Tang et al. 2010) because it can effectively arrest mitosis at the anaphase stage (Kunitake et al. 1998, Kermani et al. 2003, Nigel et al. 2007, Mohammadi et al. 2012). Till date, there are several reports on production of new hybrids or species through polyploidy breeding in crop plants (Zhu and Gao 1988, Ye et al. 2010, Lian and Chen 2013, Kanoktip et al. 2014).

Mulberry is an outstanding bioresource plant which with its holistic approach can produce high quality raw silk. Mulberry leaf is of paramount importance and metabolites play a crucial role in the sustainable growth and development of Silkworm *Bombyx mori* L. Apart from the biotic factors, abiotic factors such as temperature, humidity, rainfall, elevation, sunshine, etc., also have profound importance on the quality of mulberry leaves. Mulberry crop is grown in both temperate and tropical countries and several countries have been practicing sericulture for several decades. China is pioneer in silk production and is in driver's seat with regard to raw silk production followed by India. Tropical country like India is blessed with salubrious climatic conditions for growing mulberry and continuous silkworm rearing throughout the year, thus providing employment to rural folk. Being a labour intensive enterprise, it continues to flourish in labour abundant countries like India and China (Maribashetty

[1] Dept. of Biotechnology, Azyme Biosciences Pvt. Ltd., Jayanagar, Bengaluru-560069, Karnataka, India.
[2] Dept. of Sericulture, V.V.Pura College of Science, K.R. Road, Bengaluru-560007, Karnataka, India.
* Corresponding author: dryoganand16@gmail.com

et al. 1997). It is the agricultural cash crop with minimum investment and maximum profitability providing frequent and attractive remunerations throughout the year especially in tropical belt (Nataraja and Sampath 1988). The word SERI is fondly known as 'Self Employment Remunerative Industry' in India. About 92.20% of silk produced in the world is obtained from mulberry silkworm *Bombyx mori* L. reared solely on mulberry (*Morus* spp.) leaves and remaining 7.80% of silk is obtained from non-mulberry silkworms namely, tasar, muga and eri (Datta 1994).

Mulberry genotypes can be introduced to a region by means of importation or cultivating a selected genotype from among screened germplasm resources. The selected genotypes can be subjected to mutation breeding and/or polyploidy breeding. Careful evaluation of variants is a must for commercial exploitation. Evaluation is a process of correlating the morpho-genetic traits of a crop according to the consumer needs (Dandin and Kumar 1989). Providing the best variety over the existing ones is the aim of any crop improvement programme. Mulberry is also grown as a fodder crop despite its leaves being the sole food for silkworms. Therefore improvement of mulberry is mainly related to the quantitative increase in leaf yield per unit area. High leaf productivity brings in the reduction of cocoon production cost. Mulberry breeding aims at evolving new varieties for augmented yield, enriched quality, palatability, wide range of adaptability, good response to agronomic inputs and resistance to harmful biotic and abiotic factors. These goals can be achieved either by employing conventional methods like introduction, selection and hybridization, or other breeding methods like mutation breeding, polyploidy breeding in addition to tissue culture, protoplast fusion and genetic engineering. Cross pollination is the rule rather than exception in mulberry and this provides a high degree of heterozygosity and genetic diversity in the mulberry gene pool. These aspects make mulberry amenable for genetic experimentation and provide ample scope for evolving useful variants. Taxonomy of mulberry is highly confusing due to polygenic nature of genes. Evolving a new variety of mulberry is a herculean task and needs great degree of patience. Researchers have tried both conventional and non-conventional methods of genetic improvement in mulberry and not many beneficial varieties have been released as popular varieties although both central and state government institutions are maintaining germplasm with large number of mulberry accessions. Maintenance and evaluation of both exogenous and indigenous varieties is regularly undertaken and breeders are striving tirelessly to evolve new varieties.

Further, mulberry has capability of multiplication through vegetative propagation with which valuable and persisting variants can be multiplied, thus maintaining the parental characters. Mutations can be induced through both ionizing (X-rays, β-rays, gamma rays and fast neutrons) and non-ionizing (ultraviolet rays) radiations and treatment with chemical mutagens. Mutation breeding has been widely employed in recent times for improving vegetatively propagated crop plants. It offers the breeder a rapid method to increase genetic variation which, otherwise, would require many years to achieve through conventional breeding. Successful exploitation of various mutagenic agents for inducing aberrations has become one of the most important lines of contemporary research. Mutation induction in mulberry started towards the end of 1950s in Japan (Sugiyama and Tojyo 1962, Hazama 1967a). Katagiri (1970) has done extensive work on irradiation of mulberry by using gamma rays and obtained variations in leaf color, shape, size and internodal distance. Besides using irradiations, induction of mutation using chemicals such as Colchicine has also been found to be successful in mulberry (Das and Krishnaswami 1969, Das and Prasad 1974, Sastry et al. 1974, 1976, Dwivedi et al. 1986, 1989a, Yang and Yang 1991, Dwivedi 1994, Sikdar and Jolly 1995, Eswar Rao 1996). Colchicine is believed to interfere with spindle formation during cell division inducing polyploidy. Polyploidy has been useful in developing many new varieties of crop plants. Colchicine induced polyploids are known as Colchiploids. Colchiploids in mulberry have shown all the symptoms of polyploids such as larger leaf size, bigger stomatal size, dark green colored leaves, delayed flowering, slower growth, etc., when compared to their diploid counterparts. In commercially important plants, polyploidy is commonly induced to bring about gigas characters such as larger and thicker leaves, flowers and fruits. In mulberry, triploidy is found to be the optimum ploidy level. Generally triploids are highly vigorous, have high yield potential, good rooting ability, good regeneration of buds, high palatability and nutritive value to the silkworms.

Polyploidy in mulberry aims at improving the already existing cultivars by inducing improved morpho-economic traits either through gamma irradiation or colchicine treatment. Different treatments of these were given and the responses against each of the parameters analysed and compared with control. Qualitative and quantitative aspects of leaf production, root initiation and proliferation, sprouting, survivability and branching pattern were assessed in the selected variants for an effective evaluation. Chemo-assay and bio-assay studies were also carried out in the promising induced variants. These tests were mandatory to assess the qualitative aspects of the induced variants and to weigh their feeding efficiency in order to ascertain the commercial parameters (cocoon weight, shell weight, filament length, reelability percentage, denier, renditta and effective rate of rearing, etc.) required for silk production.

Evaluation of Popular Mulberry Genotypes in India

Different varieties of mulberry are available in nature but not all these genotypes are useful to breeder for lack of one or the other beneficial trait. The main objective of any breeding programme is to evolve a new breed having better yield parameters, resistance to pests and diseases, leaves with enriched nutrient components, survivability of new variety in different regions and seasons as well as under drought or stress conditions. Large number of mulberry varieties have been evolved but Mysore Local served as a pioneer mulberry variety which is grown in traditional areas of the southern zone of India. This mulberry variety could very well adapt to soil and climatic conditions of the southern zone. However, it was less productive and subsequently a high yielding-open pollinated variety M_5 was selected for the purpose of cultivation and exploitation in place of Mysore Local variety. Gradually M_5 gained popularity and was cultivated in different agro-climatic zones. Similarly, DD and TG varieties released into the field failed to live up to the expectations. Through mutation breeding and selection, Central Silk Research and Training Institute, Berhampore, India has evolved some promising elite varieties such as S_{30}, S_{36}, S_{41}, and S_{54} for cultivation under assured irrigation. Further, drought resistant varieties like S_{13} and S_{34} were also successfully introduced for cultivation under arid and semi-arid conditions. However, exotic varieties namely Goshoerami, Ichinose, Psukasakuwa, Italian, Philippinensis, etc., imported from other countries for commercial exploitation failed to acclimatize due to poor rooting and sprouting. Besides the introduction of elite varieties new high yielding mulberry variety 'V_1' has been introduced which is virtually ruling the mulberry gardens cultivated across the country with potential of 65–70 Tones/Hectare/Year of leaf production. S_{54} is considered more suitable for rearing chawki silkworms, S_{13} and S_{34} are drought resistant varieties and considered as popular genotypes.

Morphological Parameters of Popular Mulberry Genotypes

Morphology of Mysore Local

This mulberry variety belongs to *Morus indica* L. It is a popular tropical mulberry variety widely cultivated in Southern India and called as 'Natikaddi'. It is comparatively low yielding variety, but known for its adaptability to low agronomic inputs and poor management practices both under rain fed and irrigated conditions. Stem is woody, cylindrical, erect and clothed with many lenticels; milky latex present in the stem. Plant is branching type, erect, straight, monopodial; open type bushes, branches are simple, vertical and dark greenish. Leaves are smooth, heterophyllous (with both lobed and unlobed occurring on the same plant), alternately or spirally arranged, ovate to broadly ovate, palmately veined and coriaceous. Leaf yield is about 8000 kg and 25000 kg/ha/yr under rain fed and irrigated conditions respectively.

Morphology of M_5/Kanva-2

It is a diploid mulberry variety belongs to *Morus indica* L. selected from natural populations of Mysore local variety. It is a popular tropical mulberry variety widely cultivated in Southern India, grows vigorously and responds well to agronomic practices both under rainfed and irrigated conditions. Its acceptability and palatability by silkworms is superior to that of Mysore Local cultivar. It is being cultivated as a perennial bush and the plant has tap root system. Stem is woody, cylindrical, erect and clothed with many lenticels; milky latex present in the stem. Plant is branching type, erect, straight, monopodial; young stem is light green and mature stem is grayish brown in color. The leaves are simple, alternate, stipulate and petiolate. Stipules two, lanceolate, free, lateral, caducous. Upper surface of the leaf is green, smooth and lower surface is light green and slightly rough. Lamina unlobed, ovate, multicostate-reticulate, serrate and acuminate. Inflorescence is a catkin, cylindrical, axillary and possesses 30–35 flowers which are compactly arranged on the peduncle. Flower small, ebracteate, sessile, incomplete, actinomorphic, unisexual (female), tetramerous, cyclic and hypogynous. Female flower bears four perianth lobes which are hairy, fleshy, yellowish green in color with imbricate aestivation. Androecium is absent. Ovary is superior, bicarpellary, syncarpous, unilocular with single pendulous ovule attached to the marginal placenta. Style single, short with a feathery bifid stigma. The fruit is sorosis, green in color turns to black at maturation.

Morphology of S_{13}

Mulberry variety S_{13} is a selection from open pollinated hybrids of Kanva-2 during 1986. It is a male clone, exhibiting fast growth and quick rooting ability. Open type bushes, branches are simple, erect and greyish green in color. The variety is characterized by short internodes and having a capacity to produce large number of branches. Stem is woody, cylindrical, clothed with whitish grey lenticels; young stem greenish in color and mature stem whitish grey in color. Leaves are simple, alternate or spirally arranged, stipulate, lateral, lanceolate and caducous. Lamina unlobed with smooth surface, ovate to broadly ovate, coriaceous, multicostate-reticulate, serrate, acuminate, palmately veined, upper surface dark green and lower surface pale green. Inflorescence is catkin, axillary. This variety is recommended for rainfed areas with red loamy soils and also for water scarce areas. Because of its better rooting ability it has become popular in semi-irrigated conditions. It yields 13–15 MT of leaf/ha/year under rainfed conditions.

Morphology of S_{34}

S_{34} was evolved in 1986 from cross-pollinated hybrids of S_{30} and Ber. C_{776} varieties of *Morus indica* L. and its male clone. This variety is recommended for black cotton soils. The variety has extensive and deep rooting system and therefore grows well under moisture stress (rainfed) conditions. Branches are erect with short internodal distance. Stem woody, glabrous, greyish white and clothed with sparsely distributed lenticels. Leaves simple, smooth, unlobed, alternate, petiolate, stipulate, and lateral. Lamina cordate, multicostate-reticulate, coriaceous, serrate, acuminate, dark green in color, and glabrous. Each male flower bear 4 perianth lobes with imbricate aestivation. Stamens 4, opposite to perianth lobes, filament short, anther dithecous with longitudinal dehiscence. Pollen grains are small, and gynoecium is absent. Moisture content is 74.80% as it exhibits good moisture retention capacity. Leaf yield is about 12–15 MT/ha/year under rainfed conditions.

Morphology of S_{54}

S_{54} is a beneficial mulberry (*Morus indica*) variety evolved in 1984 by chemical mutagenesis through EMS treatment of Berhampore local variety at Central Sericulture Research and Training Institute, Mysore. It has better rooting and sprouting abilities and is capable of thriving well both in temperate and tropical climatic conditions. With yields of about 40 metric tonne herbage/hectare/year. This variety has been recommended as one of the promising high yielding variety under irrigated conditions. Leaves are most suited for rearing young age (Chawki) worms. S_{54} is a tree with drooping branches. Stem is woody, cylindrical, glabrous, with lenticels; young stem pale green and mature one is light grey in color. Leaves simple, alternate, stipulate, stipules-2, lateral, linear lanceolate, caducous, petiolate; lamina cordate, unlobed, multicostate reticulate with 3–5 prominent basal veins, serrate, acute and glabrous. Internodal distance measures 3.78–4.02 cm. Stomatal frequency is $602.41/cm^2$ and each stoma measures 17.1×3.8 μm in size. Guard cell possesses about 8 chloroplasts. The catkin bears 26–30 flowers. Flowers unisexual (male), sub-sessile, actinomorphic, incomplete. Perianth-4 petals, greenish yellow in color, succulent, imbricate aestivation. Stamens-4, opposite to perianth lobes, filament slender, inflexed in bud, anthers dithecous and dehisce longitudinal. Pollen grains small, round and viability is around 88–90%. Rudimentary ovary is present.

Morphology of Victory-1

This mulberry variety of *Morus indica* popularly known as V_1 was developed in 1997 by Central Sericultural Research and Training Institute in Mysore. Selected from controlled pollinated hybrids of S_{30} and Ber. C_{776}

the variety has been tested in seven centres in South India and recommended as resistant to diseases. V_1 variety bushes are of open type, erect with many branches, simple, vertical, rough with short internodes. Stem is hard, thick and gray in color. Leaves are shiny, thick, dark green, simple, smooth and glossy, unlobed, alternate or spirally arranged, entire, ovate to broadly ovate with truncate base, palmately veined and succulent with high moisture content. It has got good agronomic traits like high rooting ability, fast growth and high yield; performs better in wider spacing with paired row system. The average leaf yield is 70,295 kg/hectare at Thalagattapura under irrigated condition and 10,960 kg/hectare under rain-fed condition at Chamrajanagar in Karnataka using recommended package of cultural practices.

Polyploidization in Mulberry

Polyploidization is an appropriate and significant plant breeding method successfully used to enhance morphological and physiological characteristics in order to produce valuable germplasm resources in many plant species (Tang et al. 2010, Dhooghe et al. 2011). Colchicine has been extensively applied to induce plant polyploidy since it interrupts the mitosis at the anaphase stage resulting in chromosome doubling (Nigel et al. 2007, Mohammadi et al. 2012). Majority of mulberry plants are unisexual although small number of bisexual plants are also observed which are diploid in nature with chromosome number $2n = 2x = 28$. Polyploidization in plants with more than two sets of chromosome is a major mechanism for adaptation and speciation. In mulberry polyploidy is frequent and a lot of triploids have been noticed. Similarly tetraploid, pentaploid and hexaploid varieties have also been noticed. *Morus nigra* L. is dexoploid ($2n = 22x = 308$) with high chromosome number among the phanerogams. Polyploids exhibiting gigas characters in plants with increased girth of stem, leaf thickness, slow growth, decreased branching, internodal distance and decreased stomatal frequency is also associated with changes in sex expression of mulberry plants. Mostly colchicine is used in inducing polyploidy in mulberry. Triploid and polyploid mulberry varieties are found to be nutritionally better as compared to related diploids in terms of silk worm foraging and production of quantitative yield of raw silk (Tojo 1985).

Appraisal of Earlier Work on Polyploidy in Mulberry

Seki (1951) treated the seeds of open pollinated mulberry varieties Nezimeigheshi and Karyonezumigheshi in 0.1% Colchicine in aqueous solution for 24 hours. Out of the treated seeds, only 5 of them germinated and attained maturity. They secured three tetraploid plants. Hamada (1963) reported the occurrence of about 1000 mulberry varieties in Japan. Of these, 224 diploids and 126 triploids were identified by him. He opined that the triploid varieties produced higher nutritive qualities of leaves and exhibited resistance to cold and blight disease. Abdullaev (1962, 1963) reported the occurrence of spontaneous polyploidy forms of mulberry. He observed that varieties with bigger and sweeter fruits, and those yielding more leaves annually were generally polyploids. In these polyploid varieties fruits developed parthenogenetically. Further, he produced valuable polyploids using colchicine treatment or gamma irradiation. Das and Krishnaswami (1969) treated the seeds of diploid ($2n = 2x = 28$) mulberry variety CSR-I and bush malda with 0.05%, 0.1%, 0.2% and 0.4% colchicine. Based on the results obtained, they concluded that young seedlings or vegetative buds are ideal materials for the induction of polyploidy in mulberry instead of seeds. Kedarnath and Lakshmikanthan (1965) treated the vegetative buds of *Morus alba* with 0.2% and 0.5% aqueous solutions of colchicine. The resultant polyploid shoots exhibited delayed sprouting, slow growth rate, thick and dark green leaves. Sastry et al. (1976) treated the seeds of mulberry variety Kanva$_2$ with 0.1%, 0.2% and 0.4% of aqueous colchicine solution for a period of 6 hours, 12 hours and 24 hours. Treated seeds showed good germination percentage. Dzhafarov and Abbasov (1967, 1970) and Dzhafarov et al. (1985) standardized the colchicine treatment method for the production of tetraploids and higher polyploids in mulberry. They found 0.1% aqueous solution of colchicine treatment to germinated seeds, 0.04% to 0.1% colchicine treatment at the cotyledonary stage, and 0.05% to 0.5% colchicine application on growing points of adult trees very effective in the production of polyploids. They succeeded in obtaining 1000 tetraploid forms in different mulberry varieties. Alekperova (1979) induced polyploids by treating the growing stem of *Morus kagayamae* with 1% aqueous solution of colchicine.

Verma et al. (1986) treated the mulberry seedling with 0.2%, 0.4% and 0.6% colchicine solution for 2–3 days for 5 hours daily. They found that 0.4% colchicine is most effective in inducing polyploids in the genus *Morus*. Dwivedi et al. (1986) treated the sprouting axillary buds of RFS_{135} mulberry variety with 0.2%, 0.3% and 0.4% colchicine for 6 hours and 8 hours daily for three consecutive days. They obtained five tetraploid varieties in saplings treated with 0.4% colchicine for 8 hours schedule. They opined that this schedule is most effective for the induction of tetraploidy in mulberry. Dwivedi et al. (1988b) induced autotetraploids in M_5 genotype by treating the apical and axillary buds with 0.35%, 0.4% and 0.45% colchicine for 6 hours and 8 hours daily for three consecutive days. They secured six tetraploids from apical bud treatment and eight tetraploids from axillary bud treatment. Sikdar et al. (1986) treated the sprouting buds of *Morus indica* var. S_{41} with 0.4% colchicine solution prepared in 5% glycerol to induce polyploidy. They obtained one autotetraploid possessing $2n = 4x = 56$ chromosomes with improved leaf yield parameters.

Yang and Yang (1989) obtained tetraploid mulberry by treating the axillary and apical buds as well as seedlings at the cotyledonary stage with 0.1% to 0.4% colchicine. Further, Yang and Yang (1991) reviewed the research activities on polyploidy breeding in mulberry and found 'P.R. China' as the country where natural polyploid varieties of mulberry occurred most abundantly with different levels of ploidy such as triploid, tetraploid, hexaploid, octaploid and decaploids. They also reported the presence of spontaneous haploids in this variety. Sikdar and Jolly (1994) treated the apical and axillary buds with 0.2%, 0.4% and 0.6% colchicine for 6 hours and 9 hours daily for three consecutive days and recovered 14 tetraploids. Recovery of tetraploids was reported to be marginally higher in the apical bud treatment than in the axillary bud treatment.

Method of Inducing Polyploidy in Popular Mulberry Varieties in India

Colchicine has proved very potent polyploidising agent extracted from the plant *Colchicum autumnale* L. Mulberry bushes are given middle pruning to hasten the sprouting of axillary buds. Five different concentrations of aqueous colchicine ($C_{22}H_{25}NO_6$) viz., 0.1%, 0.2%, 0.3%, 0.4% and 0.5% have been generally used to treat the vegetative buds. Five buds selected per plant are earmarked by tagging for the treatment and selected buds washed thoroughly in distilled water before the application of colchicine. To buds that are covered with cotton swabs colchicine solution is applied from time to time from 8 am to 5 pm. This is done for three consecutive days (Dwivedi et al. 1986, 1988b), while simultaneously control buds are treated with distilled water. After completion of colchicine treatment and removal of the cotton swabs a thorough washing of buds in distilled water is done for allowing the treated and control plants to grow by providing the needed agricultural inputs. Stomatal frequency is determined only from the fully developed leaves (Sikdar et al. 1986).

Colchicine Induced Variations in Mulberry Variety M_5

Mulberry variety M_5 has been used for the induction of variation through colchicine treatment. Five different concentrations of aqueous colchicine [$C_{22}H_{25}NO_5$] viz., 0.1%–0.5% were used to treat the vegetative buds. It was found that sprouting percentage, rooting percentage, and plant height decreased with a linear increase in the colchicine percentage. Increase in leaf area was observed at 0.4% (166.94 cm²) and 0.5% (165.20 cm²) compared to control (155.41 cm²) plants. Thick, succulent and dark leaves (Figure 1) were found associated with the increased leaf area. In the M_2 generation, the variant exhibited decreased number of stomata/unit area (39.57) and increased number of chloroplast (18–22) compared to control (49.42 and 11–14). Yield of the treated plants (0.573 kg/plant) was noticeably increased. Plants grown to full maturity were tested for their potential use in breeding programmes oriented toward producing lines with improved quantity and quality of leaves (Ramesh et al. 2011).

Colchicine Induced Variations in Mulberry Variety Kajali

Mulberry variety Kajali is grown in tropical condition and the vegetative buds are treated with aqueous solution (0.1%–0.5%) of colchicine. Agro-botanical traits such as sprouting, rooting, survivability percentages, plant height, leaf area and intermodal distance were evaluated. The results showed decrease in the growth parameters

Figure 1. Polyploid variant of M_5 at 0.4% Colchicine.

of Kajali variety with the increase in colchicine concentration. Plants recovered at C_1 (Colchicine treated plants in F_1 generation) generation had beneficial characters only at 0.4% colchicine treatment in C_0 plants. Leaf area considerably increased to 196.11 cm^2 compared to control (178.27 cm^2). In dwarf, stout, thick, greenish, lobed and unlobed, leaves (Figure 2), yield increased to 10–11% and number of chloroplast ranged from 23–27 per cell than the control (44.00) in C_1 generation. C_1 mulberry plants require further systematic yield trials and evaluation of Colchiploids over a period of time to establish their potentiality as cultivars (Ramesh and Yogananda Murthy 2014).

Several investigators have succeeded in inducing polyploidy and utilizing these polyploids either directly for commercial purpose or for further breeding work in various crop plants (Sybenga 1992). The survival percentage of treated axillary buds of varieties M_5 and Kajali declined upon treatment with increased colchicine concentration. Similar observations were reported earlier by Tojyo (1966) in Japanese mulberry genotypes, and Dwivedi et al. (1986), Verma et al. (1986), Sikdar (1990), Eshwar Rao (1996) in tropical mulberry varieties namely, RFS$_{135}$, M_5, S_{30}, S_{36} and S_{41}. It is evident that the higher concentrations of colchicine (0.4%–0.5%) not only drastically affect the sprouting and survivability but also results in the delayed emergence of buds. It may also be due to physico-chemical disturbances of cells and reduced rate of cell division and polyploidization (Swanson 1965). Lower concentration of colchicine (0.1%, 0.2% and 0.3%) did not affect the rooting behaviour, but number of roots considerably decreased in the progeny treated with higher concentration of colchicine (0.4% and 0.5%). Dwivedi et al. (1989a) also observed poor rooting behaviour in colchicine induced autotetraploids of S_{30} and S_{36} mulberry varieties. Similar decrease in plant height was observed in M_5 and Kajali varieties when treated with high colchicine concentrations.

Figure 2. Polyploid variant of Kajali showing lobed and unlobed leaves at 0.5% Colchicine.

Scientists were of the opinion that, the reason for reduced growth in colchicine treated material is due to abnormal cytological behaviour. The stunted growth with deformities following colchicine treatment is probably due to serious hormonal imbalances resulting in physiological disorder (Bharathi Behera and Patnaik 1975) and also due to reduced rate of cell division (Swanson 1965). Colchicine treated plants, however, showed marginal reduction in the number of branches. Internodal distance is another phenotypic attribute that determines the total foliage produced. Short internodes were noticed in some populations of M_5 and Kajali treated with 0.4% and 0.5% of colchicine, respectively. Das et al. (1970) reported 0.4% and 0.6% aqueous colchicine solution is most effective in the induction of tetraploids by treating the growing apical buds of mulberry seedlings. Dwivedi et al. (1986) found that 0.3% and 0.4% colchicine applied for 6 hours, and 8 hours, respectively, for three consecutive days effected the induction of tetraploids in *Morus alba* variety RFS_{135}. Some of the irregularities observed in leaf size, shape, texture and coloration in tetraploids are attributed to the differential rate of cell division coupled with physiological disturbances in the treated buds as commonly noticed in colchicine treated populations (Dwivedi et al. 1988b, Sikdar 1990). Colchicine (0.5%) treatment of mulberry variety Kosen resulted in plants with enhanced number of chloroplasts, leaf yield and leaf weight compared to control (Ramesh and Munirajappa 2001). The beneficial variant procured from 0.4% colchicine treatment, however, had stomatal frequency and stomatal chloroplast counts that revealed it to be polyploid in nature. In this variant, decreased stomata per unit area and increased number of chloroplast per stoma were encountered. Reduced pollen fertility was also observed. Similar observations on stomatal frequency, stomatal chloroplast counts, and pollen fertility were adopted to ascertain the polyploidization in colchicine treated mulberry plant populations (Sikdar et al. 1986, Susheelamma et al. 1991).

Mutagenicity in Mulberry

A mutation is a sudden heritable change in the DNA in a living cell, not caused by genetic segregation or genetic recombination. Mutation breeding is one of the many techniques available to plant breeders for the development of superior genotypes. Mutation breeding implies the purposeful application of mutations in plant breeding (Van Harten 1998). Unlike hybridization and selection, mutation breeding has the advantage of improving a defect in an otherwise elite cultivar, without losing its agronomic and quality characteristics. Mutation breeding is the only straightforward alternative for improving seedless crop. Mutation breeding has found a niche in plant breeding because of these advantages. Plant breeding can be seen as human-guided evolution of crop plants and genetic variation is its footing. When anticipated variation is obtainable in different cultivars of a crop, its breeding objectives may be attained through cross-breeding. However, when one or more of the parent cultivars possessing the desired characters used in cross-breeding are poorly adapted, it becomes essential to implement a back-crossing approach to recover the elite type. Another problem encountered in conventional crossbreeding is the poor combining ability of some parental genotypes. For example, aromatic rice varieties have poor combining ability and cross-breeding with non-aromatic varieties will lead to a decrease in aroma and quality (Bourgis et al. 2008, Pathirana et al. 2009). Under such circumstances, mutation induction can be of benefit to produce cultivars with anticipated traits within defined germplasm pools.

Mulberry is considered as the most suitable material for mutation experiment as once a desirable mutant is identified it can be propagated successfully either by cuttings or grafts. Mulberry breeders are often faced with long juvenile phase of the plant. Heterozygocity coupled with long juvenile phase make the development of superior recombinants through conventional breeding very difficult and time consuming. One important advantage of mutation breeding over cross breeding is that the appropriate treatment may lead to change in one or few genetic characters to be improved without disturbing the overall genetic setup of the material. The main objectives of the mulberry breeders are high yield with better leaf quality, resistance to diseases and pests, adverse climatic conditions, adaptability, improved agro botanical characters, increasing leaf area etc. (Hazama 1968b, Nakajima 1973, Kukimura et al. 1975).

Induction of Mutation

As the spontaneous mutation rate is very low, it is important to increase the mutation rate through mutagens (agents capable of inducing mutation in plants).

Physical Mutagens

Physical mutagens emit irradiations highly penetrating in nature which due to excitation of electrons cause alteration in genome amongst the cells resulting in mutant formation. Physical mutagens namely X-ray, gamma, alpha and beta rays, and fast neutrons are commonly used for irradiation processes. In mulberry mutagenesis, considerable research work has been done and Fu-sang is the only mulberry variety developed by the Institute of Silkworm, Mulberry and Tree Research, Jiangxi Academy of Agriculture Science, China using gamma rays irradiation method. Due to the polygenic nature and confused taxonomy of mulberry not many varieties are released but efforts are continuing for induction of variation using physical mutagens. From extensive research on mutation breeding in mulberry in India two indigenous cultivars M_5, S_{54} and one exotic variety Kosen have been selected. Both M_5 and S_{54} showed favourable characters compared to exotic variety and substantial results were also noticed in chemo assay and bio assay studies (Das et al.1970, Rao et al. 1984, Tikader et al. 1996, Eswar Rao et al. 2004, Deshpande et al. 2010).

Method of Treatment with Physical Mutagens for Mutagenesis in Mulberry

Gamma Irradiation Studies

Juvenile twigs of mulberry taxa are selected for cutting slip preparation and only middle part of the twigs taken. Each cutting about 15 cm in length with 2–3 active vegetative buds (free from pathogens and pests) are used for irradiation purpose. Care is taken to avoid damage to the buds, and cut ends, while preparing the cutting (Hartman and Kester 1976, Bindroo et al. 1988). The cuttings are irradiated with different doses of gamma rays (1 kR to 10 kR) from Co^{60} gamma unit. Irradiated cutting slips then planted in earthen pots filled with a mixture of well dried pulverized garden soil, fine sand and well decomposed farmyard manure in the proportion of 1:1:1. Three replications having 10 cuttings each in pots are usually maintained for six months before transplanting them to the main field. Suitable practice is to arrange pots in rows giving a spacing of ½' between the pots and 1' between the rows and later. The transplanted saplings are maintained in randomized block design with 90 cm × 90 cm spacing. Necessary cultural operations such as timely irrigation, weeding, intercultivation, manuring, protection against desiccation, diseases and pests need to be ensured. Controls also are maintained in similar conditions for comparative studies. This is followed by recording data on M_1 and M_2 generations related to growth responses such as sprouting, rooting, survivability, internodal distance, branching pattern, leaf area and pollen fertility.

Appraisal of Earlier Work on Mutagenicity using Physical Mutagene in Mulberry

Hazama et al. (1968) studied the varietal differences of mulberry to radio-sensitivity and reported bud mutation. They found the inhibition of plant height, growth and induction of fasciation among 15 mulberry varieties of Japan due to subsequent gamma irradiation. They also reported that mutation rate/plant was quite high at 3 kR to 10 kR. Hazama (1968a,b) irradiated 50 plants of variety KNG (Kairyone-zumigheshi) with gamma rays (5 kR–15 kR) and secured one mutant possessing both the types of entire and dissected leaves. The mutant showed increased leaf yield over the control. The irradiated clones were subjected to pruning for 3 successive years in order to widen mutation spectrum and to increase mutation frequency. Tojyo (1966) carried out an experiment in which mulberry cuttings preserved at 3–5°C were irradiated with different doses of gamma rays (2.5 kR–10 kR). In the plants developed from irradiated cuttings, growth of flower buds was inhibited, size of the pollen grains varied and meiotic irregularities encountered. Das et al. (1970) studied the effects of gamma irradiation on seed germination and seedling development in mulberry. Seeds and seedlings of *Morus alba* Var. Ichinose were irradiated with gamma rays (2.5, 5, 7.5, 10 and 15 R). He reported that 10kR is lethal dose for growth and development of mulberry seedlings.

Nakajima (1972) recovered a beneficial mutant in mulberry by gamma irradiation. He secured a promising mutant 3183 by irradiating one year old grafts of KNG variety with 5kR gamma rays. This mutant showed changed leaf morphology, i.e., entire leaf (unlobed) in contrast to 5 lobed leaf or control. It also revealed 7% increase in leaf thickness, 20% in inter-nodal distance and 10% increase in leaf yield. Nakajima (1973) irradiated the hardwood cuttings of three mulberry cultivars namely KNG, Kenmochi and Ichinose. The experiment resulted in the production of mutants showing increased leaf yield and resistance to *Diaporthae nomurai* disease. Fujitha and Nakajima (1973) subjected the induced mutants to re-irradiation with 10 kR gamma rays. The shoots of re-irradiated plants were pruned repeatedly. They observed that mutant 3198 and IRB 240-I exhibited higher mutation frequency than the parent variety Ichinose. Further, they recorded the reversal of leaf shape trait due to re-irradiation. Katagiri (1976a) irradiated the vigorously growing shoots of mulberry Var. Ichinose with 6 kR (120 Rad/h) and 7.5 kR (150 Rad/h) gamma rays during early July, August and September months. Irradiated shoots of early September month produced deformed narrow leaves and failed to grow at higher dose, whereas the irradiated shoots of early July and August months had normal growth and development at lower doses (6 kR). These results imply that shoots exposed at higher doses 7.5 kR as LD_{50} fail to develop. Nakajima (1977) subjected the irradiated plants for repeated pruning in order to develop wholly mutated shoots from the vegetative buds. He reported that the entire leaf of mutant reverted back to normal lobed leaf due to re-irradiation. Fujitha et al. (1980) re-irradiated the mutant line 3198 of the variety Ichinose by 10 kR gamma rays. The mutant line was obtained originally by irradiating the cuttings of variety Ichinose with 10 kR gamma rays. They observed the increased frequency of mutations in the mutant (7.1%) over the control (0.5%) due to re-irradiation. Mutant forms recovered exhibited improvement in leaf and shoot characteristics. Gatin and Ogurtsov (1981) demonstrated that the wide spectrum of new mutant forms induced from among diploids allow useful mutations to be selected. They opined that the mutation spectrum was less wide among polyploids. Further, they stated that the induced mutagenesis among polyploids can diversify the gene pool in mulberry.

Fujitha and Wada (1982) also re-irradiated the induced mutant of Ichinose variety with gamma rays (10 kR). They recorded the increased mutation frequency and wider mutation spectrum due to re-irradiation. They obtained a mutant with entire leaf by irradiating a cultivar possessing 5 lobed leaf. Interestingly, Katagiri and Nakajima (1980) irradiated the mulberry trees with 5 kR gamma rays at the rate of 0.2 kR/hr. The irradiated trees were allowed to grow and subjected to pruning three times in a growing season. They obtained tetraploids and mutants. The frequency of tetraploids was more important in mutants recovered. Rao et al. (1984) studied the effects of different doses of maleic hydrazide and concentrations of gamma irradiation, respectively, on M_5 and local mulberry varieties. They recorded promising results with regard to traits such as sprouting and survival percentage, plant height, branching pattern, leaf size and total yield in M_1 generation. Based on the observed results, they concluded that gamma irradiation is a very effective tool to induce variation and to genetically improve mulberry cultivars. Jayaramaiah and Munirajappa (1987) conducted irradiation studies on 'Mysore Local' mulberry variety. They irradiated the hard wood cuttings with different dosages of gamma rays (1kR to 10kR) and studied various parameters like survival percentage, sprouting ability, vigour of plant and cytological features. They observed mutants with visible morphological traits like shortened internodes, thick, succulent and dark green leaves, opposite phyllotaxy, etc. Further, they also reported the meiotic anomalies like clumping of chromosomes, laggards and anaphase bridges in mutants. Kuchkarov and Ogurtsov (1987) studied the different forms of mulberry mutants exhibiting jointed snake like shoots, rolled leaf blades, weeping habit and other traits of varieties, e.g., var. globosa, pyramidalis and aurea.

Mutagenesis by Physical Mutagens in Popular Indian Mulberry Varieties

Gamma Induced Variations in Mulberry Variety M_5

M_5 mulberry variety cuttings were irradiated with different doses of gamma rays (1 kR–10 kR) and recorded various propagation parameters viz., sprouting percentage, rooting percentage, survivability, height of the plant, internodal distance, and leaf area. Moderate dosages of 4 kR–7 kR proved fruitful in the induction of beneficial variability. Morphology of leaf in mutants showed variation like enlarged leaf (4 kR), curled and mosaic leaf (8 kR), and biforked leaf (9 kR) in the treated population. Further in M_2 generation, important features like shortened internodal distance and increased leaf area was observed. Rooting percentage increased at 4 kR (90.03%) when compared to control (89.20%). LD_{50} of M_5 proved good between 6 kR and 7 kR, while doses 9 kR and 10 kR turned out lethal for the taxa studied. Height of plants in the irradiated population got significantly reduced with the increase in the dosages of gamma rays though slight increase in plant height was recorded in the population irradiated at 5 kR. Increased tendency in the number of branches occurred at 4kR, while higher doses of gamma rays, i.e., 8 kR to 10 kR drastically reduced the number of branches. The treated plants were grown for full maturity in order to test their potentiality in breeding programme (Ramesh et al. 2012a).

Gamma Induced Variations in Mulberry Variety Kosen

Mulberry cuttings of variety Kosen were treated with different doses of gamma radiations (1 kR–10 kR from Co^{60}). Overall results revealed a declined trend in all the growth parameters, e.g., sprouting (83.66%–18.66%), rooting (77.96%–19.59%), height of the plant (105.00 cm–58.03 cm), number of the branches (7.73–4.23), intermodal distance (3.76 cm–3.94 cm), petiole length (2.89 cm–2.59 cm) and pollen fertility (77.93%–40.66%), with deleterious effect though the response of these growth parameters against different doses also showed fluctuating behaviour. For example, sprouting percentage had decreasing trend with delayed sprouting on increase in dosage of gamma rays whereas the cuttings irradiated at low dosage (1 kR to 5 kR) took 13–15 days to sprout. Survivability of plant cuttings varied depending on the dosage of gamma rays administered. Saplings recovered from cuttings irradiated with higher doses of gamma rays (10 kR) sprouted, grew initially, but failed to develop further after 45–50 days (Figure 3). Plant height of the irradiated population of var. Kosen was significantly reduced with the increase in gamma rays administered. The number of branches and branching pattern showed variations. Internodal distance was also found less than in control plants. Leaf area was considerably decreased due to gamma irradiation and some irradiated populations at 8 kR, 9 kR and 10 kR exhibited small, crumpled, biforked and chlorophyll deficient leaves (Ramesh et al. 2012b).

Figure 3. Mutant Kosen failed to bloom at 10 kR Gamma rays.

Gamma Induced Variations in Mulberry Variety S_{54}

Gamma ray of different concentrations (1 kR–10 kR) were used to induce variability in juvenile twigs of mulberry variety S_{54} for various agro-botanical characters viz., sprouting, rooting, internodal distance, leaf area, plant height, etc., leaves were subjected to biochemical analysis. Mulberry variety S_{54} showed linear decrease in growth parameters with the increased gamma ray dosage. However, plants obtained from cuttings treated with dosage 7KR exhibited variability with increased rooting (81.33%), plant height (147.86 cm) and leaf area (146.22 cm^2) compared to control. Treated variants exhibit notably boat shaped leaves at 8 kR irradiation (Figure 4). Mutants showing desired characters were grown upto M_2 generation as the progeny showed marked improvement in growth and yield parameters. Biochemical constituents of leaves in S_{54} mutant also had increased proteins, amino acid, chlorophyll a and b content. Mutants of M_2 generation thus demonstrated marked improvement consistently in growth, yield and bio-chemical parameters (Ramesh et al. 2014).

Figure 4. Mutant S_{54} showing boat shaped leaves at 8 kR Gamma rays.

Causes of Variations Affected by Gamma Rays

(a) Decrease in sprouting has been attributed to the destruction of auxin (Skoog 1935) or due to inhibition of auxin synthesis (Gordon 1957). It may be also due to variation in temperature, water content and oxygen level at the time of treatment (Nybom et al. 1952, Rao et al. 1984, Jayaramaiah and Munirajappa 1987, Tikader et al. 1996). Sprouting is adversely affected by higher doses of gamma rays. Eswar Rao et al. (2004) and Deshpande et al. (2010) reported that, gamma rays being more potent and highly penetrating in nature might have impacted cells which were undergoing meiotic division in the bud region. Katagiri (1970) reported that, decrease in sprouting percentage with the increase in gamma rays dosage is due to partial cell death. Similar reduction in sprouting and survival percentage of vegetatively propagated crops was reported (Banerji and Datta 1991, Hemalatha 1998). According to Kaicker (1992) reduction in sprouting may be due to the toxic effect of higher doses of gamma rays whereas the same at lower levels hastened the metabolic activity. Fujitha and Wada (1982) reported relatively low percentage of rooting (50%) in mutant strains of Ichinose over its control. They reported that spontaneous mutant of variety KNG also produced low percentage of rooting (16%–32%), whereas normal KNG variety showed high percentage (96%) of rooting. Kukimura et al. (1975) too reported the increased rooting ability of mulberry mutants. It has been concluded that, lower doses of gamma irradiation could be safely used as an effective tool to induce variations towards improving mulberry cultivars (Nakajima 1972, Das et al. 1987, Jayaramaiah and Munirajappa 1987, Hyadalu Lingappa Ramesh et al. 2013). They also opined that the increased dosages of gamma rays (9 kR and 10 kR) produced semi lethality to complete lethality. Skoog (1935), Smith and Kersten (1942) opined that the decrease in survival percentage after radiation treatment was attributed to the destruction of auxins. Sastry et al. (1974)

reported that survivability in mulberry varieties S_{30} and K_2 showed that the injury was directly proportional to the concentration of mutagen. The authors also suggested that disturbances caused at the physicochemical level in cells or acute chromosomal damage are due to the combined effect of both. Gray (1990) observed that decrease in the survivability of irradiated plant material may be related to a series of events occurring at the cellular level that affect the vital macromolecules and result in physiological imbalance. Survivability indicates the capacity of treated population to withstand even the severe dosage of gamma rays. Deshpande et al. (2010) reported the survival percentage in all the irradiated mulberry varieties showed a continuously decreasing trend. Sensitivity of the plant material depends on the genetic constitution, dose employed, DNA content, its replication time at initial stages, moisture content, development stage and the genotype.

(b) Height of the irradiated population is significantly reduced with the increase in gamma rays administered. Mulberry is basically polygenic and plant height is a quantitative trait predominantly controlled by polygenes. Each gene contributes to small effects, which is called genetic additive effect (Thohirah Lee Abdullah et al. 2009). Anon (1977), Kearsey and Pooni (1996) were of the view that in some cases mutation are not stable because they will undergo recombination during meiosis. Rao et al. (1984), based on his studies on irradiation effects in M_5 and local mulberry varieties, reported that higher doses of gamma rays irradiation (8 kR to 10 kR) drastically reduce not only the height but the number of branches in the irradiated progeny leading to semi lethality to complete lethality. Mutagens can have direct negative effect on plant tissue due to the fact that primary injuries cause inhibition of cell division or cell death affecting the growth habit and change in plant morphology. Contrarily if doses are too low, a low mutation frequency occurs that results in small mutated sector (Nazir et al. 1998). Katagiri (1976a) observed leaf deformation and growth inhibition at higher doses of gamma rays in mulberry var. Ichinose. These effects are cytological such as chromosomal damages, inhibited mitotic division, degeneration of nuclei, cell enlargement, etc. (Sparrow et al. 1952, Pollard 1964, Karpate and Chaudhuri 1997). Further leaf parameters such as leaf area and length of internodes, which are desirable morpho-economic traits in mulberry, were also affected due to irradiation (Hazama 1968b, Katagiri 1970, Kuchkarov and Ogurtsov 1987). Dandin et al. (1996) registered several spontaneous leaf mutants with abnormal traits like venosa, wrinkled leaf, glossy leaves, etc. The abnormalities in leaves are attributed to disturbances in phytochromes, chromosomal aberrations, and mitotic inhibition (Moh 1962, Mickaelsen et al. 1968, Abraham and Ninan 1968). Internodal length variation was found to be affected by cell number and cell size or both in mutants of Mysore Local variety raised after gamma irradiation (Jayaramaiah and Munirajappa 1987).

Chemical Mutagens

Recent decade has witnessed intensive work on the induction of mutations in mulberry by chemical mutagenic agents (Appa Rao and Jana 1976, Wang 2017). The frequency of induced mutations is almost manifold than those naturally occurring and they have been looked as a powerful tool for the development of new cultivars. EMS is a potent mutagen that has been extensively used in genetic research. Of all the chemical mutagens available today DES and EMS are reported to be more effective and could be safely used for mulberry (Sastry et al. 1983, Rao et al. 1984, Yang and Yang 1991, Santhosh Lal and Pavithran 1997, Deshpande et al. 2010). In higher organisms, there is clear-cut evidence that, EMS is able to break chromosomes, although the mechanism involved is not well understood, but there is some evidence that EMS can cause base-pair insertion or deletions as well as more extensive intragenic deletions (Gary and Sega 1984). The main advantage of this method is for mutation breeding in vegetatively propagated crops as it has the ability to change one or few traits of outstanding cultivars with little alteration in their genotype. Since mulberry is a vegetatively as well as sexually propagated plant exhibiting a high degree of heterozygosity, its improvement needs chemical mutation induction. Chemical mutagen Ethane-Methane-Sulfonate (EMS) induces varied morphological changes in mulberry although reversal of characters were observed in C_2 generation due to the polygenic nature of mulberry. Procedure employed in chemical mutation is very simple.

Method of EMS Induced Mutagenesis in Mulberry

Using Seeds

In India, prominent mutants evolved at Central Silk Research and Training Institute, at Berhampore, Mysore were obtained by treatment of seeds with EMS at concentrations of 0.1% to 0.6% for 6, 12 and 24 hours, respectively. Mutants such as S_{30}, S_{36}, S_{41} and S_{54} were isolated. S_{36} exhibited increased leaf yield (more than 25%) over the standard variety $Kanva_2$. S_{54} has been found more suitable for raring Chawki silkworms.

Using Stem Cuttings

Indigenous mulberry variety M_5 was chosen for investigation. Stem cuttings were procured from healthy mulberry plantation aged about 8 to 10 years. The stem cuttings were planted in earthen pots and filled with standard soil mixture (Red earth, sand and farm yard manure) in the proportion of 1:1:1. A week after planting, the vegetative buds were treated with varying concentration of EMS (0.1%–0.5%) from dawn to dusk (8 am to 6 pm) via cotton plugs placed on the buds (Broertjes and Van Harten 1998) intermittently at every one hour. Ten replicas were maintained for each concentration and each replica consisted of three stem cuttings. The earthen pots were placed under optimum sunlight and periodically watered. After treatment till the end of the eighth months observations were made regularly from the day of sprouting to record the variations in the vegetative characteristics (Dandin and Jolly 1986, Dandin and Kumar 1989).

Appraisal of Earlier Work on Chemical Mutagenicity in Mulberry

Sastry et al. (1974) treated the seeds of Berhampore mulberry variety with aqueous solution of EMS in different concentrations (0.1%, 0.15%, 0.3%, 0.45% and 0.6%) for 6, 12 and 24 hours at 5°C and 20°C. They reported that maximum survival rate at 12 hours treatment duration at 5°C. Beneficial mutants S_{30}, S_{41} and S_{54}, superior to $Kanva_2$, were recovered. Aliev (1977) used chemical mutagens Ethyl methane sulphonate, Diethyl methane sulphonate and N-nitrosomethyl urea in combination with hybridization techniques in few mulberry varieties to evolve beneficial genotypes. Mulov and Kovalik (1977) treated hybrids produced after interspecific hybridization among open pollinated mulberry genotypes with chemical mutagens for production of mutants. Kozikova (1980) obtained useful mutants of mulberry genotypes for breeding programme at M_2 generation showing significant morphological deviations with chemical mutagens. Sastry et al. (1983) studied the different sensitivity of chemical mutagens on sprouting, survival and injury in three mulberry varieties S_{30}, S_{36}, and $Kanva_2$, by treating the buds with aqueous solutions of sodium azide (0.005%. 0.01%, 0.015%), diethyl sulphate (250, 500, 750 ppm) and compared to effects of 5, 10 and 15 kR gamma rays. Results showed that the injury was directly proportional to the concentration and doses of mutagens used. Thus, it is evident that the higher concentrations of mutagens are lethal and injurious, affecting the survivability. Rao et al. (1984) studied survival, sprouting, plant height, branches and number of nodes, leaf size and leaf yield in *Morus indica* var. M_5 as well as local mulberry varieties, by treating the cuttings both with different doses of gamma rays and maleic hydrazide. They opined that both as effective tools in inducing beneficial mutants in mulberry.

EMS Induced Variations in Mulberry Variety M_5

Mulberry variety M_5 was used for the investigation of mutagenic effect of Ethyl Methane Sulphonate (EMS). The EMS (0.1%, 0.2%, 0.3%, 0.4% and 0.5%) treated M_5 mulberry cuttings effected variation in sprouting percentage. At 0.4% concentration, higher sprouting of 75.71% was recorded and the least sprouting (44.44%) occurred at 0.1% concentration when compared to control (89.66%). Similarly, rooting percentage (78.71%) was noticed at 0.4% concentration and least rooting percentage (46.27%) at 0.2% concentration as compared to control (91.04%). Height of the plant and leaf area increased considerably at 0.4% EMS treatment. Stem dichotomy, fusion of leaves (Figure 5), increase in thickness, change in leaf texture (Figure 6) and occurrence of albino xantha were

Figure 5. M_5 showing fusion of leaves at 0.4% EMS.

Figure 6. M_5 showing leathery and wrinkled texture leaves at 0.3% EMS.

common in 0.3% EMS treated plants. M_5 variety clones treated with 0.3% EMS also showed significant decrease in internodal distance (0.66 cm) and plant height as compared to control (1.12 cm). However, more number of leaves/plant (12.33) occurred at 0.3% EMS concentration, less (8.64) at 0.1% EMS, as against the control (8.73), while at the same time, more number of inflorescence and fruits (4.98) were recorded at 0.1% EMS concentration, followed by 4.62 at 0.4% EMS and least number of inflorescence and fruits (3.84) at 0.5% EMS in comparison to control (3.70). Leaf area (166.23 cm) recorded 2.4 fold increase at 0.4% EMS concentration compared to control (69.25 cm) which is approximately 2 fold increases (81.04 cm and 83.40 cm, respectively) than 0.1% and 0.2% EMS concentrations (Yogananda Murthy et al. 2011).

Mulberry genotypes show variation in sprouting ability affected by mutagen treatment (Agastian et al. 1995, Eswar Rao et al. 2004, Hardansau et al. 1995). Decrease in sprouting percentage has been attributed to destruction of auxin or due to the inhibition of auxin synthesis, variation in temperature, water content and oxygen tension at the time of treatment (Gordon 1957). Shirasawa et al. (2016) reported stunted growth as a result of hydroxyl amine synthesis on account of EMS treatment in *Lycopersicon*. With increase in chemical concentration the rate of mutation also increased leading to variations in rooting. Delay in mitosis coupled with other physiological factors could be the result of dwarfing in some of the EMS treated plants (Chakraborthi et al. 1998). The irregularities observed in plant leaves by the earlier workers has been attributed to various causes such as phytochrome disturbance chromosome aberrations, disruption of auxin synthesis, DNA synthesis, mineral deficiencies. Enlargement of palisade, spongy and mesophyll tissues observed despite these irregularities (Karpate and Choudhary 1997, Mikaelsen et al. 1968, Ahloowalia and Maluszynski 2001) may, however, result in increased leaf area (Dwivedi et al. 1989a, Singh et al. 1999). Waghmare and Mehra (2000) also recorded increased plant height and number of branches in *Lathyrus sativus* L. on treatment with 0.5% EMS. Raisinghani and Mahna (1994) noticed growth abnormalities including variation in height in gamma ray irradiated and alkylating agent treated vegetative shoots in *Vigna mungo*. Konzak et al. (1961), Rahman and Soriano (1973) have interpreted, like earlier reports, the reduced growth in treated materials due to abnormal cytological behaviour, irregular cell enlargement and degeneration of nuclei. Increase in leaf area in EMS treated population is quite important as it reflects on the biomass of the plant (Datta et al. 1978,

Dwivedi et al. 1988b, Sarinee Chaicharoen et al. 1995). Brunner (1995) and Hotta (1954) opined that leaf area increase may be due to enlargement in palisade and spongy layers both in length and breadth and also due abnormal chromosome number. The reduction in size of the inflorescence due to EMS treatment have also been observed. Such floral mutants were reported in EMS treated *Brassica juncea* L. (Bhat et al. 2001).

Conclusion

Evolution of new variety should satisfy criterion like yield, resistance to pests and diseases, nutritional quality of leaves, regional and seasonal specific conditions. Breeder employs both conventional and non-conventional methods of breeding. The progeny selected should exhibit traits better than mat of parent including combination of characters which may suffice the needs of the end user. Mulberry breeding assumed importance in India since several decades. In the earlier days, breeding was solely dependent on selection and hybridization techniques which required minimum knowledge and scientific temperament. Advances in breeding technology made breeders to adopt advanced methods like mutation and polyploidy breeding with noteworthy contributions.

Mulberry breeding is a herculean task due to its heterozygous, cross pollinated, polygenic nature and confused taxonomy. Though several physical mutagens like X-rays, α-rays, β-rays and gamma rays as well as fast neutrons are available, most of the mulberry breeders depend much on gamma irradiation using source Co^{60} due to reported success stories in breeding and possible observation of variable characters in mulberry. Seeds, seedlings and cuttings of mulberry are exposed to gamma ray irradiation at various dosages (1 kR–10 kR). The rays penetrate deep into the tissues causing instability in cells due to excitation of electrons. Addition, deletion or frame shift mutations change the genomic and morphological characteristics. Only such plants show beneficial traits over the parents and are selected for growing a series of generations to release a variety.

Polyploidy in mulberry is induced by colchicine as it inhibits spindle formation in cells during mitosis. Polyploids of mulberry exhibit stout stem, thick green leaves, reduced number of stomata, increased guard cell size, etc. Silkworms prefer both polyploids for its foraging activity due to enriched nutritional quality of leaves.

Chemical mutagens also induce variations in mulberry and some of the commonly used mutagens include EMS, MMS, DES, Mustard gas, Sodium azide, etc. In mulberry breeding, EMS is extensively used to evolve beneficial varieties. S_{30}, S_{36}, S_{41} and S_{54} varieties were released after treating Berhampore local variety with different EMS concentrations (0.1%–0.6%). Both mulberry seeds and cuttings can be used for chemical mutagen treatment.

Mulberry breeding is a tough job and breeder confronts with several technical problems. Varieties showing beneficial traits in F_1 and F_2 generations suddenly revert back to the parental traits in next generation due to polygenic nature of plants. Evolving a new mulberry variety is cumbersome and time consuming process which requires patience. Not many varieties, therefore, could be released and even if new variety is evolved, it lacks in one or the other beneficial trait required for silkworm rearing. Research institutes at both central and state governments of India, and Universities strive hard to produce varieties which can meet the required goals for the enhancement of silk production in the country and contributing substantially to the national exchequer.

References

Abdullaev, I.K. 1962. Utilisation of triploidy in the development of high yielding forms of mulberry for the silkworm. Agrobiologiya (Moscow) 6: 861–865.

Abdullaev, I.K. 1963. Polyploidy and selection of mulberry. (Rus) Akad. Nauk. Azerb. SSR. Dok. 19(1): 49–53.

Abraham, A. and C.A. Ninan. 1968. Genetic improvement of the coconut palm: Some problems and possibilities. Indian J. Genet. and Plant Breeding 28A: 142–153.

Agastian Sim Yan Theoder, P., D. Dorcus and M. Vivekanandan. 1995. Screening of mulberry varieties for saline tolerance. Sericologia 35(2): 487–492.

Ahloowalia, B.S. 1967. Colchicine-induced polyploids in ryegrass. Euphytica 16(1): 49–60. doi.10.1007/BF00034098.

Ahloowalia, B.S. and M. Maluszynski. 2001. Induced mutations—A new paradigm in plant breeding. Euphytica. 118(2): 167–173. doi.10.1023/A:1004162323428.

Alekperova, O.R. 1979. A useful autotetraploid form of mulberry. Geneti Selektsiya v Azerid zhane 3: 97–103 (Ru). From Referativny Zhurnal 1980. 5. 65. 636.

Aliev, M.O. 1977. Use of chemical mutagens combined with hybridization of mulberry forms differing in ploidy. Sheik 1: 7–8.

Anon. 1977. Manual on mutation breeding. Technical report series No.119, IAEA, Vienna. 1974. pp. 169–192.

Appa Rao, S. and M.K. Jana. 1976. Leaf mutations induced in black gram by X-rays and EMS. Environmental and Experimental Botany. 16(2-3): 151–154. doi.10.1016/0098-8472(76)90007-1.

Banerji, B.K. and S.K. Datta. 1991. Induction of somatic mutation in chrysanthemum cultivar 'Anupam'. Journal of Nuclear Agriculture Biology 19: 252–256.

Bharathi Behera and S.N. Patnaik. 1975. Cytotaxonomy studies in the family Amaranthaceae. Cytologia. 39: 121–131.

Bhat, S.R., A. Haque and V.L. Chopra. 2001. Induction of mutants for cytoplasmic male sterility and some rare phenotypes in Indian mustard (*Brassica Juncea* L). Indian J. Genet. 61(4): 335–340.

Bindroo, B.B., A.K. Tiku and R.K. Pandit. 1988. Response of Japanese mulberry varieties propagated through cuttings under Kashmir eco-climate. Geobios 7(1): 26–39.

Bourgis, F., R. Guyot, H. Gherbi, E. Tailliez, I. Amabile and J. Salse. 2008. Characterization of the major fragrance gene from anaromatic japonica rice and analysis of its diversity in Asian cultivated rice. Theoretical and Applied Genetics 117: 353–68.

Broertjes, C. and A.M. Van Harten. 1988. Developments in Crop Science. Vol. 12. Applied Mutation Breeding for Vegetatively Propagated Crops. [Elsevier Science Publication. New York, U.S.A]. pp. 3–23. ISBN:9781483289991.

Brunner, H. 1995. Radiation induced mutations for plant selection. Applied Radiation and Isotopes 46(6-7): 589–594. doi. org/10.1016/0969-8043(95)00096-8.

Chakraborthi, S.P., K. Vijayan, B.N. Roy and S.M.H. Quadri. 1998. *In vitro* induction of tetraploidy in mulberry (*Morus alba* L.) Plant Cell Rep. 17: 779–803.

Dandin, S.B. and M.S. Jolly. 1986. Mulberry descriptor. Sericologia 26(4): 465–475.

Dandin, S.B. and R. Kumar. 1989. Evaluation of mulberry genotypes for different growth and yield parameters. *In*: Genetic resources of mulberry and utilization. Ed. By K. Sengupta and S.B. Dandin, C.S.R & T.I, Mysore. pp. 143–151.

Dandin, S.B., M.V. Rajan and R.S. Mallikarjunappa. 1996. Mutant forms in mulberry (*Morus* spp.). Sericologia 36(2): 353–358.

Das, B.C. and S. Krishnaswami. 1969. Estimation of components of variation of leaf yield and its traits in mulberry. J. Seric. Sci. Japan. 38: 242–248.

Das, B.C., D.N. Prasad and A.K. Sikdar. 1970. Colchicine induced tetraploids of mulberry. Caryologia 23: 283–293.

Das, B.C. and D.N. Prasad. 1974. Evaluation of some tetraploid and triploid mulberry varieties through chemical analysis and feeding experiments. Indian J. Seric. 13(1): 17–22.

Das, B.C., B.B. Bindroo, A.K. Tiku and R.K. Pandit. 1987. Propagation of mulberry through cuttings. Indian Silk 26(1): 12–13.

Datta, R.K., K. Sengupta and S.K. Das. 1978. Induction of dominant lethals with ethyl methane sulfonate in male germ cells of mulberry silkworm *Bombyx mori* L. Mutation Res. 56: 299–304.

Datta, R.K. 1994. Production and demand of silk in India. Global Silk Scenario – 2001. Proc. Inter. Natl. Conf. Seri. October, 25–29, C.S.R & T.I, Mysore, India. pp. 44–54.

Deshpande, K.N., S.S. Mehetre and S.D. Pingle. 2010. Effect of different mutagens for induction of mutations in mulberry. Asian J. Exp. Biol. Sci. Spl. 104–108.

Dhooghe, E., L.K.Van, T. Eeckhaut, L. Leus and H.J. Van. 2011. Mitotic chromosome doubling of plant tissues *in vitro*. Plant Cell Tiss Org. 104: 359–373.

Dwivedi, N.K., A.K. Sikdar, S.B. Dandin, C.R. Sastry and M.S. Jolly. 1986. Induced tetraploidy in mulberry I. Morphological, anatomical and cytological investigations in cultivars RFS$_{135}$. Cytologia. 51: 393–401.

Dwivedi, N.K., A.K. Sikdar, M.S. Jolly, B.N. Susheelamma and N. Suryanrayana. 1988b. Colchicine induced monoecious mutants in mulberry. Curr Sci. 57(4): 208–210.

Dwivedi, N.K., B.N. Susheelamma, A.K. Sikdar, N. Suryanarayana, M.S. Jolly and K. Sengupta. 1989a. Induced tetraploidy in mulberry III. Morphological and hybridization studies in cultivars S$_{30}$ and S$_{36}$. Indian J. Seric. 28(2): 131–138.

Dwivedi, N.K. 1994. Induced auto triploid in mulberry. Adv. Plant Sci. 7(1): 35–40.

Dzhafarov, N.A. and S.N. Abbasov. 1967. The hybridization of mulberry forms and the study of their progeny for seed purposes. Trans. Az. Seric. Res. Inst. Silk Prod. 6: 67–84 (Ru).

Dzhafarov, N.A. and S.N. Abbasov. 1970. Hybridization of polyploids and its importance in seed production in mulberry. Polyploidiya u Shelkovitsy, (Moscow). 168–173.

Dzhafarov, N.A., L.V. Turachinova, O.R. Alekperova and L.A. Shirieva. 1985. The new triploid variety AzN II Sh9. Shelk. 3: 4–5.

Eswar Rao, M.S. 1996. Improvement of mulberry through polyploidy breeding. Ph.D. Thesis, Bangalore University, Bangalore, India.

Eswar Rao, M.S., S.B. Dandin, R.S. Mallikarjunappa, H.V. Venkateshaiah and U.D. Bongale. 2004. Evaluation of induced tetraploid and evolved triploid mulberry genotypes for propagation, growth and yield parameters. Indian J. Seric. 43(1): 88–90.

Fujita, H. and K. Nakajima. 1973. Retreatment of induced mulberry mutants with gamma rays. Gamma Field Sym. 2: 49–61.

Fujita, H., T. Yokoyama and K. Nakajima. 1980. Re-treatment of induced mulberry mutants with gamma-rays. Technical News 23: 28–29.

Fujitha, H. and M. Wada. 1982. Studies on mutation breeding in mulberry (*Morus* spp.) *In*: Induced mutation in vegetatively propagated plants. IAEA. Vienna pp. 249–279.

Gary, A. and Sega. 1984. A review of the genetic effects of ethyl methane sulfonate. Reviews in Genetic Toxicology 134(2): 113–142.

Gatin, F.G. and K.S. Ogurtsov. 1981. Mutagenic effect of N-nitroso-N-methyl urea in the MV2 of mulberry. Sheik. 6: 6–7.

Gordon, S.A. 1957. The effect of ionizing radiation on plants, biochemical and physiological aspects. Quant. Rev. Biol. 32: 3–14.

Gray, E. 1990. Evidence of phenotypic plasticity in mulberry (*Morus* L.). Castanea. 55(4): 272–281.

Hamada, S. 1963. Polyploid mulberry trees in practice. Indian J. Seri. 1(3): 3–4.

Hardhansau, P.K. Sahu, B.R. Dayakar Yadav and B. Saratchandra, 1995. Evaluation of mulberry (*Morus* spp.) genetic resources-I sprouting, survival and rooting ability. J. Environ. Res. 3(1): 11–13.

Hartman, H.T. and D.E. Kester. 1976. Plant propagation-Principles and Practices. Prentice Hall of India. pp. 120–135.

Hazama, K. 1967a. Induced mutations and plant breeding methods in vegetatively propagated species. J. Seric. Sci. Jpn. 36(4): 346–352.

Hazama, K., K. Katagiri and S. Takato. 1968. Varietal difference of radiosensitivity and bud mutation of mulberry tree in the gamma irradiation field. J. Seric. Sci. Jpn. 37(5): 427–433.

Hazama, K. 1968a. Breeding of mulberry tree. JARQ 3: 15–19.

Hazama, K. 1968b. Adaptability of mutant in mulberry tree. Gamma Field Symp. 7: 79–85.

Hemalatha, K. 1998. Induction of mutation in carnation (*Dianthus caryophyllus* L.) through gamma rays and EMS. Ph.D. Thesis, University of Agricultural Sciences, Bangalore, India.

Hieter, P. and T. Griffiths. 1999. Polyploidy—More is more or less. Science 285: 210–211.

Hotta, T. 1954. Taxonomical study on cultivated mulberry in Japan, Kyoto, Japan. pp. 125.

Hyadalu Lingappa Ramesh, Veerapura Narayanappa Yogananda Murthy and Munirajappa. 2013. Gamma ray induced radio sensitivity in three different mulberry (*Morus*) genotypes. American Journal of Plant Sciences 4(7): 1351–1358. doi:10.4236/ajps.2013.47165.

Jayaramaiah, V.C. and Munirajappa. 1987. Induction of mutations in mulberry variety 'Mysore Local' by gamma-irradiation. Sericologia 27(2): 199–204.

Kaicker, U.S. 1992. Rose breeding in India and Cytology of induced mutation of H.T.Cv, 'Folklore'. Acta Horticulture 320: 105–112.

Kanoktip, P., S. Ratchada, N. Ikuo, M. Masahiro and S. Kanyaratt. 2014. Tetraploid induction of *Mitracarpus hirtus* L. by colchicine and its characterization including antibacterial activity. Plant Cell Tissue and Organ Culture 117(3): 381–391. doi:10.1007/s11240-014-0447-y.

Karpate, R.R. and A.D. Choudhary. 1997. Induced mutation in *Linum usitatissimum* L. J. Cytol. Genet. 32(1): 41–48.

Katagiri, K. 1970. Varietal differences of mutations rate and mutation spectrum after acute gamma ray irradiation in mulberry. J. Seric. Sci. Jpn. 39(3): 194–200.

Katagiri, K. 1976a. Polyploidy induction in mulberry by gamma radiations. Mut. Breed News Letter 8: 11–12.

Katagiri, K. and K. Nakajima. 1980. Tetraploid induction by gamma rays irradiation in mulberry. In Induced mutations in vegetatively propagated plants. Proceedings of the final Research co-ordination meeting Coimbatore 11–15 Feb. 1980. pp. 235–248.

Kearsey, M.J. and H.S. Pooni. 1996. The genetic analysis of quantitative traits. Plant genetic group school of Biological Science, University of Birmingham, U.K. Chapman and Hall, ISBN: 0412609800. pp. 381–395.

Kedarnath, S. and D. Lakshmikanthan. 1965. Induction of polyploidy in mulberry (*Morus alba* L.). Indian Forester 91(9): 682–683.

Kermani, M.J., V. Sarasan, A.V. Roberts, K. Yokoya, J. Wentworth and V.K. Sieber. 2003. Oryzalin-induced chromosome doubling in Rosa and its effect on plant morphology and pollen viability. Theor. Appl. Genet. 107: 1195–1200.

Konzak, C.F., R.A. Nilan, R.R. Ligault and R.J. Foster. 1961. Modification of induced genetic damage in seeds. pp. 155–169. *In*: Proc. Symp. on effects of ionizing radiations on seeds. IAEA. Vienna.

Kozikova, R.K. 1980. The nature of mutagenic stimulation in mulberry. Shovkivnit Stvo. 13: 3–6.

Kuchkarov, U. and K.U. Ogurtsov. 1987. Spontaneous mutants of mulberry. Shelk. 6: 3–4.

Kukimura, H., F. Ikeda, H. Fujitha, T. Maeta, K. Nakajima, K. Katagiri, K. Nakahira and M. Somegou. 1975. Genetical, cytological and physiological studies on the induced mutants with special reference to effective methods for obtaining useful mutants in perennial woody plants. pp. 83–104. *In*: Improvement of Vegetatively Propagated Plants through Induced Mutations, Tokai, IAEA, Vienna.

Kunitake, H., T. Nakashima, K. Mori and M. Tanaka. 1998. Somaclonal and chromosomal effects of genotype, ploidy and culture duration in *Asparagus officinalis* L. Euphytica. 102: 309–316.

Lian, L.K. and Q.F. Chen. 2013. A comparative study of seed protein content and seed flavonoid content between diploid and tetraploid tartary buck-wheat. Seed. 32: 1–5.

Majdi, M., G. Karimzadeh, M.A. Malboobi, R. Omidbaigi and G. Mirzaghaderi. 2010. Induction of tetraploidy to feverfew (*Tanacetum parthenium* Schulz-Bip.): morphological, physiological, cytological and phytochemical changes. Hort Science 45: 16–21.

Maribashetty, V.G., M.V. Chandrakala and K.C. Narayanaswamy. 1997. Health hazards in Sericulture. Indian Silk 36(6): 38–40.

Mickaelsen, K., G. Ahnstrom and W.C. Li. 1968. Genetic effects of alkylating agent in barley. Influence of past-storage, metabolic state and pH of mutagen solution. Heriditas 59: 353–374.

Moh, C.C. 1962. The use of radiation induced mutations in crop breeding in Latin America and some biological effects of radiation in Coffee. Int. Jour. Appl. Radiation and Isotopes 13: 467–475.

Mohammadi, P.P., A. Moieni, A. Ebrahimi and F. Javidfar. 2012. Doubled haploid plants following colchicine treatment of microspore-derived embryos of oilseed rape (*Brassica napus* L.). Plant Cell Tissue and Organ Culture 108: 251–256. doi:10.1007/s11240-011-0036-2.

Mulov, G.B. and A.I. Kovalik. 1977. New Ukranian mulberry varieties and methods of breeding them. In Tr Vses Seminara pogenet.i Shelkovitsy tutovogo Shelkopryada i Shelkovitsy, Tashkent, Uzbek. SSR. pp. 127–133(Ru).

Nakajima, K. 1972. Induction of useful mutation in mulberry by gamma irradiation. J.A.R.Q. 6(4): 195–198.

Nakajima, K. 1973. Induction of useful mutations of mulberry and roses by gamma rays. In Induced mutation in vegetatively propagated plants. IAEA, Vienna. pp. 105–117.

Nakajima, K. 1977. Studies on effective methods for induction of mutation of vegetatively propagated plants using a gamma field. Acta. Rad. Bot. Genet. Jpn. 4: 103.

Nataraja, N. and J. Sampath. 1988. Indian Silk Industry. Silk in India: Statistical Biennial. Central Silk Board, Bangalore.

Nazir, M.B., O. Mohamad, A.A. Affida and A. Sakinah. 1998. Research highlights on the use of induced mutations for plant improvement in Malaysia. Malaysian Institute for Nuclear Technology Research (MINT), Bangi, Kajang, Selangor, Malaysia.

Nigel, A.R.U., H. Jennie and M. Therese. 2007. Generation and characterization of colchicine-induced autotetraploid *Lavandula angustifolia*. Euphytica 156: 257–266.

Nybom, N., A. Gustafsson and L. Ehrenberg. 1952. On the injurious action of ionizing radiation in plants. Bot. Notiser. 105: 343–365.

Pathirana, R., T. Vitiyala and N.S. Gunaratne. 2009. Use of induced mutations to adopt aromatic rice to low country conditions of Sri Lanka. pp. 388–390. *In*: Induced Plant Mutations in the Genomics Era. Proceedings of an International Joint FAO/IAEA Symposium. International Atomic Energy Agency, Vienna, Austria.

Pollard, E. 1964. Ionizing radiation: Effect on genetic transcription. Science 146: 927–929.

Raisinghani, G. and S.K. Mahna. 1994. Mutants of *Vigna mungo* L. induced by gamma rays and two alkylating agents. J. Cytol. Genet. 29: 137–141.

Rahman, N.M. and J.D. Soriano. 1973. Studies on the mutagenic effect of some monofunctional alkylating agents in Rice. Rad. Bot. 12(4): 291–295.

Ramesh, H.L. and Munirajappa. 2001. Colchicine induced variability in mulberry variety Kosen. Bull. Ind. Acad. Seri. 5(2): 34–41.

Ramesh, H.L., V.N. Yogananda Murthy and Munirajappa. 2011. Colchicine induced morphological variations in mulberry variety M_5. The Bioscan. 6(1): 115–118.

Ramesh, H.L., V.N. Yogananda Murthy and Munirajappa. 2012b. Effect of different doses of gamma radiation on growth parameters of mulberry (*Morus*) variety Kosen. Journal of Applied and Natural Science 4(1): 10–15. doi.org/10.31018/jans.v4i1.214.

Ramesh, H.L., V.N. Yogananda Murthy and Munirajappa. 2012a. Effect of gamma irradiation on morphological and agro-botanical parameters of mulberry variety M_5. International Journal of Science and Nature 3(2): 447–452.

Ramesh, H.L. and V.N. Yogananda Murthy. 2014. Induction of colchiploids in mulberry (*Morus*) variety Kajali in C_1 generation. International Journal of Advanced Research 2(4): 468–473.

Ramesh, H.L., V.N. Yogananda Murthy and Munirajappa. 2014. Induction of useful mutation in mulberry (*Morus*) variety S_{54} by gamma irradiation in M_1 generation. American Journal of Experimental Agriculture 4(1): 48-57. doi:10.9734/AJEA/2014/5517.

Rao, P., J.M.M. Rao and N.L. Sarojini. 1984. Mutation breeding in mulberry *Morus indica* L. Indian J. Bot. 7(1): 106–111.

Santhoshlal, P.S. and K. Pavithran. 1997. Genetics of EMS induced recessive tall mutation in rice. Indian J. Genet. 57(2): 210–213.

Sarinee Chaicharoen, Ariya Satrabhandhu and Maleeya Kruatrachue. 1995. *In-vitro* induction of polyploidy in white mulberry (*Morus alba* Var.S_{54}) by colchicine treatment. J. Sci. Soc. Thailand. 21: 229–242.

Sastry, C.R., C.V. Venkataramu, Azeez Khan and J.V. Krishna Rao. 1974. Chemical mutagenesis for productive breeding in mulberry. Paper presented at the seminar organised in commemoration of silver jubilee of Central Silk Board, India.

Sastry, C.R., C.V. Venkataramu and A. Khan. 1976. Evaluation of mulberry strains obtained by chemical mutagen. L.C. Dunn. Th. Dobzhonsky memorial symposium on Genetics, December, 16–18. pp. 57.

Sastry, C.R., R. Kumar, S.B. Dandin and N.K. Dwivedi. 1983. Effect of physical and chemical mutagens on sprouting, survival and injury in three varieties of mulberry. National Seminar on silk Research and Development, Bangalore, March 10–13, 1983. PP.54.

Seki. 1951. Studies on polyploid mulberry trees (I). Tetraploid mulberry trees induce by colchicine. Research Reports in Textile Sericulture 3: 11–17.

Shaked, H., K. Kashkush, H. Ozkan, M. Feldman and A.A. Levy. 2001. Sequence elimination and cytosine methylation are rapid and reproducible responses of the genome to wide hybridization and allopolyploidy in wheat. The Plant Cell 13: 1749–1759.

Shirasawa, K., H. Hirakawa, T. Nunome, S. Tabata and S. Isobe. 2016. Genome-wide survey of artificial mutations induced by ethyl methane sulfonate and gamma rays in tomato. Plant Biotechnol. J. 14: 51–60.

Sikdar, A.K., N.K. Dwivedi, S.B. Dandin, R. Kumar and K. Giridhar. 1986. Stomatal chloroplast count technique as a tool to ascertain different ploidy in mulberry. Indian J. Seric. 25(2): 88–90.

Sikdar, A.K. 1990. Qualitative and Quantitative improvement of mulberry (*Morus* spp.) by induction of polyploidy. Ph.D. Thesis, University of Mysore, Karnataka, India.

Sikdar, A.K. and M.S. Jolly. 1994. Induced polyploidy in mulberry (*Morus* spp.): Induction of tetraploids. Sericologia. 34: 105–116.

Sikdar, A.K. and M.S. Jolly. 1995. Induced polyploidy in mulberry (*Morus* spp.) II. Production of triploids and their yield evaluation. Bull. Sericult. Res. 6: 39–46.

Singh, V.P., Man Singh and J.P. Pal. 1999. Mutagenic effects of gamma rays and EMS on frequency and spectrum of chlorophyll and macro mutations in urdbean (*Vigna mungo* L. Hepper). Indian J. Genet. 59(2): 203–210.

Skoog, F. 1935. The effect of radiation on auxin and plant growth. J. Cell. Comp. Physiol. 7: 227–270.

Smith, G.F. and H. Kersten. 1942. Auxin in seedlings from x-rayed seeds. Amer. J. Bot. 29: 785–819.

Sparrow, A.H., M.J. Moses and R.J. Dubow. 1952. Relationship between ionizing radiation, chromosome breakage and certain other nuclear disturbances. Exptl. Cell Research Suppl. 2: 245–267. PBA 23: 2345 [Tradescantia; Trillium].

Sugiyama, T. and I. Tojyo. 1962. Studies on the effect of irradiation on bud of mulberry cutting in the hybridization. Bull. Seri. Exp. Sta. Tokyo. 18(2): 115–132.

Sugiyama, S.I. 2005. Polyploidy and cellular mechanisms changing leaf size: Comparison of diploid and autotetraploid populations in two species of Lolium. Annals of Botany 96: 931–938.

Susheelamma, B.N., T. Mogili, M.N. Padma, K. Sengupta and N. Suryanarayana. 1991. Comparative morphology of autotetraploids and crossability studies in mulberry. Mysore J. Agric. Sci. 25: 469–473.

Swanson, C.P. 1965. Cytology and Cytogenetics, Prentice Hall Inc., Englewood Cliffs, U.S.A.

Sybenga, J. 1992. Cytogenetics in plant breeding. Monographs on Theoretical and Applied Genetics. Springer-Verlag, Berlin and Heidelberg GmbH & Co.KG, New York, Alemanha/USA.

Tang, Z.Q., D.L. Chen, Z.J. Song, Y.C. He and D.T. Cai. 2010. *In vitro* induction and identification of tetraploid plants of *Paulownia tomentosa*. Plant Cell, Tissue and Organ Culture 102(2): 213–220. doi.org/10.1007/s11240-010-9724-6.

Thohirah Lee Abdullah, Johari Endan and B. Mohd Nazir. 2009. Changes in flower development, chlorophyll mutation and alteration in plant morphology of *curcuma alismatifolia* by gamma irradiation. American J. Applied Sciences 6(7): 1436–1439.

Tikader, A., K. Vijayan, B.N. Roy and T. Pavankumar. 1996. Studies on propagation efficiency of mulberry [*Morus* spp.] at ploidy level. Sericologia. 36(2): 345–349.

Tojyo, I. 1966. Studies on polyploidy mulberry tree I. Breeding of artificial autotetraploids. Bull. Seri. Expt. Stn. 20(3): 187–207.

Tojo, I. 1985. Research of polyploidy and its application in *Morus*. JARQ 18: 222–228.

Van Harten, A.M. 1998. Mutation Breeding: Theory and Practical Applications. Cambridge University Press, Cambridge, UK. pp. 35–48.

Verma, R.C., A. Sarkar and S. Sarkar. 1986. Induced amphiploids in mulberry. Curr. Sci. 55: 1203–1204.

Waghmare, V.N. and R.B. Mehra. 2000. Induced genetic variability for quantitative characters in Grasspea (*Lathyrus sativus* L.). Indian J. Genet. 60(1): 81–87.

Wang, L.J., M.Y. Sheng, P.C. Wen and J.Y. Du. 2017. Morphological, physiological, cytological and phytochemical studies in diploid and colchicine-induced tetraploid plants of *Fagopyrum tataricum* (L.). Gaertn. Bot. Studies. 58: 2. pp. 1–12. doi:10.1186/s40529-016-0157-3.

Xing, S.C., Y.H. Cai and K.D. Zhou. 2010. A new approach for obtaining rapid uniformity in rice (*Oryza sativa* L.) via a 3x × 2x cross. Genetics and Molecular Biology 33: 325–327.

Yang, J.H. and X.H. Yang. 1989. Breeding of artificial triploids in mulberry. Seric. Sci. Jpn. 15: 65–70.

Yang, J.H. and X.H. Yang. 1991. Research on polyploidy and polyploidy breeding in mulberry in China. Sericologia 31(4): 625–630.

Ye, M.Y., J. Tong, X.P. Shi, W. Yuan and G.R. Li. 2010. Morphological and cytological studies of diploid and colchicine-induced tetraploid lines of crape myrtle (*Lagerstroemia indica* L.). Sci Hortic. 124: 95–101.

Yogananda Murthy, V.N., H.L. Ramesh and Munirajappa. 2011. Ethyl methane sulphonate induced morphological variations in mulberry (*Morus*) variety M_5. Journal of Applied and Natural Science 3(1): 114–118. doi.10.31018/jans.v3i1.167.

Zhu, B.C. and L.R. Gao. 1988. Study on autotetraploid common buckwheat I. comparison of morphology and cytology between autotetraploid and diploid common buckwheat. Hereditas 10: 6–8.

Chapter 8

Plant Tissue Culture in Mulberry Improvement

Anupama Razdan Tiku,[1] *T Dennis Thomas*[2] *and MK Razdan*[3,*,®]

INTRODUCTION

Science of plant tissue culture is linked to Henry-Louis Duhamel du Monceau's (1756) pioneering experiments on wound healing in plants. He demonstrated that spontaneous callus formation on decorticated regions of elm plants resulted in their wound healing. This discovery *in situ* paved way for application of plant tissue culture studies *in vitro*. Plant tissue culture is blanket term for protoplast, cell, and tissue and organ culture under aseptic conditions (Bhojwani and Razdan 1983). An important revelation of plant tissue culture has been that living cells in a plant body can potentially give rise to whole plants, thus substantiating the concept of cellular potency (Street 1977, Thorpe 2007). With advances made in the development of protocols defining various physical and chemical factors that enable organogenesis and whole plant regeneration *in vitro* from cells, tissues and organs of different plant species; plant tissue culture is in essence the non-conventional procedure of plant growth and development with potential applications in clonal propagation, genetic improvement via protoplast culture and fusion, gene transformation programmes, production of pathogen free plants, pharmaceuticals and other secondary metabolites of industrial value (Stewart 2008, Trigiano and Gray 2011, Fasella and Hussain 2014). Further the plant tissue culture has proven to have substantial role in conservation of biological diversity. With advances made in these areas plant tissue culture has become an integral part of agricultural biotechnology (George et al. 2008, Razdan 2003, 2016).

Mulberry (*Morus* spp.) is an important horticultural plant in sericulture industry belonging to the family Moraceae. The leaf of mulberry is used to feed the silkworm *Bombyx mori* L. for production of silk. Apart from it, mulberry is also used as a fodder and has pharmaceutical including antioxidant value (Vijayan et al. 2011, 2014). The mulberry fruit in addition to medicinal uses is also consumed for human consumption as jam, jelly, marmalade, ice cream, frozen desserts, juices, etc. (Arfan et al. 2012). Due to high economic, horticultural and industrial importance, mulberry is cultivated in large areas across the globe. It is estimated that about 60% of the total cost of silk cocoon production is dependent on production and maintenance of mulberry trees. This attracts a large attention for improvement in the quality and quantity of mulberry trees considering the challenges especially under ongoing climate change. Plant tissue culture techniques which are known to produce genetically improved plants (Trigiano and Gray 2011, Kumar and Loh 2012, Marthe 2018) could thus be applied in the production and

[1] Department of Botany, Ramjas College, University of Delhi, Delhi, India.
[2] Department of Plant Science, Central University of Kerala, Kasargod, Kerala, India.
[3] Shyam Lal College, University of Delhi, Shahdara, Delhi, India.
* Corresponding author: maharaj_razdan@yahoo.co.uk
® Retired as Principal of this Institute

multiplication of new genotypes of mulberry. This chapter presents a general overview on these techniques of tissue culture that have been used for producing genetically improved mulberry trees.

Basic Aspects

Explant Source

Explants source with minimal or optimal size is one of the important factors that determine the potential for cell and tissue differentiation in cultures. Small pieces of explants isolated from shoot, leaf, roots, flower or inflorescences proliferate to form callus which undergoes organogenesis and subsequent differentiation into shoots on suitable media *in vitro*. Pieces thicker than 20 mm are found to observe proliferation; however, the optimum size of explants is specific to a plant species (Khayat 2012). In attempts to recover pathogen free plants through tissue culture, explants used for initiating cultures are designated as 'shoot tip', 'tip', meristem or 'meristem tip' (Baker and Phillips 1962, cutter 1965). For micropropagation the most suitable explants are nodal cuttings. In some horticultural plants, like ornamentals, terminal buds, or apical shoots, are used to initiate shoot regeneration in cultures. Axillary buds or meristems in leaf axils are also used but it may be noted that on source mother plant their capacity to develop shoot *in vivo* is limited due to apical dominance. Chaturvedi et al. (2004) obtained *in vitro* clonal propagation of an adult tree of neem (*Azadirachta indica* A. Juss) by forced axillary branching. In monocots, leaf- base and scale-base explants are highly meristematic and cultures are initiated using explants joined with small pieces of basal plate. Immature zygotic embryos or nucellar explants are highly used for establishing embryogenic cultures. These cultures have been found most suitable for propagation of woody and other plant species that initially failed to show growth response in tissue culture (for details see Bhojwani and Razdan 1996).

Age, physiological state, genotype and orientations of the explant, while in contact with medium, are other factors that are known to influence the induction of organogenesis, shoot regeneration with or without callus and embryogenesis (Mantell et al. 1985, Merkle et al. 1995, Thorpe 2007).

In mulberry, Ohyama (1970) using axillary buds of *Morus alba* demonstrated for the first time that complete plants can be regenerated from these explants *in vitro* on a suitable medium. Later, Oka and Ohyama (1986) were successful to induce adventitious buds from nodal, shoot-tip and leaf explants of mulberry without the formation of callus. Embryos, hypocotyls, cotyledons and leaf tips are, however, equally suitable explants for micropropagation of mulberry (Thomas 2003). Additionally, anther, ovule, and protoplasts isolated from mulberry (*Morus* spp.) have also been cultured to obtain genetically improved plants.

Development of Media

One of the important factors for advances made in tissue culture technology relates to overall composition of culture media. The principal components of most plant tissue culture media are inorganic nutrients (macronutrients and micronutrients), carbon source(s), organic supplements, growth regulators and a gelling agent. Some tissues grow on simple media containing inorganic salts and carbon source, e.g., Knop's mineral solution (Nobécourt 1937, Gautheret 1942), but most explants require medium composed of organic substances like vitamins, amino acids and growth substances in addition to inorganic nutrients with sugar as carbon source. A number of media have been devised for specific plant tissues and organs. Of these MS (Murashige and Skoog 1962), LS (Linsmaeir and Skoog 1965) media are widely used to induce organogenesis and plant regeneration in cultured tissues. For protoplast culture and subsequent regeneration into whole plants B5 medium has proved valuable. Other media Nitsch and Nitsch (1969) and N6 (Chu 1978) were developed particularly for anther culture. The need to develop various media depends on the fact that ratio as well as the concentration of nutrients in the medium should nearly match the optimum requirement with regard to the growth and differentiation of respective cell or tissue system (Steward et al. 1952, George et al. 2008).

Callus Induction and Multiplication

Various media have been tried in mulberry for *in vitro* callus induction as well as multiplication from various explants and for undergoing organogenesis. Of these MS (Murashige and Skoog 1962) medium has been of most common use with 2, 4-D as plant growth hormone. The addition of Kn, IAA or NAA into the medium together with 2, 4-D enhances the multiplication of callus; the ratio of their combination, however, depending on specific mulberry species (Jain and Datta 1992, Zaki et al. 2011). Supplementing Abscisic acid (ABA) in the medium has been observed to sustain callus viability for longer periods (Ohnishi et al. 1986). Subculturing callus after 4 weeks interval prevents blackening of callus and promotes better growth (Shajahan et al. 1997).

Shoot-bud differentiation and Shoot/Plant Regeneration

Organogenesis by differentiation of shoot-bud in callus is dependent on auxin/cytokinin ratio in the media. According to Skoog and Miller (1957), the organ formation in callus or cultured tissue is controlled by quantitative interaction (ratio rather than absolute concentration) of growth regulators. They observed that high concentration of cytokinin (e.g., kinetin) plus casein hydrolysate neutralises the inhibitory effect of auxin (e.g., IAA) in White's medium and promotes shoot bud and shoot regeneration on the tobacco callus, whereas increasing concentrations of auxin (IAA) favours cell proliferation and root differentiation. Similar effects of auxin/cytokinin ratio in media were observed subsequently on the callus cultures of various other plant species. In mulberry, shoot bud (*Morus bombycis*) or plantlet regeneration (*Morus alba*) was obtained from the callus cultured in medium supplemented with higher concentration of cytokinin (BA) and less auxin (IAA or NAA); however, LS medium was found suitable for *M. alba* Narayan et al. (1993) and MS medium for *M. bombycis* (Jain and Datta 1992). The addition of gibberellic acid (GA3) and dithiothreitol (DTT) to culture medium reportedly breaks apparent dormancy in the stored callus of mulberry and initiates its regenerative capacity (Yasukura and Onishi 1990). Direct plant regeneration from explants without callus induction is also reported in *Morus* spp., which may have advantage of generating genetically homogenous populations. The success achieved on Organogenic response of different explants of various mulberry species in culture media is mentioned in detail by Vijayan et al. (2011).

Acclimatisation of Tissue Culture Raised Plants

Plants regenerated from *in vitro* tissue cultures are transplanted to soil in pots. The potted plants are ultimately transferred to greenhouses or growth cabinets under controlled conditions of light, temperature and humidity for maintenance and further growth. However, these *in vitro* plants show poor photosynthetic efficiency, heterotrophic mode of nutrition, marked decrease in epicuticular wax, poor mechanism of water control loss and hypolignified stem. These factors make these plants vulnerable to transplantation shocks. Moreover, there is change of substrate which is rich in organic nutrients *in vitro* to a soil substrate composed of mostly inorganic nutrients. Acclimatisation of transplanted plants under controlled conditions of humidity, light and temperature inside growth cabinet or greenhouse becomes necessary as it leads to hardening of shoot system, thereby inducing autotrophism and reduction of transpiration level through the development of surface wax on leaves which enhances the photosynthetic activity of these plants.

Built up of high humidity initially around transplanted plants is prerequisite for hardening. This is generally achieved by covering these plants in pots with clean transparent polythene bags with a small hole for air circulation and maintaining them inside growth cabinets or in greenhouse under controlled conditions. Alternatively, fogging system or misting in small potting room adjacent to tissue culture laboratory can be arranged where transplanted plants in pots are kept under the cover of plastic bags. The humidity is then gradually decreased to ambient level with concomitant increase in light intensity before final transfer to greenhouse or field (Razdan 2016). In mulberry, Patnaik and Chand (1997) transferred plantlets of *Morus cathayama*, *M. lhou* and *M. serrata* with well-developed roots to plastic pots (5 cm diameter) filled with autoclaved vermi-compost, covered them with plastic bags, and maintained inside the growth chamber under 80–85% RH (relative humidity), $25 \pm 1^{\circ}C$ temperature, and 50 μE $m^{-2} s^{-1}$ light intensity generated by cool white fluorescent tubes. Similar procedure was adopted for *Morus alba* plants but in place of pots, plastic cups (5 cm Diameter) were used for transplantation (Chakraborti et al. 1998). The plantlets were moistened with ½-strength MS basal liquid every 4 days or by normal water for a period of two weeks. Co-cultivation of vesicular arbuscular mycorrhiza (VAM) with plants in pots has also been found suitable for hardening of several woody plants (Mathur et al. 2008). For *Morus alba* plants, Kashyap and Sharma (2006) used arbuscular mycorrhiza (AM) and *Azotobacter chroococcum* during acclimatisation and achieved at least 50% survival rate. More information on impact of AM in growth and quality of mulberry is provided in Chapter 14.

Applications to Mulberry

Improvement in basic aspects of *in vitro* plant regeneration followed the development of various efficient methods of plant tissue culture that provide a wide scope for production of new genotypes which can be ultimately micropropagated and also conserved *ex situ*. Especially genetically variable new plants raised through these methods can be investigated for their success in surviving unfavourable environmental conditions, such as climate change. Indiscriminate clearing of forests and conversion of agricultural land for industrialisation, urbanisation and other measures of economic development have adversely affected the ecosystem. Additionally, environmental hazards *in situ* such as wind, storm, tornedos, tsunamis, hurricanes, volcanic eruptions, and cyclones, etc., coupled with global warming, have considerably contributed to the loss of genetic resources or gene pools world over. Depletion of naturally occurring plant genetic resources in these processes are reported to the extent that nearly 2000–60,000 higher plant species have become endangered, rare, or on threshold of extinction; many of these species being the agricultural crops (FAO 2009).

In consideration of the various ecological disturbances as mentioned above the application of undermentioned plant cell and tissue culture methods seem appropriate as an alternative for production of plants with new improved new genotypes in *Mulberry* spp. that can have potential to survive the impact of climate change. Further the germplasm of these improved plant varieties require to be preserved for shorter or longer periods so that they are released for growth and development at appropriate climatic conditions.

Haploid Production in Mulberry

*E*ver since A.D. Bergner (1921) discovered haploid plants in *Datura stramonium*, plant breeders have worked intensively to obtain haploids (sporophytes of higher plants with gametophytic chromosome constitution) *in vivo*. According to Pierik (1989), the *in vivo* techniques employed to induce haploid production include gynogenesis, ovule androgenesis, genome elimination by distant hybridisation, semigamy, chemical treatment, temperature shocks, and irradiation effects. Haploids also arise as a result of parthenogenesis in nature, but these plants rarely produce characters of male parent. Moreover, haploid yield produced *in situ* or *in vivo* are of low frequency, therefore, attempts were made to obtain haploid plants *in vitro*. The first report showing that isolated anthers of *Datura innoxia* were able to form haploid embryos *in vitro* was published by Guha and Maheshwari (1964, 1966). Subsequently, Bourgin and Nitsch (1967) obtained first haploid plants from anther cultures of *Nicotiana*. Since then much progress has been made in anther culture of rice, wheat, maize, brassica, pepper and other crop species (Dwivedi et al. 2015). There are reports on haploids being produced by anther culture (androgenesis) or pollen culture from about 250 species and hybrids distributed in 40 families of higher plants (Razdan et al. 2008, Mishra and Goswami 2014).

Haploid plants can also be produced by *in vitro* gynogenesis. San Noeum (1976) was first to demonstrate that gynogenesis (production of a haploid individual by the development of an unfertilised egg-cell as a result of delayed pollination), essentially an *in vivo* phenomenon, can be also be induced under *in vitro* conditions. She obtained gynogenic haploids from unfertilised ovary culture of *Hordeum vulgare*. Subsequently, haploids have been obtained from *in vitro* cultures of unfertilised ovary or ovule cultures of different species of ornamental, food, and tree crops (Doi et al. 2013).

For mulberry species attempts have been made to obtain haploids by *in vitro* androgenesis as well as *in vitro* gynogenesis. The objective of obtaining haploid plants is to induce of double haploid lines by duplicating chromosome complement of haploids by colchicine or other chemical treatment (Murovec and Bohanec 2012).

The homozygous (inbred) lines thus generated ensure stability of new genotypes developed by haploidisation in shortest time. In mulberry, the development of inbred lines through traditional methods of breeding is a difficult task because it takes longer period to flower which extends to several years like in other trees. More importantly, mulberry is dioecious and a natural outbreeder. As a result it has developed inbreeding depression and phenotypes developed are heterozygous. Lin et al. (1987) made first attempts to generate haploids from (*Morus*) anther culture, whereas Katagiri (1989) induced callus from mulberry pollen cultures. Subsequently, Katagiri and Venkateswaralu (1993) reported organ-like structures in callus induced from isolated pollen cultures of *Morus australis*. Earlier Sethi et al. (1992) observed embryo differentiation from mulberry anther cultures maintained in the dark at $26 \pm 1°C$ for a period of 15 days. Tewary et al. (1989) and Chakraborti et al. (1999) also found globular and heart-shaped embryo formation in cultures of pollen isolated from anthers of mulberry on MS media supplemented with glutamine, CM and 2,4-D. Jain and Datta (1992) and Jain et al. (1996) reported the development of rooted embryoids on callus induced from uninucleate microspore cultures of *M. indica* cv. RFS 135. In all these studies, however, reproducibility of haploid plant production has been negligible (Sarkar et al. 2018).

Success achieved in production of haploid plants of mulberry has been through *in vitro* gynogenesis. Gynogenic plants from ovary cultures of *Morus indica* without callus phase were obtained by Lakshmi Sita and Ravindran (1991), but subsequently Thomas et al. (1999) developed reproducible protocol for production of haploid plants of *Morus alba* from unpollinated ovary cultures on MS medium supplemented with 2,4-D, glycine and proline. Production of homozygous lines based on double haploidy of *in vitro* raised haploid plants in mulberry remains to be investigated.

Triploid Production in Mulberry

In the majority of angiosperms there occurs a unique phenomenon of double fertilisation. This results in formation of two fusion products namely zygote (diploid arising from fusion of an egg and one of the male gametes) and primary endosperm (which is triploid formed by fusion of second male gamete with two polar nuclei) in the ovule. Zygote divides which leads to development of embryo (diploid tissue), whereas primary endosperm divides giving rise to endosperm (triploid tissue). Endosperm is important as it serves the nutrient requirements of growing zygotic embryo. Non development or dysfunction of endosperm causes embryo abortion in the seed. Considering the fact that endosperm is triploid tissue and lacks organogenic or vascular differentiation it has been cultured *in vitro* to obtain triploid plants. Triploids generally show enhanced vegetative characteristics than their related diploids and are of significant value to woody plants. For example, the triploids of *Populus tremuloides* have better quality pulpwood—a characteristic important in forest industry. Further various trisomics from among triploids may be useful in gene mapping. Triploid production by endosperm culture though initially successful in a limited number of species (Chaturvedi et al. 2003) has been applied with success for genetic improvement of other economically important plant species (Razdan-Tiku et al. 2014).

Various natural and *in vivo* induced triploids have been reported in mulberry (Dwivedi et al. 1989). Triploid plants are sterile and do not set seeds, yet triploid mulberry plants have been observed with improved traits in terms of leaf yield and nutritive quality in comparison to their related diploid. Triploid production in trees like mulberry through conventional breeding being a tedious and lengthy process (Seki and Oshikane 1959), endosperm culture can be used an alternative to triploid plant regeneration in mulberry. Thomas et al. (2000) were successful in raising triploid plants of mulberry (*Morus alba* L.) by endosperm culture via callus phase with improved traits (For more details refer to Chapter 9). Further, tetraploid mulberry plant production *in vitro* has also been reported by Chakraborty et al. (1998) but in terms of commercial cultivation these genotypes did not find favour with respect to quality of traits (Sarkar et al. 2018).

Protoplast Culture and Somatic Hybridisation

In plant breeding programmes many desirable combinations of phenotypic traits cannot be transmitted in hybrids obtained through conventional methods due to either incompatibility barriers or flowers produced bear sterile seeds. Further some plants (e.g., banana) are only vegetatively propagated and combination of good quality traits of related crop species is not possible. So "another process" other than sexual breeding has become available particularly for higher plants that can lead to genetic recombination leading to plant improvement (Cocking 1979). This non-conventional genetic procedure involves fusion between isolated somatic protoplasts (wall-less naked cells) of two different species under *in vitro* conditions and subsequently developing their product, hybrid cell (hetrokaryon), to a hybrid plant. This process named as somatic hybridisation thus provides opportunities for genetic improvement of plants first by exploring genetic variations among existing crops and then attempting to transfer available genetic information from one species to another through fusion of protoplasts isolated from the somatic tissues of the desired crop species (Razdan and Cocking 1981).

In mulberry, flowering period differs in relation to its geographic origin (Doss et al. 1998, Tikader and Dandin 2005). Tropical mulberry genotypes flower in the end of winter and beginning of spring, while temperate mulberry genotypes flower in the middle of spring to first half of the summer. This non-synchronised flowering pattern is an impediment for initiating hybridisation experiments in mulberry through conventional procedures. Somatic hybridisation appears plausible option to obtain new improved mulberry genotypes combining some of the tropical and temperate traits. Attempts therefore were made to regenerate protoplasts isolated from callus (Ohnishi and Kiyama 1987) and leaf mesophyll cells (Umate et al. 2005, Mallick et al. 2016) of mulberry species using standard procedures of protoplast isolation and culture. Whole plant regeneration could be obtained only from callus originating from mesophyll protoplasts of *Morus alba* (Ming et al. 1992) and *Morus indica* (Umate et al. 2005), but attempts at somatic hybridisation in regeneration of somatic hybrid plants of mulberry seem to be on the anvil. For more information see Chapter 10.

Somaclonal Variant Selection for Stress Tolerance

In plant tissue culture studies, the genetic variations are observed to occur in undifferentiated cells, protoplasts, and calli. Morphological traits of plants regenerated from these cells or tissues *in vitro* may vary due to cytological heterogeneity and new irregularities brought about by cultural environment. Superior variant traits observed in regenerated plants can thus be selected and transmitted to the progeny through sexual (lettuce, tobacco) or vegetative (sugarcane, potato) propagation. Variants selected from callus cultures are referred to as 'calliclones' (Skirvin 1978) and from protoplast cultures 'protoclones' (Shepard et al. 1980). A general term 'somaclonal variation' was, however, coined by Larkin and Scowcroft (1981) for plant variants derived from callus or protoplast cultures. Studies on sugarcane, potato, tomato, banana, some ornamentals like geranium, cereals and grasses, etc., highlighted the fact that somaclonal variation may be an additional tool for crop improvement (Baer et al. 2007).

One of the factors affected by climate change could be variation in soil characteristics leading to abiotic stress caused by drought, cold, salinity and alkalinity. To achieve abiotic stress tolerance, plants evolve adaptive mechanisms through morphological and developmental changes mediated by alterations in physiological, biochemical and molecular processes (Sarkar et al. 2016, 2017). Somaclonal variation emanating from tissue culture could affect genetic changes efficiently to screen and identify saline, alkaline and drought tolerant genotypes of mulberry (Tewary et al. 2000). Somaclonal variants selected in mulberry (*M. alba* cv. S14) exhibited higher leaf yield with more number of branches than the mother plant (Narayan et al. 1993); subsequently, somaclonal variant with shorter intermodal distance, thicker leaves, higher chlorophyll, and better moisture content was obtained by Susheelamma et al. (1996) from leaf-derived callus of mulberry. Attempts have also been made for *in vitro* selection of $NaHCO_3$ tolerant cultivars of *Morus alba* (Ahmad et al. 2007). More details on stress tolerance in mulberry species are given in Chapter 13.

Transgenic Plant Production

One of the most current areas of plant cell and tissue culture developed for genetic improvement of plants has been the transfer and expression of foreign genes into plant cells or tissues. Such cells or tissues with foreign gene regenerate *in vitro* plants called transgenic or GM (genetically modified) plants since the genome of these plants has become transformed or modified to express desired trait of incorporated gene. This biotechnological process, also referred to as genetic engineering technology, has been particularly used to explore potential benefits in agriculture. Engineering plants for improvement of nutritional value, production of biofuels, pharmaceuticals, and abiotic stress tolerance are some of the aspects where this transformation technology has proved successful (Harfouche et al. 2012). Availability of restriction endonucleases, development of recombinant DNA technology, marker, promotor and terminator genes, made it possible to develop plasmid constructs as vectors carrying the desired gene for incorporation into host plant cells and tissues. Cells and tissues transformed by incorporated desired gene express the new trait not only at regenerated whole plant level but also in successive generations of these transgenic plants (Smith 2013, Fasella and Hussain 2014).

In mulberry, genetic transformation has been attempted by incorporation of vectors carrying exogenous DNA into cells and tissues using particle bombardment, electroporation, *Agrobacterium rhizogenes*-mediated transformation and in planta procedures (Bhatnagar et al. 2002, Bhatnagar and Khurana 2003). These attempts initially did not yield success in transgenic plant regeneration although Bhatnagar et al. (2003) reported stable incorporation of GUS gene in transgenic plants of mulberry. Further attempts were made to develop efficient protocols using different explants and genotypes of mulberry, heterologous genes, and different transformation procedure. Lal et al. (2008) and Checker et al. (2012) used hypocotyl, cotyledon explant of *Morus indica* cv. K2 and transgene *Hval* to obtain transgenic mulberry plant through *Agrobacterium tumefaciens*-mediated transformation procedure which showed tolerance to drought, cold and salinity stress. Following the same protocol using leaf-derived callus, hypocotyl, cotyledon explants and transgene *Osmotin* (Das et al. 2011), *bch l* (Saeed et al. 2015), success was achieved in production of transgenic plants of mulberry *M. indica* cv. K2 which were tolerant to drought, salinity stress, high temperature, irradiance stress and resistance to fungi. In another mulberry genotype hypocotyl, cotyledon explants of *M. indica* cv. M5 transformed by transgene SHN *I* produced transgenic plants with enhanced leaf moisture retention capacity (Sajeevan et al. 2011). These studies suggest that protocol for transgenic plant production in mulberry has been apparently standardised and opens avenues for developing genetically improved mulberry transgenic plants with capability of tolerance to climatic stress. For further information regarding genetic engineering in mulberry refer to Chapter 11.

Secondary Metabolite Production

Application of plant tissue culture for production of compounds valuable to industry has been subject of considerable interest for biotechnologists. Success achieved in the growth and maintenance of microorganisms in cultures for industrial level production of drugs, breweries, pharmaceuticals and milk products is well established. Higher plants too are valuable sources of important natural products and quite a few of them have been produced on large-scale at industrial level from cells and tissues cultured in specially designed vessels called bioreactors (Bisaria and Panda 1991). Examples of successful industrial level production of secondary metabolites from higher plant cell and tissue cultures are shikonin obtained from root cultures of *Lithospermum erythrorhizon*, ginseng saponins extracted from roots of *Panax ginseng*, berberine from cell cultures of *Coptix japonica*, and taxol from suspension cultures of *Taxus cuspidata* (Trigiano and Gray 2011). These compounds mostly belong to a metabolic group collectively referred to as secondary metabolites as they do not participate in vital metabolic functions of host-plant tissues in the same manner as amino acids, nucleic acids, or other primary metabolites. They appear to serve as chemical interface between host plant and surrounding environment presumably to attract pollinators, ward off predators, and to some extent help in combating infectious diseases (Wink et al. 2005, Sarkar et al. 2018).

Production of secondary metabolite production using plant cell and tissue cultures has an advantage over *in vivo* synthesis of these compounds in naturally growing plants especially in the circumstances when climatic and geographic factors are unfavourable for cultivation of plantation crops. *In vitro* cultures on the other hand can be maintained at any place with proper laboratory infrastructure facilities ensuring secondary metabolite production all through the year. Mulberry also produces important medicinally and economically secondary metabolites such as rutin, mulberroside A, morusin, cyclomorusin, auercetin, etc. Attempts have made to produce mulberroside A from cell suspension cultures of *Morus alba* by immobilisation of free cells in liquid medium with calcium alginate and subsequent elicitation with yeast extract and methyl jasmonate to enhance two-fold production of this metabolite (Inayai et al. 2019). Earlier Lee et al. (2011) studied the influence of phytohormones and nitrogen in the culture medium on production of rutin from callus and adventitious roots of white mulberry tree (*M. alba*). El-

Mawla and his co-workers (El-Mawla et al. 2011) were able to trigger synthesis of active flavonoids in cell cultures of *Morus nigra* and tested their hypoglycaemic efficiency. Commercial production of these medicinal secondary metabolites of mulberry if achieved would add value to sericultural industry.

Micropropagation of Improved Mulberry Genotypes

In nature, mulberry species are heterozygous, cross-pollinated, dioecious, perennial and woody trees. The progeny derived from sexually propagated plants demonstrate a high amount of heterogeneity since seed progeny are not true to type. Thus the new improved trait expressed in genetically improved mulberry plants raised through cell/tissue culture techniques when grown in the field shows segregation of this trait in the seed progeny due to cross pollination. As a result these plants may or may not express the improved trait. Stem-cuttings, layering, grafting and budding could be the possible *in vivo* methods of producing identical improved mulberry plants. However, the *in vivo* method of vegetative multiplication of plants (called clonal propagation) is found not only difficult, expensive, and time consuming, but even unsuccessful for mulberry. According to Murashige (1974), micropropagation (clonal propagation through tissue culture) offers only alternative means of plant propagation that can be achieved in a short time and space through which there will be no segregation of genetically improved trait because progeny produced comprise of identical plants. Morel (1960) found micropropagation as the only commercially viable approach for orchid propagation. Since then several crop species have been micropropagated and recipes are available which growers can use for commercial production (see Bhojwani and Razdan 1996, Razdan 2003, 2016).

Micropropagation of mulberry was first time demonstrated by Ohyama (1970) who was able to multiply whole plants developing from axillary buds of *Morus alba*. Axillary buds were cultured in MS (Murashige and Skoog 1962) medium supplemented with growth regulators. Murashige (1978) proposed principles of micropropagation comprising of four distinct stages that can be adopted for plant production at commercial scale. Stages I-III is followed under *in vitro* conditions, whereas stage IV is accomplished in the greenhouse environment. In all stages there is no possibility of cross pollination and micropropagated plants are genetically identical. An additional Stage 0 is also suggested for various micropropagating systems (Debergh and Maene 1981). Stage 0 is meant for selection and maintenance of stock plants for culture initiation under controlled conditions. The initiation and establishment of aseptic cultures comprises Stage I. The main steps of this stage involve explant isolation, surface sterilisation, washing and establishment of suitable culture medium. Multiplication of shoots developed from explants directly or explant derived callus, or somatic embryo, using defined culture medium is essential feature of Stage II. Rooting of regenerated shoots and, or germination of embryos, constitutes Stage III. The last Stage IV includes transfer of plantlets to sterilised soil in pots and hardening under greenhouse environment.

In micropropagation of perennial species including mulberry trees apparently the composition of defined media in various stages is essential depending on genotype and explant used. Attempts have been made to micropropagate *Morus alba*, *M.* australis, *M. bombycis*, *M. cathyana*, *M. nigra*, *M. laevigata*, *M. latifolia*, *M. lohu*, *M. indica*, *M. serrata*, and *Morus* spp. The information about various mulberry explants (e.g., winter buds, axillary buds, shoot tips, young leaves, hypocotyl, cotyledon, epicotyl, nodal segments, etc.) and genotypes used by various workers; media devised along with plant growth regulator combinations at different stages such as shoot initiation, shoot multiplication and rooting has been comprehensively summarised by Vijayan et al. (2011). Although a number of reports are available on micropropagation of mulberry, but traits in genetically improved micropropagated mulberry plants (Gogoi et al. 2017) need further studies to use them for production on commercial scale.

Conservation of Mulberry Germplasm

Conservation of forest genetic resources was taught and decreed in India and China as far back as 700 B.C. (Swaminathan 1983). The primitive cultivars, their wild relatives, and *in vitro* regenerated new genotypes constitute a pool of genetic diversity called 'germplasm' or 'gene bank' which has great value in breeding programmes, forestry, multiplication of genetically improved plants, and production of pharmaceuticals. More importantly, germplasm conservation ensures minimal loss of plant genetic diversity against natural calamities as well as the climate change enabling preservation of particular plant or genetic stock for future use (Manek and Hank 2014). Initially, farmers, breeders and researchers would follow *in situ* mode of conservation of plant genetic resources. The *in situ* conservation of habitats as well new genotypes received high priority in the world conservation strategy programmes launched since 1980 and according to UNEP (United Nations Environment Programme 1995) even threatened plant species are conserved internationally mainly in botanical gardens and national biodiversity parks. This mode of conservation, however, has limitations as there is risk of the material being lost due environmental hazards, pests and diseases besides the cost of maintaining a large proportion of available genotypes in nurseries or fields being cost prohibitive. *Ex situ* conservation is now chief mode of conservation of genetic diversity which includes wild, cultivated and biotechnologically produced plant materials. Generally seeds or *in vitro* maintained plant cells, tissues and organs are preserved *ex situ* under appropriate conditions for long-term storage as gene banks (Razdan and Cocking 1997, 2000).

Germplasm conservation in the form of seeds is most convenient and economical since seeds occupy a relatively small space, but owing to high heterozygosity the conservation of mulberry germplasm through seeds is not practiced. Cryopreservation seems to be ideal method for long-term preservation of plant materials of improved mulberry genotypes. The principle underlying cryopreservation involves bringing *in vitro* grown plant cells or tissues to a non-dividing or zero metabolism state by subjecting them to super low temperature in presence or absence of cryoprotectants (Sakai 1997). In this technique the plant material is frozen and maintained at the temperature of liquid nitrogen (LN), which is around $-196°C$ or $-150°C$ in the vapour phase. Although various protocols for cryopreservation have been devised since sensitivity of cells or tissues to low temperatures varies according to the species or genotypes (Finkel and Ulrich 1983) the general practice followed is to suspend the material in the culture medium and after treating with cryoprotectant transfer it to polypropylene ampoules with screw caps and then freeze it using any of these methods: (a) slow-freezing, (b) rapid-freezing, (c) step-wise freezing , (d) dry freezing or (e) vitrification (For details see Bhojwani and Razdan 1996, Razdan 2003). Success has been achieved in cryopreservation of several accessions of mulberry genotypes belonging to *Morus alba, M. australis, M. cathayana, M. bombycis, M. indica, M. laevigata, M. latifolia, M. multicaulis, M. nigra, M. rotundioba and M. sinensis* (Rao et al. 2007, 2009; Chapter 17).

It is essential that viability of cryopreserved germplasm be tested periodically on some samples and properly recorded. To enable this process of thawing and reculture of sample materials is needed. For thawing cryopreserved materials are plunged into warm water at 37–40°C which gives a rapid thawing rate of 500–750°C min^{-1}. After thawing about 90s the material is transferred to either an ice bath or water bath at ambient temperature until ready for reculture. Soon after thawing cryoprotectants are removed by washing thawed materials several times to avoid any deplasmolytic injury to cells and tissues. Washed materials are then recultured in a fresh culture medium which has been found suitable for *in vitro* regeneration of particular thawed cells and tissues into plants. Thus the principle objective underlying cryopreservation in context of mulberry is to conserve germplasm of somaclonal variants, maintenance of cell lines producing medicines, preserving tissues or propagules of genetically engineered as well as somatic hybrid plant materials, storage of meristematic buds for micropropagation, and storage of pollen for sexual hybridisation especially of mulberry species that have different seasons of flowering due to their occurrence in different habitats. Encapsulated shoot tips (*Morus alba*) with calcium alginate coating can also be cryopreserved (Padro et al. 2012). On the whole cryopreservation has been quite successful in mulberry as is evident from the experiment that its dormant winter buds could be cryopreserved for 11 years without loss of viability (Fukui et al. 2011).

Concluding Comments

Plant tissue culture is essentially a non-conventional procedure of plant growth and development having potential applications in clonal propagation, genetic improvement via protoplast culture and fusion, genetic transformation, production of pathogen free plants, pharmaceuticals, and other secondary metabolites of industrial value. In mulberry (*Morus* spp.), the basic aspects of plant tissue culture such as explant source, media composition for callus induction and multiplication, shoot-bud differentiation up to whole plant regeneration, acclimatisation and hardening of *in vitro* raised plants after transplantation into the soil have been standardised. Further improvisation on these aspects followed the development of various efficient methods of plant tissue culture which can be applied to mulberry with specific perspective of achieving improved traits. Attempts have been made to produce genetically improved trees expressing desired traits in different species of mulberry through haploid and triploid production, protoplast culture and somatic hybridisation, somaclonal variant selection, transgenic plant production, and secondary metabolite production for medicinal use. *In vitro* raised these mulberry plants with improved traits can thus be micropropagated as well as maintained under controlled environmental conditions to withstand impact of climate change. Further the germplasm of these plants can be conserved on long-term basis by cryopreservation till favourable conditions occur for their *in situ* growth and development. Integration of genetic engineering and molecular techniques with methods of tissue culture is much desired and concerted efforts are being made to raise genetically improved traits in mulberry especially for stress tolerance, quality of foliage, fruits, and production of pharmaceuticals on commercial scale.

References

Ahmad, P., S. Sharma and P.S. Sharma. 2007. *In vitro* selection of $NaHCO_3$-tolerant cultivars of *Morus alba* (Local and Sujanpuri) in response to morphological and biochemical parameters. Horticultural Science (Prague) 34: 114–122.

Arfan, M., R. Khan and A. Rybarczyk. 2012. Antioxidant activity of mulberry fruit extracts. Int. J. Mol. Sci. 13: 2472–2480.

Baer, G.Y., A.I. Yemets, N.A. Stadnichuk, D.B. Rakhmetov and Y.B. Blume. 2007. Somaclonal variability as a source for creation of new cultivars of finger millet (*Eleusine coracana* Gaertn.). Cytol. Genet. 56: 145–150.

Baker, R. and D.J. Phillips. 1962. Obtaining pathogen free stock by shoot-tip culture. Phytopathology 52: 1242–1244.

Bergner, A.D. 1921. (Cited by Hu Han. 1987. Application of pollen-derived plants in crop improvement. pp. 155–168. *In*: S. Natesh et al. (eds.). Biotechnology in Agriculture. Oxford & IBH Publishing Co. Pvt. Ltd., New Delhi.

Bhatnagar, S., A. Kapur and P. Khurana. 2002. Evaluation of parameters for high efficiency gene transfer via particle bombardment in Indian mulberry, *Morus indica* cv. K-2. Indian J. Exp. Biol. 40: 1387–1392.

Bhatnagar, S. and P. Khurana. 2003. *Agrobacterium tumefaciens*—mediated transformation of Indian mulberry, *Morus indica* cv. K-2: A time phased screening strategy. Plant Cell Rep. 21(7): 669–675.

Bhatnagar, A. Kapur and P. Khurana. 2003. Evaluation of parameters for high efficiency gene transfer via *Agrobacterium tumefaciens* and production of transformants in Indian mulberry, *Morus indica* cv. K-2. Plant Biotechnol. 21: 1–8.

Bhojwani, S.S. and M.K. Razdan. 1983. Plant Tissue Culture: Theory and Practice. Elsevier Science Publishers, Amsterdam 502 pp.

Bhojwani, S.S. and M.K. Razdan. 1996. Plant Tissue Culture: Theory and Practice, A Revised Edition. Elsevier, Amsterdam 467 pp.

Bisaria, V. and A. Panda. 1991. Large-scale plant cell culture: methods, applications and products. Current Opinion in Biotechnol. 2: 370–374.

Bourgin, J.P. and J.P. Nitsch. 1967. Obtention de Nicotiana haploides à partir d'étamines cultivées in vitro. Ann. Physiol. Veg. 9: 377–382.

Chakraborti, S.P., K. Vijayan, B.N. Roy and S.M.H. Quadri. 1998. *In vitro* induction in tetraploidy in mulberry (*Morus alba* L.). Plant Cell Rep. 17: 794–803.

Chakraborti, S.P., K. Vijayan and B.N. Roy. 1999. Isolated microspore culture in mulberry (*Morus* spp.). Sericologia 39: 541–549.

Chaturvedi, R., M.K. Razdan and S. Bhojwani. 2003. An efficient protocol for production of triploid plants from endosperm callus of neem, Azadirachta indica A. Juss. J. Plant Physiol. 160: 557–564.

Chaturvedi, R., M.K. Razdan and S.S. Bhojwani. 2004. *In vitro* clonal propagation of an adult tree of neem (*Azadirachta indica* A. Juss.) by forced axillary branching. Plant Sci. 166: 501–506.

Checker, V.G., A.K. Chhibbar and P. Khurana. 2012. Stress-inducible expression of barley Hva1 gene in transgenic mulberry displays enhanced tolerance against drought, salinity and cold stress. Transgenic Res. 21: 939–957.

Chu, C.C. 1978. The N_6 medium and its applications to anther culture of cereal crops. *In*: Proc. Symp. Tiss. Cult. Science Press, Peking, pp. 43–50.

Cocking, E.C. 1979. Parasexual reproduction in flowering plants. New Zealand J. Bot. 17: 665–671.

Cutter, E.G. 1965. Recent experimental studies of the shoot apex and shoot morphogenesis. Bot. Rev. 31: 7–113.

Das, M., H. Chauhan, A. Chibbar, Q.M.R. Haq and P. Khurana. 2011. High efficiency transformation and selective tolerance against biotic and abiotic stress in mulberry, *Morus indica* cv. K-2, by constitutive and inducible expression of tobacco Osmotin. Transgenic Res. 20(2): 231–246.

Debergh, P.C. and L.J. Maene. 1981. A scheme for commercial propagation of ornamental plants by tissue culture. Sci. Hort. 14: 335–345.

Doss, S.G., K. Vijayan, S.P. Chakraborti and B.N. Roy. 1998. Studies on flowering time and its relation with geographic origin in mulberry. Ind. J. Forestry 24(2): 203–205.

Doi, H., N. Hoshi, E. Yamada, S. Yokoi, M. Nishihara, T. Hikage and Y. Takhata. 2013. Efficient haploid and doubled haploid production from unpollinated ovule culture of gentianas (*Gentiana* spp.). Breed Sci. 63(4): 400–406.

Duhamel du Monceau, H.L. 1756. La Physique des Arbes, ou II Est trait ê de I'Anatomie des plantes et de l'economie Végétale pour Servir d'Introduction au Trait ê complet des Bois et des Forêsts. P. H. L. Guêrin pub. (Cited by Gautheret 1985, p. 52).

Dwivedi, N.K., N. Suryanarayan, A.K. Sikdar, B.N. Susheelamma and M.S. Jolly. 1989. Cytomorphological studies in triploid mulberry evolved by dipfoidization of female gamete cells. Cytologia 54: 1319.

Dwivedi, S.L., A.B. Britt, L. Tripathi, S. Sharma, H.D. Upadhaya and R. Ortiz. 2015. Haploids: constraints and opportunities in plant breeding. Biotechnol. Adv. 33: 812–829.

El-Mawla, A.A.M., K.M. Mohamed and A.M. Mustafa. 2011. Induction of Biologically active flavonoids in cell cultures of *Morus nigra* and testing their hypoglycaemic efficacy. Sci. Pharmaceut. 79(4): 951–956.

FAO. 2009. Draft second report on World's plant genetic resources for food and agriculture. Food and Agriculture Organization of United Nations, Rome, pp. 112.

Fasella, F. and A. Hussain. 2014. Plant Biotechnology. Scientific Int. Pvt. Ltd., New Delhi, 499 pp.

Finkel, B. and J.M. Ulrich. 1983. Protocols of cryopreservation. pp. 806–815. *In*: Evans D.A. et al. (eds.). Handbook of Plant Cell Culture, Vol. 1: Techniques for Propagation and Breeding, Macmillan Publishing Co. N.Y.

Fukui, K., K. Shirata, T. Niino and I.M. Kashif. 2011. Cryopreservation of mulberry winter buds in Japan. Acta. Hort. 908: 483–488.

Gautheret, R.J. 1942. Manual Technique de Culture de Tissus Végétaux. Masson Publishers, Paris.

George, E.F., M.A. Hall and G. J. Deklerk (eds.). 2008. Plant Propagation by Tissue Culture 3rd Edition, Vol.1. The background. Springer, the Netherlands, 501 pp.

Gogoi, G., P.K. Borua and J.M. Al-Khayri. 2017. Improved micropropagation and *in vitro* fruiting of Morus indica L. (K-2 cultivar). J. Genet. Engineer. Biotechnol. 15: 249–256.

Guha, S. and S.C. Maheshwari. 1964. *In vitro* production of embryos from anthers of Datura. Nature (London) 204: 497.

Guha, S. and S.C. Maheshwari. 1966. Cell division and differentiation of embryos in pollen grains of Datura *in vitro*. Nature 212: 97–98.

Harfouche, A., R. Meilan, K. Grant and V.K. Shier. 2012. Intellectual property rights of biotechnologically improved plants. *In*: A. Altman and P.M. Hasegawa (eds.). Plant Biotechnology and Agriculture for 21st Century, Elsevier Inc., Oxford. U.K., pp. 525–539.

Inayi, C., P. Boonsnongcheep, J. Kamaikul, B. Sritularak, H. Tanaka and W. Putalun. 2019. Alginate immobilization of *Morus alba* L. Cell suspension cultures improved the accumulation and secretion of stilbenoids. Bioprocess Biosyst. Eng. 42(1): 131–141.

Jain, A.K. and S.K. Datta. 1992. Shoot organogenesis and plant regeneration in mulberry (Morus bombycis Koidz): Factors influencing morphogenetic potential in callus cultures. Plant Cell Tiss. Org. Cult. 29: 43–50.

Jain, A.K., A. Sarkar and R.K. Datta. 1996. Induction of haploid callus and embryogenesis in *in vitro* cultured callus of mulberry (*Morus indica*). Plant Cell Tiss. Org. Cult. 44: 143–147.

Kashyap, S. and S. Sharma. 2006. *In vitro* selection of salt tolerant *Morus alba* and its field performance with bioinoculants. Hort. Sci. 33: 77–86.

Katagiri, K. 1989. Colony formation in culture of mulberry mesophyll protoplasts. J. Sericult. Sci. Japan 58: 267–268.

Katagiri, K. and M. Venkateswaralu. 1993. Induction and calli and organ-like structures in isolated pollen cultures of mulberry, *Morus australis*. J. Sericult. Sci. Japan 60: 514: 516.

Khayat, E. 2012. An engineering view to micropropagation and generation of true to type and pathogen free plants. *In*: A. Altman and P. M. Hasegawa (eds.). Plant Biotechnology and Agriculture for 21st Century. Elsevier, London, pp. 229–241.

Kumar, P.P. and C.D. Loh. 2012. Plant tissue culture for biotechnology. In: A. Altman and P.M. Hasegawa (eds.), Plant Biotechnology and Agriculture for 21st Century. Elsevier, London, pp. 131–138.

Lakshmi Sita, G. and S. Ravindran. 1991. Gynogenic plants from ovary cultures of mulberry (Morus indica). *In*: J. Prakash and K.L.M. Pierik (eds.). Horticulture New Techniques and Applications, Kluwer Academic Publishers, London, pp. 225–229.

Larkin, P.J. and W.R. Scowcroft. 1981. Somaclonal variation—a novel source of variability from cell cultures for plant improvement. Theor. Appl. Genet. 67: 443–455.

Lal, S., V. Gulyani and P. Khurana. 2008. Over expression of hva 1 gene from barley generates tolerance to salinity and water stress in transgenic mulberry (*Morus indica*). Transgen. Res. 17: 651–663.

Lee, Y.., D.-E. Lee, H.S. Lee, S.-K. Kim, W. S. Lee, S.-H. Kim and M.-W. Kim. 2011. Influence of auxins, cytokinins, and nitrogen on production of rutin from callus and adventitious roots of the white mulberry (*Morus alba* L.). Plant Cell Org. tiss. Cult. 105(1): 9–19.

Lin, S., D. Ji and J. Qui. 1987. *In vitro* production of haploid plants from mulberry (*Morus*) anther culture. Sci. Sin. 30: 853–863.

Linsmaier, E.M. and F. Skoog. 1965. Organic growth factor requirements for tobacco tissue cultures. Physiol. Plant. 18: 100–127.

Mallick, P., S. Ghosh, S. Chattaraj and S.R. Sikdar. 2016. Isolation of mesophyll protoplasts from Indian mulberry (*Morus alba* L.) cv. S. 1635. J. Environ. Sociobiol. 13(2): 217–222.

Manek, H. and M. Hanke. 2014. 10 years of fruit gene bank at Dresten-Pillmitz under Federal responsibility. J. Kulturpflanzen. 66: 117–129.

Mantell, S.H., J.A. Matthews and R.A. McKee. 1985. Principles of Biotechnology. Blackwell Scientific Publications, Oxford, U.K., 269 pp.

Marthe, F. 2018. Tissue culture approaches in relation to medicinal plant improvement. pp. 487–497. *In*: Gosal S.S. and S.H. Wani (eds.). Biotechnologies of Crop Improvement. Vol.1, Springer International Publishing AG.

Mathur, A., A.K. Mathur, P. Verma, S. Yadav, M.L. Gupta and M.P. Darukar. 2008. Biological hardening and genetic fidelity testing of micro-cloned progeny of *Chlorophytum borivilianum* Sant. Et Fernand. Afric. J. Biotechnol. 7: 1046–1053.

Merkle, S.A., W.A. Parrot and B.S. Flinn. 1995. Morphogenetic aspects of somatic embryogenesis. pp. 155–205. *In*: Thorpe T.A. (ed.). *In Vitro* Embryogenesis in Plants. Kluwer, Dordrecht.

Ming, W.Z., Z.H. Xu, N. Xu and H.M. Ren. 1992. Plant regeneration from mesophyll protoplasts of *Morus alba*. Plant Physiol. Communic. 28: 248–249.

Mishra, V. K. and R. Goswami. 2014. Haploid production in higher plants. ICJCBS Rev. paper 1(1): 1–21.

Morel, G. 1960. Producing virus-free *Cymbicium*. Am. Orchid Soc. Bull. 29: 495497.

Murashige, T. and F. Skoog. 1962. A revised medium for rapid growth and bioassays with tobacco tissue cultures. Physiol. Plant. 15: 473–497.

Murashige, T. 1974. Plant propagation through tissue culture. Annu. Rev. Plant Physiol. 25: 135–166.

Murashige, T. 1978. Principles of rapid propagation. *In*: K.W. Hughes et al. (eds.). Propaation of Higher Plants through Tissue Culture. US. Deptt. of Energy, Washington, D.C., pp. 1424.

Murovec, J. and B. Bohonec. 2012. Haploids and doubled haploids in breeding. pp. 88–106. *In*: Abdurakhmonov, I. (ed.). Plant Breeding. In Tech. Europe, Croatia.

Narayan, S.P. Chakraborti, B.N. Roy and S.S. Sinha. 1993. *In vitro* regeneration of plant from intermodal callus of *Morus alba* L. and isolation of genetic variant. Seminar Plant Cytogenet., India, Univ. of Calcutta, March 4–5, 1993, pp. 188–192.

Nitsch, J.P. and C. Nitsch. 1969. Haploid plants from pollen grains. Science 163: 85–87.

Nobécourt, P. 1937. Cultures en série de tissus Végétaux sur milieu artificial. C. R. Hebd. Seances Acad. Sci. 200: 521–523.

Ohyama, K. 1970. Tissue culture in mulberry tree. Jap. Agric. Res. Quart. 5: 30–34.

Oka, S. and K. Ohyama. 1986. Mulberry (*Morus alba* L.). pp. 384–392. *In*: Bajaj, Y.P.S. (ed.). Biotechnology in Agriculture and Forestry, Vol.1, Springer – Verlag, Berlin.

Ohnishi, T. and S. Kiyama. 1987. Effects of change in temperature, pH, and Ca ion concentration in solution used for protoplast fusion on the improvement of fusion ability of mulberry protoplasts. J. Sericult. Sci. Jap. 56: 418–421.

Ohnishi, T, F. Yasukura and J. Tan. 1986. Preservation of mulberry callus by the addition of both abscisic acid (ABA) and pamino benzoate (PABA) in culture medium. J. Seric. Sci. Japan 55: 252255.

Padro, M.D.A., A. Frattareli, A. Sgueglia, E. Condello, C. Damiano and E. Caboni. 2012. Cryopreservation of white mulberry (*Morus alba* L.) by encapsulation-dehydration and vitrification. Plant Cell Tiss. Org. Cult. 108(1): 167–172.

Pattnaik, S.K. and P.K. Chand. 1997. Rapid clonal propagation of three mulberries, *Morus cathayana* Hemsl., *M. ilhou* Koiz. And *M. serrata* Roxb. Through *in vitro* culture of apical shoot buds and nodal explants from mature trees. Plant Cell Rep. 16: 503–508.

Pierik, R.L.M. 1989. *In Vitro* Cultures of Higher plants. Martinus Nijhoff Publishers, Dordrecht, The Netherlands, 334 pp.

Rao, A.A., R. Chaudhary, S. Kumar, D. Velu, R.P. Saraswat and C.K. Kamble. 2007. Cryopreservation of mulberry germplasm core collection and assessment of genetic stability through ISSR markers. Intl. J. Indus. Entomol. 15: 23–33.

Rao, A.A., R. Chaudhary, S. Kumar, R. Ramachandra and S.M.H. Qadri. 2009. Mulberry biodiversity conservation through cryopreservation. *In Vitro* Cell. Dev. Biol. Plant 45: 639–649.

Razdan, A., M.K. Razdan, M.V. Rajam and S.N. Raina. 2008. Efficient protocol for *in vitro* production of androgenic haploids of Phlox drummondii. Plant Cell Tiss. Organ Cult. 95: 245–250.

Razdan, M.K. and E.C. Cocking. 1997. Conservation of Plant Genetic Resources *In Vitro*, Vol. 1: General Aspects. Science Publishers Inc., Enfield, USA, 314 pp.

Razdan, M.K. and E.C. Cocking. 1981. Improvement of legumes by exploring extra-specific genetic variation. Euphytica 30: 818–833.

Razdan, M.K. and E.C. Cocking. 2000. Conservation of Plant Genetic Resources *In Vitro*, Vol. 2: Applications and Limitations. Science Publishers Inc., Enfield, USA, 315 pp.

Razdan, M.K. 2003. Introduction to Plant Tissue Culture (2nd edition). Science Publishers, Enfield, USA, 375 pp.

Razdan, M.K. 2016. Introduction to Plant Tissue Culture (3rd edition), Oxford & IBH Publishing Co. Pvt. Ltd., New Delhi, 400 pp.

Razdan-Tiku, A., M.K. Razdan and S.N. Raina. 2014. Production of triploid plants from endosperm cultures of *Phlox drummondii*. Biologia Plantarum 58: 153–158.

Saeed, B.M. Das, Q.M.R. Haq and P. Khurana. 2015. Over expression of beta carotene hydroxylase-I (bch I) in mulberry, *Morus indica* cv. K-2, confers tolerance against high-temperature and high irradiance stress induced damage. Plant Cell Tiss. Organ Cult. 120(3): 1003–1015.

Sakai, A. 1997. Potentially valuable cryogenic procedures for cryopreservation of cultured meristems. *In*: M.K. Razdan and E.C. Cocking (eds.). Conservation of plant genetic Resources *In Vitro*. Vol. 1: General Aspects. Science Publishers Inc., Enfield, USA, pp. 53–66.

San Noeum, L.H. 1976. Haploides *d'Hordeum vulgare* L. par culture in vitro non fecondes. Ann. Amelior. Plantes. 26: 751–754.

Sanjeevan, R.S., S. Jeba Singh, K.N. Nataraja and M.B. Shivanna. 2011. An efficient *in vitro* protocol for multiple shoot induction in mulberry, *Morus alba* L var. VI. Intl. Res. J. Plant Sci. 2: 254–261.

Sarkar, T., T. Radhakrishnan, A. Kumar, G.P. Mishra and J.R. Dobaria. 2016. Stress inducible expression of AtDREBIA transcription factor in transgenic peanut (*Arachis hypogaea* L.) crop conferred tolerance to soil-moisture deficit stress. Front. Plant Sci. 7: 935.

Sarkar, T., T. Mogili and V. Sivaprasad. 2017. Improvement of abiotic stress adaptive traits in mulberry (*Morus* spp.): An update on biotechnological interventions. 3 Biotech. 7: 214.

Sarkar, T., T. Mogili, S.G. Doss and V. Sivaprasad. 2018. Tissue culture in mulberry (*Morus* spp.) intending genetic improvement, micropropagation and secondary metabolite production: A review of current status and future prospects. *In*: Kumar, N. (ed.). Biotechnological Approaches for Medicinal and Aromatic Plants. Springer Nature Singapore Pte Ltd., pp. 467–487.

Seki, H. and K. Oshikane. 1959. Studies in polypoid mulberry trees III. The valuation of breeded polypoid mulberry leaves and the results of feeding silkworms on them. Res. Rep. Fac. Text Seric. Shinshu Univ. 9: 6–15.

Sethi, M., S. Bose, A. Kapur and N.S. Rangaswamy. 1992. Embryo differentiation in anther culture of mulberry. Ind. J. Exp. Biol. 30: 1146–1148.

Shajahan, A., K. Kathiravan and A. Ganapathi. 1997. Selection of salt tolerant mulberry callus tissue culture from cultured hypocotyl segments. pp. 311–313. *In*: Khan, A.I. (ed.). Frontiers in Plant Science. The Book Syndicate, Hyderabad.

Shepard, J.F., D. Bidney and E. Shahin. 1980. Potato protoplasts in crop improvement. 1980. Potato protoplasts in crop improvement. Science (N.Y.) 208: 17–24.

Skirvin, R.M. 1978. Natural and induced variation in tissue culture. Euphytica 27: 241–266.

Skoog, F. and C.O. Miller. 1957. Chemical regulation of growth and organ formation in plant tissue cultured *in vitro*. Symp. Soc. Exp. Biol. 11: 118–131.

Smith, R.H. 2013. Plant Tissue Culture: Techniques and Experiments. Elsevier, New Delhi, 188 pp.

Steward, F.C., M. Caplin and F.K. Millar. 1952. Investigations on growth and metabolism of plant cells. Ann. Bot. (London) 16: 57–77.

Stewart, Jr., C.N. 2008. Plant Biotechnology and Genetics: Principles, Techniques and Applications. John Wiley and sons Inc., Hobokeu, New Jersey, 377 pp.

Street, H.E. 1977. Cell (suspension) cultures. pp. 61–102. *In*: Street, H.E. (ed.). Plant Tissue and Cell Culture. Univ. California Press, Berkeley, Calif.

Susheelamma, B.N., K.R. Shekhar, A. Sarkar, M.R. Rao and R.K. Datta. 1996. Genotypes and hormonal effects on callus formation and regeneration in mulberry. Euphytica 90: 25–29.

Swaminathan, M.S. 1983. Genetic conservation: Microbes to man. Presidential Address, XV Intl. Congr. Genet., December 12–21, 1983, New Delhi, pp. 332.

Tewary, P.K., B.K. Gupta and G.S. Rao. 1989. *In vitro* studies on the growth rate of callus of mulberry (*Morus alba* L.). Ind. J. Forestry 12: 34–35.

Tewary, B.K., A. Sharma, M.K. Raghunath and A. Sarkar. 2000. *In vitro* response of promising mulberry (*Morus* sp.) genotype for tolerance to salt and osmotic stress. 2000. Plant Growth Regulation. 30(1): 17–21.

Thomas, T.D., A.K. Bhatnagar, M.K. Razdan and S.S. Bhojwani. 1999. A reproducible protocol for production of gynogenic haploids of mulberry (*Morus alba* L.). Euphytica 110: 169–173.

Thomas, T.D., A.K. Bhatnagar and S.S. Bhojwani. 2000. Production of triploid plants of mulberry (*Morus alba L.*) by endosperm culture. Plant Cell Reports 19: 395399.

Thomas, T.D. 2003. Thidiazuron induced multiple shoot induction and plant regeneration from cotyledonary explants of mulberry. Biologia Plantarum 46(4): 529–533.

Thorpe, T.A. 2007. History of plant tissue culture. Mol. Biotechnol. 37: 169–180.

Tikader, A. and S.B. Dandin. 2005. Biodiversity, geographical distribution, utilization and conservation of wild mulberry *Morus serrata* Roxb. Caspian J. Env. Sci. 3: 179–186.

Trigiano, R.N. and D.J. Gray (eds.). 2011. Plant Tissue Culture, Development and Biotechnology. CRC Press, Boca Raton, 608 pp.

Umate, P., V.K. Rao, K. Kiranmayee, T. Jayashree and A. Sadanandam. 2005. Plant regeneration of mulberry (*Morus indica*) from mesophyll-derived protoplasts. Plant Cell Tiss. Organ. Cult. 82: 289–293.

UNEP. 1995. Global diversity Assessment. Cambridge Univ. Press, 457–475.

Vijayan, K., A. Tikader and J.A. Teixeira da Silva. 2011. Application of tissue culture techniques for propagation and crop improvement in mulberry (*Morus* spp.). Tree and Forestry Science and Biotechnology 5(1): 1–13.

Vijayan, K., P. Jayarama Raju and B. Saratchandra. 2014. Biotechnology of mulberry (*Morus* L.)—A Review. Emir. J. Food Agric. 26(6): 472–496.

Wink, A., A.W. Alfermann, R. Franke, B. Wetteraner, M. Distl, J. Windhovel, O. Krahn, E. Fuss, H. Garden, A. Mohagzadeh, E. Wildi and P. Ripplinger. 2005. Sustainable bioproduction of phytochemicals by plants *in vitro* cultures: anticancer agents. Plant Genetic Resour. Charact. Util. 3: 90–100.

Yasukura, F. and T. Ohnishi. 1990. Effects of regeneration period of mulberry cultures on their growth after their storage and after the pseudo dormancy induced by storage. J. Sericult. Sci. Jap. 59: 255–258.

Zaki, M., Z.A. Kaloo and M. Sofi. 2011. Micropropagation of mulberry *Morus nigra* L. from nodal segments with axillary buds. World J. Agric. Sci. 7: 496–503.

CHAPTER 9

In vitro Production of Haploids and Triploids in Mulberry

Anju M,[1] *Razdan MK*[2] *and T Dennis Thomas*[1,]*

INTRODUCTION

Mulberry is a fast growing economically important deciduous woody perennial tree mainly distributed in Asia. Mulberry belongs to the genus *Morus* of the family Moraceae. The genus *Morus* consists of several species. Among them four species predominantly represent the genus *Morus* in India which includes *M. indica* L., *M. laevigata* Wall., *M. alba* L. and *M. serrata* Roxb (Tikader and Kamble 2008). Mulberry leaves are economically important and widely used for the rearing of silkworm (*Bombyx mori* L.) larva to produce silk yarn. The proteins present in mulberry leaves are transformed into silk proteins such as fibroin and sericin by the silk worm larva (Ghosh et al. 2017). The Mulberry silk constitute about 90% of the total raw silk production in the world that is used for the production of top quality silk garments (Ghosh et al. 2017). Mulberry leaves also form major part of the fodder for livestock, whereas delicious fruits are used in the preparation of several commercial products such as jam, frozen desserts, ice cream, marmalade, paste and wine for human consumption (Vijayan et al. 2014).

The medicinal properties of mulberry are also noteworthy. Several studies showed that mulberry has high pharmaceutical and pharmacological activities. The leaf extract of mulberry showed significant inhibitory action towards the amyloid β-peptide fibril formation which has a significant role in etiology of Alzheimer's disease and neurotoxicity (Niidome et al. 2007). Mulberry (*M. indica*) leaf extract is known to evince anti-oxidant and anti-hyperglycemic properties (Andallu and Varadachargulu 2003) and for anti-inflammatory and anti-pyretic activities (Chatterjee et al. 1983). According to their study the root extract of *M. indica* has major anti-inflammatory effect in exudative, proliferative, and chronic phases of inflammation. The antipyretic activities of *M. indica* were also reported by Ghosh et al. (2017) and Niidome et al. (2007). *M. alba* is generally known as white mulberry having cardioprotective, anti-ulcer, neuroprotective, and skin whitening properties. It is also used as liver and kidney tonic (Zafar et al. 2013). Ayoola et al. (2011) reported the anti-microbial and anti-fungal activities of *M. alba*. Kuwanon - G, an anti-bacterial agent, was isolated from *M. alba* (Park et al. 2003), which inactivated *Streptococcus* mutants responsible for tooth decay. Bioactive molecules such as sanggenon - B and D, morusin and kuwanon - C isolated from mulberry also have the anti-microbial properties (Zafar et al. 2013, Nomura et al. 1988), whereas some

[1] Department of Plant Science, Central University of Kerala, Tejaswini Hills, Periye P.O, Pin-671320, Kasaragod, Kerala, India.
[2] Shyam Lal College, University of Delhi, Shahdara, Delhi, India (Retd.).
* Corresponding author: den_thuruthiyil@yahoo.com

phytochemicals like morusin, morusin 4-glucoside and kuwanon - H isolated from mulberry demonstrate anti HIV activity (Zafar et al. 2013, Shi-De et al. 1995).

Biotechnological Approaches in Mulberry Improvement

Although sexual and asexual reproduction is possible in mulberry, the propagation through seeds is not practiced due to the high heterozygosity of parental lines and prolonged juvenile period (Vijayan et al. 2011). Usually stem cuttings and bud graftings are employed for vegetative propagation of mulberry. However, success and propagation rate are unpredictable. The success of vegetative propagation in mulberry is dependent on age, genotype, physiological condition of the plant, environmental factors, climatic conditions and cultural practices (Vijayan et al. 2014). The tissue culture techniques have been applied in mulberry as an efficient and rapid method for the production of numerous true-to-type plants with desirable characters within a short span of time irrespective of seasonal variations under suitable conditions (Anis et al. 2003).

As reported in Chapter 8 the micropropagation of mulberry was initiated by Ohyama (1970) from axillary buds of *M. alba*. Since then large number of researchers micropropagated various species of mulberry using different types of explants such as shoot tips, nodes, cotyledons, leaves, hypocotyls, axillary buds, etc. MS (Murashige and Skoog 1962) medium with various concentrations of different hormones was used by majority of the researchers for nodal segment, apical and axillary bud culture (Thomas 2002). It is interesting to note that MS medium supplemented with various concentrations of thidiazuron (TDZ) developed more shoots than same concentrations of BAP (6-benzylaminopurine) for same aged explants (Thomas 2003). Thomas (2004) found that it is possible to modify the sex expression of dioecious plant mulberry using MS medium with 1.13 mg/l BAP, 2000 µg/l ethrel (2-Chloroethylphosphonic acid) and 2500 µg/l silver nitrate. Induction of male, female and mixed inflorescences was produced using this technique. Somatic embryogenesis was also reported in mulberry, but it was not very successful as in other crop plants due to various reasons (Vijayan et al. 2014). Similarly, somatic hybridization had been reported in mulberry by using chemical fusogen (Ohnishi and Kiyama 1987) and electro-fusion (Ohnishi and Tanabe 1989). This technique also met with several problems and hence the development of somatic hybrid in mulberry did not yield promising results (Vijayan et al. 2011). The genetic engineering technique via *Agrobacterium tumefaciens*-mediated genetic transformation has been applied for mulberry transformation (Bhatnagar and Khurana 2003). According to this study, *A. tumefaciens* strain LBA—4404 was more infective than other strains GV2260 and A281. More studies on *Morus* sp. regarding these aspects are described in various chapters of this book.

Importance of Haploids

Haploids are plants having gametophytic number of chromosomes (i.e., one set of chromosomes) in the sporophyte. Angiosperms are generally diploid plants having sporophytic plant body. Haploid cells are generated from sporophytic diploid plants during the gamete (i.e., pollen grains and egg) formation by meiosis, in angiosperm plants. Fertilization of male and female gametes leads to development of diploid sporophytic phase. Haploid production is the rapid and most feasible way of obtaining 100% pure homozygous lines after chromosome duplication using suitable chemicals. The first natural haploids were reported in *Datura stramonium* and *Nicotiana* sp. which led to the recognition and significance of haploids in genetic improvement and plant breeding research (Blakeslee et al. 1922, Blakeslee and Belling 1924, Kostoff 1929). Subsequently it was found that haploids have immense use in many areas including genetic analysis, in vitro selection, genetic transformation, mutation breeding and formation of inbred lines to utilize hybrid vigor. The doubling of chromosome number of haploids by applying colchicine is a routine procedure these days to get homozygous diploid lines in crop plants as well as the economically important trees which have long reproductive cycle, self-incompatibility, and high degree of heterozygosity. It is really a Herculean task to obtain pure lines in woody tree species through conventional inbreeding methods and therefore, the research on genetic studies apparently become difficult.

Haploids may be formed spontaneously through the production of embryo from haploid cell other than egg or from unfertilized egg (Germana 2006). Haploid plants can be produced by inducing microspore, egg or any other cell in the embryo sac to form sporophyte without the fertilization. Wide or distant hybridization is another technique for the production of haploid plants which leads to the selective elimination of one set of chromosome during early embryogenesis after fertilization. Embryos formed as a result of this process are thus rescued and maintained *in vitro* to grow into a haploid plant (Bhojwani and Dantu 2010). The conventional method of homozygous plant production is achieved through self-pollination of an angiosperm in several generations which is a time consuming process. In this context doubled haploid plant production *in vitro* has advantages over the conventional method of producing pure lines. These include direct expression of recessive character at the phenotypic level and elimination of lethal mutation or gene defects in plants (Murovec and Bohanec 2012). In some cases, spontaneous duplication of chromosomes may occur in pollen derived plants (Bhojwani and Razdan 1996). Chase (1952) had induced haploid plants *in vivo* through parthenogenesis and applied chromosome doubling treatment to induce homozygous diploid lines of maize (Wędzony et al. 2009). Pollen irradiation, sparse pollination, alien cytoplasm, selecting seed with twin embryos, wide hybridization are other available methods employed for increasing the frequency of parthenogenesis in various systems (Kasha and Maluszynski 2003). Crossing of plants with different ploidy levels will give rise to haploid plants such as crossing of diploid and triploid plants (Mishra and Goswami 2014). However, number of plants produced through this technique is very less. Therefore, efficient protocols of plant tissue culture techniques could be used for the production of homozygous lines from gametic cell-raised haploid plants in short period of time as compared to conventional plant breeding methods (Morrison and Evans 1988).

Haploids of Mulberry Produced *in vitro*

Androgenesis

Anther is the male reproductive structure in plants and contains diploid sporogenous tissue. The micropore mother cell undergoes meiosis to form haploid microspores. These microspores are transformed into pollen grains by certain morphological changes. Large number of haploid pollen grains will be formed in a single anther. The production of haploid plants from the uninucleate microspores *in vitro* is known as androgenesis. The immature pollen grains (microspores) are induced to form sporophytic mode of generation in media by various physical and chemical stimuli (Bhojwani and Dantu 2010). For anther culture unopened flower buds with microspores at the uninucleate stage are used. Further, they are surface sterilized and subsequently cultured on specific medium.

Anther culture in mulberry was first reported by Shoukang et al. (1987). Types of medium, plant growth regulators, stage of anther at the time of culture are important factors that influence the induction of androgenic embryos from anthers in mulberry. MS medium supplemented with 0.99 or 2.0 mg/l BA and 0.99 or 1.99 mg/l Indole-3-butyric acid (IBA) were more effective in embryoid induction than other treatments and anthers harvested at the mid-uninucleate stage developed maximum rate of induction. Generally higher concentration of cytokinin and lower concentration of auxin was found favorable for differentiation of embryoids. The anther derived plants thus obtained were having haploid, diploid and heteroploid chromosome numbers. Interestingly, majority of the cells were haploid.

Katagiri and Modala (1991) investigated the influence of various sugars on division of pollen grain in mulberry *in vitro*. From their study it is confirmed that disaccharides play a crucial role in binucleated pollen division than monosaccharides. The calli thus obtained, later, produced embryo and organ like structures.

Sethi et al. (1992) studied the influence of dark period on androgenic embryo induction in three cultivars of mulberry, i.e., 'RFS-175', 'RFS-135' and 'Goshoerami'. MS medium supplemented with 1.01 to 2.03 mg/l IBA and 1.13 to 3.38 mg/l BA was employed for culture and the cultures were incubated in dark at 26 ± 1°C for 15 days. Cultivar 'RFS-175' showed 1.87% androgenic response while 'RFS-135' and 'Goshoerami' showed 0.23% and 0.39% respectively.

Jain et al. (1996) studied the effect of various temperature treatments on cultured anthers in mulberry (*Morus indica*). The catkins of mulberry were treated with cold shock (4°C for 0, 24, 48, 72 and 96 h) or high temperature (36°C for 0.5, 1,0, 2,0, 4,0 and 6.0 h) followed by incubating at 4°C. Of the various treatments, cold shock at 4°C for 24 h was most effective whereas high temperature treatment was detrimental. The pretreated anthers were cultured on Modified Bourgin (MB) medium (Qian et al. 1982) supplemented with various concentrations of NAA

(0.0, 0.5, 1.0, 1.5, or 2.0 mg/l) or BA (0.0, 0.5, 1.0, 1.5 or 2.0 mg/l) alone or in combination. Medium supplemented with a combination of 0.5 mg/l NAA and 1.0 mg/l BA with 80 g/l sucrose was found most effective for callus induction. The anthers split open and produced two types of calli, (a) pale, yellow, soft, friable and yellow-white, and (b) compact globular calli. Both calli exhibited embryogenic potential but globular white callus responded better than the other. Root induction from calli were very common and the frequency of rooting increased with the increase in concentration of sucrose up to 80 g/l. Cytological analysis of the roots showed majority of cells had haploid chromosome number. Histological analysis confirmed the presence of meristematic zones in the callus cells with dense cytoplasm and dark stained nuclei. MB medium supplemented with various types of hormones such as 2,4-D (2,4-Dichlorophenoxyacetic acid; 0.25–2.0 mg/l), NAA (0.1–1.0 mg/l), IBA (0.25–2.0 mg/l), BA (0.5–2.0 mg/l), kinetin (0.1–1.0 mg/l), zeatin (0.05–0.25 mg/l) and GA_3 (Gibberellic acid; 0.25–1.0 mg/l) were used for the subculture of anther derived calli to induce embryogenesis. Embryoid induction was noticed in 80% cultures on medium supplemented with NAA (0.5 mg/l), BA (0.5 mg/l) and 2,4-D (1.0 mg/l). The embryoids further developed roots when 2,4-D was removed from medium and developed elongated shoots without forming cotyledons (Jain et al. 1996). Although the authors obtained roots and 'shoot like' structures from the embryoids of the calli, there was no complete plant formation.

In another related report, Tewary et al. (1994) cultured mulberry pollen *in vitro*. The inflorescences of mulberry were subjected to starvation for a period of 24 h to 10 days before isolating the pollen grains. Pollen suspension was cultured in MS medium for development. Enlarged pollen with large nucleus and increased cytoplasmic contents were noticed in all six genotypes cultured and S-1 cultivar gave optimum result. However, authors failed to get any plants from the cultured pollen grains.

Gynogenesis

Pistil is the female reproductive structure which consists of 3 parts, i.e., stigma, style and ovary. Ovary produces ovules within it which are diploid in nature. The megaspore mother cell in ovule undergoes meiosis and forms 4 megaspores, of which three degenerate and the remaining one divides to form the embryo sac. Usually embryo sac contains one egg cell and two synergids on micropylar region, three antipodals on the chalazal region, and two central cells located at the central part of the embryo sac. Production of haploid plants from haploid cells of the embryo sac is known as gynogenesis. Usually haploid embryos or calli will be formed from the unfertilized egg cells as reported in several systems such as *Beta vulgaris* (Ferrant and Bouharmont 1994), *Helianthus annuus* (Gelebart and San 1987), *Hevea brasiliensis* (Guo et al. 1982), *Hordeum vulgare* (Huang et al. 1982) and *Melandrium album* (Mol 1992). However, there are reports of origin of haploid plants from other embryo sac cells like antipodal or synergid cells such as in *Hordeum vulgare* (San-Noeum 1976) and *Oryza sativa* (Zhou et al. 1986). Gynogenic plants emerge either through direct embryogenesis in *Allium cepa* (Keller and Korzun 1996) or through callus in *Gerbera* (Tosca et al. 1990) and *Lilium davidii* (Gu and Cheng 1983). Direct embryogenesis is preferred since it reduces somaclonal variation (Bhojwani and Thomas 2001). The development of embryos or plants from egg cell without fertilization is similar to pathenogenesis (Wędzony et al. 2009). Gynogeneis was first reported in barley by San-Noeum in 1976 (Bhojwani and Razdan 1996). Usually gynogenesis is successfully applied in systems where androgenesis is not effective or shows poor response. The limitation of gynogenesis is that only 8 haploid cells in the embryo sac can develop plants whereas in androgenesis thousands of pollen grains present in a single pollen can have the capacity to develop into plants (Wędzony et al. 2009). Various factors affecting gynogenesis include genotype, explant type, pretreatment and culture medium (Bhojwani and Thomas 2001). Of these, stress treatments like low temperature and high illumination (Puddephat et al. 1999), cold treatments (Sibi et al. 2001) and heat treatments (Gemes-Juhasz et al. 2002) are considered as the most common triggering factors of gynogenesis in various systems.

Gynogenic plants in mulberry were first reported by Lakshmi Sita and Ravindran (1991). They used individual ovaries from female inflorescences before or after fusion to form sorosis (fleshy multiple fruit) for the experiment. These ovaries were cultured on MS medium supplemented with plant growth regulators. They were able to produce 65 plants directly from the cultured ovaries without the formation of callus. Four plants from a single cultured ovary were noticed on MS medium supplemented with 1.0 mg/l BAP and 1.0 mg/l kinetin within three weeks.

Complete plants that emerged from ovaries had well-developed tap root system, suggesting direct embryogenesis. However, the growth of the plants was slow on induction medium and it was overcome by subculturing the plants on MS medium supplemented with 1.0 mg/l GA_3 and 0.3 mg/l Kinetin. The plants were further multiplied by nodal segment culture on MS medium supplemented with 0.1 mg/l BA and 0.2 mg/l kinetin. On this medium 3–4 multiple shoots were obtained from a single explant. Abundant rooting of shoots was found on half strength MS medium with IBA (0.5–1.0 mg/l). However, cytological preparations from root tip cells showed both haploid (n = 14) and diploid (2n = 28) number of chromosomes. Haploid number of chromosome was mostly observed in cells of first formed roots and on further transfer to soil they showed diploid number of chromosome. Authors assume that spontaneous diploidization may occur during embryonic development (Lakshmi Sita and Ravindran 1991). Diploidization of haploid plants appears higher in many systems during gynogenesis as compared to androgenesis (San and Gelebart 1986).

Thomas et al. (1999) reported the production of gynogenic haploids of female clones of mulberry (*Morus alba* L. Cv. K2) by using unpollinated ovary. They adopted a slightly different method by using *in vitro* derived female inflorescence of female clone K-2 for inducing gynogenic haploids in mulberry. Single node cuttings of mulberry from experimental plant were placed on MS medium supplemented with various concentrations of BAP (0.23, 0.56, 0.13, 1.58, 2.25 mg/l) for *in vitro* inflorescence induction. The nodal cuttings showed bud break after two weeks and inflorescence from axillary buds were formed in the axile of leaves after 3 weeks. Optimum concentration of BAP for the axillary bud break and formation of inflorescence was 1.13 mg/l. Ninety-five percent cuttings produced shoots which developed inflorescences and all inflorescences were female. The inflorescence segments having 4–5 florets were placed on MS medium with BAP, 2ip (6-Dimethylallylaminopurine), Kn (Kinetin), TDZ, 2,4-D, IBA, IAA (Indole-3-acetic acid), NAA (1-Naphthaleneacetic acid) and GA_3 alone or in various combinations. MS medium with combinations of BAP (1.01 mg/l) and 2,4-D (0.99 mg/l) was found optimum for ovary growth. On this medium 85% cultured ovaries exhibited a size above 4 × 6 mm after 3 weeks. In second stage, the individual ovaries were excised from the inflorescence segments and transferred to fresh medium. Maximum gynogenic plants were obtained when *in vitro* derived inflorescence segments were cultured on MS medium supplemented with 1.91 mg/l of BAP and 0.99 mg/l 2,4-D and individual ovaries from such cultures were transferred to MS medium with 0.99 mg/l 2,4-D, 499.5 mg/l glycine and 199.87 mg/l proline. In this combination 16% ovaries produced gynogenic plants. Cytological study revealed the haploid chromosome number (n = 14) in the root tip cells of 12 plants and another 8 plants as aneuploid with 13–17 chromosomes.

Ontogeny of gynogenic embryo was analyzed by Thomas (2004). Two cultivars of mulberry (K-2 and S-1) were used for the gynogenic haploid plant production using three-step protocol which includes *in vitro* flowering, inflorescence segment culture and isolated ovary culture. MS medium with 1.13 mg/l BAP was optimum for inflorescence development, while 1.58 mg/l BAP induced maximum number of flowers per inflorescence. Histological studies of cultured ovaries revealed the embryo sac with only the egg cell since other cells in embryo sac degenerated before the division of egg cell. Therefore, Thomas (2004) confirmed that the gynogenic plants in mulberry developed from the unfertilized eggs of mulberry plant. Histological studies on induction of gynogenesis in the inflorescence segment culture revealed that embryos could not grow beyond globular stage and they degenerated after prolonged culture. However, when each floret from inflorescence segment was excised and cultured on MS medium supplemented with 0.99 mg/l 2,4-D, 500 mg/l glycine and 200 mg/l proline, gynogenic plants developed from isolated ovaries of each floret. Further cultivar K-2 developed more gynogenic plants than S-1.

Importance of Triploid Plants

Endosperm is formed as a result of double fertilization and triple fusion. During fertilization, one of the two male gametes fuses with egg cell and the other one fuses with polar nuclei resulting in the formation of embryo and endosperm, respectively. Endosperm is present in all angiosperm families except Orchidaceae, Podostemaceae, and Trapaceae (Hoshino et al. 2011). The main function of endosperm is to provide nutrients to growing embryo. The embryo will be aborted in the absence of endosperm (Bhojwani and Razdan 1996). In some plants like cereals and coffee endosperm is present in mature seeds (albuminous seeds) and in other plants like legumes and cucurbits the endosperm is consumed by the growing embryo (exalbuminous seeds). Usually triploid plants are seed sterile.

Therefore, triploid production is not useful in plants which are commercially used for their seeds (Thomas and Chaturvedi 2008). However, triploid plants are useful if seedlessness is considered as an added advantage as in the case of some fruits such as apple, banana, papaya and grapes. Moreover, in plants where their vegetative parts are economically important, triploid plants are of great use because vegetative growth of triploid plants is more vigorous than their diploid counterparts. For example, triploid plants of *Populus tremuloids* showed superior pulp quality as compared to its diploid plants, similarly taste and size of triploid tomato was found better than its diploid plants (Hamada 1963, Bhojwani and Razdan 1996, Thomas and Chaturvedi 2008). Since triploid mulberry has more vegetative growth, it can feed more silk worm larva as compared to diploid mulberry. Conventionally, triploid plants are produced by crossing diploid plant with induced tetraploid plant (Hoshino et al. 2011). For this purpose tetraploid plants will be produced by treating the diploid plants with colchicine for chromosome duplication. Normally seeds, seedlings or vegetative buds can be treated with colchicine for producing tetraploid mulberry plants (Thomas and Chaturvedi 2008). In this method of tetraploid plant production, the induction rate is too low. Moreover, the entire protocol is lengthy and arduous. Secondly, crossing the tetraploid with diploid is not successful in most of the cases. Problems like low seed set, low seed germination and less survival rate are very common.

Lampe and Mills (1933) attempted endosperm culture for the first time in maize. Since then, many researchers experimented endosperm culture, but they failed to induce root/shoot regeneration from cultured endosperm probably due to the reasons such as wrong selection of nutrient media, incorrect stage of explants, absence of proper additives and growth regulators (Hoshino et al. 2011). Selection of endosperm at proper stage of development is the first step of endosperm culture. Usually the days after pollination (DAP) is considered as an important factor that determines the success of endosperm culture and it may vary in different systems. For example, in *Morus alba* the endosperm of seeds formed 17–20 DAP gave best results (Thomas et al. 2000) whereas in *Oryza sativa* seeds of 4–7 DAP was enough for optimum response (Nakano et al. 1975). Other significant factors which influence the endosperm culture include proper selection of medium, suitable growth regulators and extra additives. MS medium is exclusively used by most workers for endosperm culture. Main extra additives used in endosperm culture are tomato juice (TJ), yeast extract (YE), casein hydrolysate (CH), coconut milk (CM) potato extract (PE), grape juice (GJ) and cow's milk. TJ was initially found superior among other additives due to their cytokinin like activity (Sternheimer 1954). However, the cytokinin like activity of tomato juice depends on the age of the fruit (La Rue 1944, Bottomley et al. 1963). Physical factors such as light, temperature and pH have great influence on endosperm proliferation. There are conflicting reports on the effect of light in various systems. For example, *Ricinus* endosperm requires continuous light (Srivastava 1971) whereas corn endosperm needs dark period for optimum response (Straus and La Rue 1954). From the available literature it is clear that the optimum temperature for endosperm culture was 25°C and pH may vary from 5.5 to 5.8.

Triploids of Mulberry Produced *in vitro*

Triploid plants of mulberry have more biomass and in particular larger leaves which can feed more number of silk worm larva than their diploid counterparts. Further, the fruits of triploid mulberry plants being seedless are considered superior in quality. Triploid mulberry plants *in vitro* were produced from the endoperm by Thomas et al. (2000). Since mature seeds of mulberry are exalbuminous, immature endosperm from 17–20 DAP was ideal for culture initiation. Further, the seeds at this stage are having soft seed coat and hence easy to dissect.

The endosperm was cultured with or without the embryo for callus induction on MS medium supplemented with various plant growth regulators such as BAP, Kn, TDZ, 2, 4-D, IBA, NAA, GA_3 and extra additives like tomato juice (TJ), yeast extract (YE), casein hydrolysate (CH) and coconut milk (CM). Individual application of extra additives TJ, YE, CH and CM did not yield any result. However, they acted synergistically with BAP and NAA and the optimum callus induction (70–72%) was observed on MS medium supplemented with BAP (1.13 mg/l), NAA (0.19 mg/l) and CM (15%) or YE (1000 mg/l). An initial association of embryo with endosperm was found effective in inducing callus. Such embryos are distinct being green in colour and removed after 7 days. The endosperm is then incubated in dark for proliferation and the calli induced appear brown, friable and unorganized.

The highest shoot induction rate (75%) from endosperm calli could be achieved on MS medium supplemented with BAP (1.13 mg/l) and NAA (0.19 mg/l), followed by 0.22 mg/l TDZ. However, in terms of number of shoots (18) per callus was optimum in the latter medium. The brown calli formed green loci and from these regions shoot buds emerged after 30 days of culture. The regeneration potential of the endosperm derived calli was lost after 3–4 subcultures of 6 weeks each. For rooting of the shoots ½ MS medium supplemented with 1.42 mg/l IBA was optimum. On this medium 89% shoots developed 4–5 roots 30 days after culture. The rooted plants were then transplanted into soil with 71% success. Cytological analysis of the root tips of regenerated plants showed triploid number of chromosome (2n = 3x = 42). Additionally, the triploid nature was further confirmed by stomatal analysis of the leaves of endosperm derived plants. The leaves of endosperm derived plants exhibit larger stomata and its guard cells contain more number of chloroplasts as compared to diploid counterparts.

Khan and Naik (2017) attempted to produce triploid plants of mulberry by endosperm culture. The endosperm derived calli were subcultured on MS medium supplemented with BAP or TDZ alone or combination with NAA for shoot induction. However, the details of the result or the concentrations of growth regulators were not mentioned in this report.

Concluding Comments

Through various biotechnological approaches attempts have been made to genetically improve mulberry trees. Production of haploids and triploids *in vitro* of mulberry in various studies show haploid plants can be obtained through gynogensis whereas success achieved in triploid plant production has been limited. Further investigations are needed to find out the possibility of double haploids, and triploids, sustain rigours of climate change.

References

Andallu, B. and N.C. Varadacharyulu. 2003. Antioxidant role of mulberry (*Morus indica* L. cv. Anantha) leaves in streptozotocin-diabetic rats. Clin. Chim. Acta 338: 3–10.

Anis, M., M. Faisal and S.K. Singh. 2003. Micropropagation of mulberry (*Morus alba* L.) through *in vitro* culture of shoot tip and nodal explants. Plant Tiss. Cult. 13: 47–51.

Ayoola, O.A., R.A. Baiyewu, J.N. Ekunola, B.A. Olajire, J.A. Egunjobi, E.O. Ayeni and O.O. Ayodele. 2011. Phytoconstituent screening and antimicrobial principles of leaf extracts of two variants of *Morus alba* (S30 and S54). Afr. J. Pharm. Pharmacol. 5: 2161–2165.

Bhatnagar, S. and P. Khurana. 2003. *Agrobacterium tumefaciens*-mediated transformation of Indian mulberry, *Morus indica* cv. K2: a time-phased screening strategy. Plant Cell Rep. 21: 669–675.

Bhojwani, S.S. and M.K. Razdan. 1996. Plant tissue culture: theory and practice. A revised edition. Elsevier, Amsterdam, pp. 215–230.

Bhojwani, S.S. and T.D. Thomas. 2001. *In vitro* gynogenesis. pp. 489–507. *In*: Bhojwani, S.S. and W.Y. Soh (eds.). Current Trends in the Embryology of Angiosperms Kluwer Academic Publishers, Dordrecht, The Netherlands.

Bhojwani, S.S. and P.K. Dantu. 2010. Haploid plants. pp. 61–76. *In*: Davey, M.R. and P. Anthony (eds.). Plant Cell Culture: Essential Methods Wiley-Blackwell, Chichester UK.

Blakeslee, A.F., J. Belling, M.E. Farnham and A.D. Bergner. 1922. A haploid mutant in the Jimson weed, *Datura stramonium*. Science 55: 646–647.

Blakeslee, A.F. and J. Belling. 1924. Chromosomal mutations in the Jimson weed, *Datura stramonium*. J. Hered 15: 195–206.

Bottomley, W., N.P. Kefford, J.A. Zwar and P.L. Goldacre. 1963. Kinin activity from plant extracts. Aust. J. Biol. Sci. 16: 395–406.

Chase, S.S. 1952. Production of homozygous diploids of maize from monoploids. Agron. J. 44: 263–267.

Chatterjee, G.K., T.K. Burman, A.K. Nagchaudhuri and S.P. Pal. 1983. Antiinflammatory and antipyretic activities of *Morus indica*. Planta Med. 48: 116–119.

Ferrant, V. and J. Bouharmont. 1994. Origin of gynogenetic embryos of *Beta vulgaris* L. Sex. Plant Rep. 7: 12–16.

Gelebart, P. and L.H. San. 1987. Production of haploid plants of sunflower (*Helianthus annuus* L.) by *in vitro* culture of non-fertilized ovaries and ovules. Agronomie 7: 81–86.

Gemes-Juhasz, A., P. Balogh, A. Ferenczy and Z. Kristof. 2002. Effect of optimal stage of female gametophyte and heat treatment on *in vitro* gynogenesis induction in cucumber (*Cucumis sativus* L.). Plant Cell Rep. 21: 105–111.

Germana, M.A. 2006. Doubled haploid production in fruit crops. Plant Cell Tiss. Org. Cult. 86: 131–146.

Ghosh, A., Gangopadhyay, D. and T. Chowdhury. 2017. Economical and environmental importance of mulberry: a review. Int. J. Plant Environ. 3: 51–58.

Gu, Z.P. and K.C. Cheng. 1983. *In vitro* induction of haploid plantlets from unpollinated ovaries of lily and embryological observations. Acta Bot. Sin. 25: 24–28.

Guo, G., X.J. Jia and L. Chen. 1982. Induction of plantlets from ovules *in vitro* of *Hevea brasiliensis*. Hereditas (Beijing) 4: 27–28.

Hamada, S. 1963. Polyploid mulberry trees in practice. Indian J. Seric. 1: 3–6.

Hoshino, Y., T. Miyashita and T.D. Thomas. 2011. *In vitro* culture of endosperm and its application in plant breeding: Approaches to polyploidy breeding. Sci. Hortic. 130: 1–8.

Huang, Q.F., H.Y. Yang and C. Zhou. 1982. Embryological observations on ovary culture of unpollinated young flowers in *Hordeum vulgare* L. Acta Bot. Sin. 25: 24–28.

Jain, A.K., A. Sarkar and R.K. Datta. 1996. Induction of haploid callus and embryogenesis in *in vitro* cultured anthers of mulberry (*Morus indica*). Plant Cell Tiss Org. Cult. 44: 143–147.

Kasha, K.J. and M. Maluszynski. 2003. Production of doubled haploids in crop plants—an introduction. pp. 1–4. *In*: Maluszynski, M., K.J. Kasha, B.P. Forster and I. Szarejko (eds.). Doubled Haploid Production in Crop Plants, A Manual. Kluwer, Dordrecht/Boston/London.

Katagiri, K. and V. Modala. 1991. Effect of sugar and sugar alcohols on the division of mulberry pollen in tissue culture. J. Ser. Sci. Jpn. 60: 514–516.

Keller, E.R.J. and L. Korzun. 1996. Ovary and ovule culture for haploid production. pp. 217–235. *In*: Jain, S.M., S.K. Sopory and R.E. Veilleux (eds.). *In Vitro* Haploid Production in Higher Plants. Kluwer Academic Publishers, Dordrecht.

Khan, M.M. and S.S. Naik. 2017. Implementation of tissue culture technique for the production of triploid plants of mulberry (*Morus* L.). Int. J. Sci. Eng. Res. 5: 445–447.

Kostoff, D. 1929. An androgenic *Nicotiana* haploid. Z Zellforsch Mikrosk Anat. 9: 640–642.

La Rue, C.D. 1944. Regeneration of endosperm of gymnosperms and angiosperms. Am. J. Bot. 36: 798.

Lampe, L. and C.O. Mills. 1933. Growth and development of isolated endosperm and embryo of maize. Abs Papers Bot. Soc. Boston.

Lakshmi Sita, G. and S. Ravindran. 1991. Gynogenetic plants from ovary culture of mulberry (*Morus indica*). pp. 225–229. *In*: Prakash, J. and R.L.M. Pierik (eds.). Horticulture-New Technologies and Applications Kluwer Academic Publishers, Dordrecht.

Mishra, V.K. and R. Goswami. 2014. Haploid production in higher plant. Int. J. Chem. Biol. Sci. 1: 26–45.

Mol, R. 1992. *In vitro* gynogenesis in *Melandrium album* from partenogenetic embryos to mixoploid plants. Plant Sci. 8: 261–269.

Morrison, R.A. and D.A. Evans. 1988. Haploid plants from tissue culture: new plant varieties in a shortened time frame. Bio/tech. 6: 684–690.

Murashige, T. and F. Skoog. 1962. A revised medium for rapid growth and bioassays with tobacco tissue culture. Physiol. Plant 15: 473–497.

Murovec, J. and B. Bohanec. 2012. Haploids and doubled haploids in plant breeding. Intech. Open. 87–106.

Nakano, H., T. Tashiro and E. Maeda. 1975. Plant differentiation in callus tissue induced from immature endosperm of *Oryza sativa* (L.). Z. Pflanzenphysiol. 76: 444–449.

Niidome, T., K. Takahashi, Y. Goto, S. Goh, N. Tanaka, K. Kamei, M. Ichida, S. Hara, A. Akaike, T. Kihara and H. Sugimoto. 2007. Mulberry leaf extract prevents amyloid beta-peptide fibril formation and neurotoxicity. Neuroreport. 18: 813–816.

Nomura, T., T. Fukai, Y. Hano, S. Yoshizawa, M. Suganuma and H. Fujiki. 1988. Chemistry and anti-tumor promoting activity of *Morus* flavonoids. Prog. Clin. Biol. Res. 280: 267–281.

Ohnishi, T. and S. Kiyama. 1987. Effects of change in temperature, pH, Ca ion concentration in the solution used for protoplast fusion on the improvement of the fusion ability of mulberry protoplasts. J. Seric. Sci. Japan 56: 418–421.

Ohnishi, T. and K. Tanabe. 1989. On the protoplast fusion of mulberry and paper mulberry by electrofusion method. J. Seric. Sci. Japan 58: 353–354.

Ohyama, K. 1970. Tissue culture in mulberry tree. Jpn. Agric. Res. Quar. 5: 30–34.

Park, K.M., J.S. You, H.Y. Lee, N.I. Baek and J.K. Hwang. 2003. Kuwanon G: an antibacterial agent from the root bark of *Morus alba* against oral pathogens. J. Ethnopharmacol. 84: 181–185.

Puddephat, I.J., H.T. Robinson, B.M. Smith and J. Lynn. 1999. Influence of stock plant pretreatment on gynogenic embryo induction from flower buds of onion. Plant Cell Tiss. Org. Cult. 57: 145–148.

Qian, C., Z. Chen, M. Cen, M. Lin and X. Xu. 1982. Improvement of dedifferentiation medium in anther culture of *Hevea brasiliensis* Muell. Arg. Annu. Rep. Inst. Genet. Acad. Sin. 1–19.

San, L.H. and P. Gelebart. 1986. Production of gynogenetic haploids. pp. 305–322. *In*: Vasil, J.K. (ed.). Cell Culture and Somatic Cell Genetics of Plants Academic Press Inc Orlando, Florida.

San-Noeum, L.H. 1976. Haploides *d'Hordeum vulgare* L. par culture in vitro d'ovairies non fecondes. Ann. Amelior Plant 26: 751–754.

Sethi, M., S. Bose, A. Kapoor and N.S. Rangaswamy. 1992. Embryo differentiation in anther culture of mulberry. Ind. J. Exp. Bot. 30: 1146–1148.

Shi-De, L., J. Nemec and B.M. Ning. 1995. Anti-HIV flavanoids from *Morus alba*. Acta Bot. Yunnanica 17: 89–95.

Shoukang, L., J. Dongfeng and A.Q. Jun. 1987. *In vitro* production of haploid plants from mulberry (*Morus*) anther culture. Sci. Sinica (series B) 30: 853–863.

Sibi, M.L., A. Kobaissi and A. Shekafandeh. 2001. Green haploid plants from unpollinated ovary culture in tetraploid wheat (*Triticum durum* Defs.). Euphytica 122: 351–359.

Srivastava, P.S. 1971. *In vitro* growth requirements of mature endosperm of *Ricinus communis*. Curr. Sci. 40: 337–339.

Sternheimer, E.P. 1954. Method of culture and growth of maize endosperm *in vitro*. Bul. Torrey Bot. Club 81: 111–113.

Straus, J. and C.D. La Rue. 1954. Maize endosperm tissue grown *in vitro* I. Culture requirements. Am. J. Bot. 41: 687–694.

Tewary, P.K., S.P. Chakravarty, S.S. Ninha and R.K. Datta. 1994. *In vitro* study on pollen culture in mulberry. Acta Bot. Ind. 22: 186–190.

Thomas, T.D., A.K. Bhatnagar, M.K. Razdan and S.S. Bhojwani. 1999. A reproducible protocol for the production of gynogenic haploids of mulberry, *Morus alba* L. Euphytica 110: 169–173.

Thomas, T.D., A.K. Bhatnagar and S.S. Bhojwani. 2000. Production of triploid plants of mulberry (*Morus alba* L.) by endosperm culture. Plant Cell Rep. 19: 395–399.

Thomas, T.D. 2002. Advances in mulberry tissue culture. J. Plant Biol. 45: 7–21.

Thomas, T.D. 2003. Thidiazuron induced multiple shoot induction and plant regeneration from cotyledonary explants of mulberry. Biol. Plant 46: 529–533.

Thomas, T.D. 2004. *In vitro* modification of sex expression in mulberry (*Morus alba*) by ethrel and silver nitrate. Plant Cell Tiss. Org. Cult. 77: 277–281.

Thomas, T.D. and R. Chaturvedi. 2008. Endosperm culture: a novel method for triploid plant production. Plant Cell Tiss. Org. Cult. 93: 1–14.

Tikader, A. and C.K. Kamble. 2008. Mulberry wild species in India and their use in crop improvement a review. Aust. J. Crop Sci. 2: 64–72.

Tosca, A., R. Pandolfi, S. Citterio, A. Fasoli and S. Sgorbati. 1990. Determination by flow cytometry of the chromosome doubling capacity of colchicine and oryzalin in gynogenetic haploids of *Gerbera*. Plant Cell Rep. 14: 455–458.

Vijayan, K., A. Tikader and J.A.T. da Silva. 2011. Application of tissue culture techniques for propagation and crop improvement in mulberry (*Morus* spp.). Tree Forest Sci. Biotech. 5: 1–13.

Vijayan, K., P.J. Raju, A. Tikader and B. Saratchnadra B. 2014. Biotechnology of mulberry (*Morus* L.)—A review. Emir. J. Food Agric. 26: 472–496.

Wędzony, M., B.P. Forster, I. Żur, E. Golemiec, M. Szechyńska-Hebda, E. Dubas and G. Gotębiowska. 2009. Progress in doubled haploid technology in higher plants. pp. 1–33. *In*: Touraev, A., B.P. Forster and S.M. Jain (eds.). Advances in Haploid Production in Higher Plants Springer, Berlin.

Zafar, M.S., F. Muhammad, I. Javed, M. Akhtar, T. Khaliq, B. Aslam, A. Waheed, R. Yasmin and H. Zafar. 2013. White mulberry (*Morus alba*): A brief phytochemical and pharmacological evaluations account. Int. J. Agric Biol. 15: 612–620.

Zhou, C., H. Yang, H. Tian, Z. Liu and H. Yan. 1986. *In vitro* culture of unpollinated ovaries in *Oryza sativa* L. pp. 165–181. *In*: Hu, H. and H. Yang (eds.). Haploids of Higher Plants *In Vitro*, China Academic Publishers, Beijing- Springer-Verlag, Berlin.

CHAPTER 10

Improvement of Mulberry (*Morus* spp.) by Protoplast Culture

Gulab Khan Rohela,[1,]* *Pawan Shukla,*[1] *Pawan Saini,*[1] *Rajesh Kumar,*[1]
Kunjupillai Vijayan[2] *and Sukhen Roy Chowdhury*[1]

INTRODUCTION

Mulberry (*Morus* spp.) of Moraceae family is a deciduous and highly heterozygous tree species with wide distribution across the continents of Asia, Australia, Europe, North America, South America and some parts of Africa (Ercisli and Orhan 2007, Perez-Gregorio et al. 2011). Various species of mulberry are globally recognized as economically important with major utilization in sericulture industry, food industry, cosmetic industry and pharmaceutical industry (Bao et al. 2016, Shukla et al. 2016, Xiao et al. 2017, Chauhan et al. 2018). Mulberry is also known for its role in environmental safety and protection through eco-restoration, afforestation, carbon sequestration, bioremediation of toxic environmental pollutants like heavy metals, preventing soil erosion and conservation of water by increasing the water holding capacity of soil through their dense network of secondary and tertiary roots, which are long and deeply rooted (Du et al. 2001, Han 2007, Jothimani et al. 2013, Zhou et al. 2017).

Considering the importance of mulberry in economic empowerment, and environmental safety, it is highly exploited across the countries for revenue generation and sustainable development (Jian et al. 2012). Therefore, every nation will be looking forward to have a mulberry variety/cultivar with suitable characters which can be grown laboriously and are able to withstand the global climatic changes over long duration of years. With rapid changes in global atmospheric and local environmental conditions, mulberry plants have to sustain under both biotic and abiotic stresses which in turn adversely impact the leaf quality and yield (Nicky and Peter 2012). In mulberry, the quantity and quality of leaf as well as fruit plays an important role in generation of economy across various industrial sectors (Shivakumar et al. 1995, Shukla et al. 2019). Hence genetic improvement of mulberry is needed in order to cope up with adverse effects of environmental stresses so that leaf and fruit yield are not compromised.

Even though mulberry is genetically better equipped to withstand to some extent the adverse environmental conditions, yet there is a need to genetically improve the mulberry species for better leaf characteristics (a prime

[1] Central Sericultural Research & Training Institute, Central Silk Board, Ministry of Textiles, Government of India, Pampore -192 121, Jammu and Kashmir, INDIA.
[2] Central Silk Board, BTM Layout, Bangalore - 560 068, Karnataka, INDIA.
* Corresponding author: gulab_biotech@yahoo.co.in

requirement of sericulture industry), fruit characteristics, improved fruit production for beverage/pharmaceutical industries, better adaptability to harsh conditions of drought, salinity, heavy metal, other pollutant contaminated sites and also utilizing it as a carbon-sequestering plant for the increased uptake of air pollutants across the metropolitan cities (Lu et al. 2004, Vijayan et al. 2003, Ghosh et al. 2017).

Genetic variability among the mulberry genotypes is an essential requirement in context of the environmental changes which are occurring at a rapid rate in the modern era of human domination (Rohela et al. 2016). This variability can contribute to genetic improvement of mulberry which may be possible either through raising the hybrid plants by conventional breeding or through the advanced biotechnological methods (Vijayan and Tsou 2010b). Conventional breeding has several limitations such as unmatched breeding season and floral asynchrony among the parental lines of mulberry especially in temperate regions (Shabnam et al. 2018), require to raise several generations (F_8-F_{12}) to have homozygous lines with regard to various improved characters. And also it is a time consuming process to have trait specific improvement of mulberry through conventional breeding as it takes several decades to have a trait specific improvement in hybrid plants.

Biotechnology, on the other hand, offers plant improvement with desired traits in short time. Marker assisted selection through molecular breeding, somaclonal variations, haploid production through androgenesis and gynogenesis, *in vitro* mutagenesis and protoplast based *in vitro* techniques can produce improved lines of mulberry (Tikader et al. 2014) (Figure 1). Resistant/tolerant genes offering resistance against various stresses (Hammond Kosack and Jones 1997, Parmar et al. 2017) have been identified in different plant species. These genes can be introduced to mulberry for the genetic improvement coding for biotic stress (viral, fungal and bacterial) and abiotic stress (drought, herbicide, pesticide, salinity and cold) tolerance. Genome editing is another biotechnological method (Chen et al. 2019, Wolter et al. 2019) by which important genes of mulberry can be edited to have quality leaves with increased synthesis of proteins and other useful phytochemicals.

Among the various biotechnological tools, protoplast technology is an advanced and emerging technology which can have several advantages for the improvement of mulberry over the conventional methods of plant breeding (Tikader et al. 2014). Through protoplast fusion, hybrids can be raised even from the sexually incompatible parents, as protoplasts with no cell wall can easily undergo fusion even if they are derived from somatic cells of incompatible species (Rohela et al. 2016). During protoplast fusion, protoplasts isolated from somatic cells of two parental lines are fused and both nuclear as well as cytoplasmic (chloroplast and mitochondria) genes incorporated in heterokaryon (somatic hybrid cell) which can regenerate into somatic hybrid plants. It is highly advantageous to have chloroplast and mitochondrial genes of two parental lines in a single hybrid plant, as the chloroplast and mitochondrial DNA possess genes for herbicide tolerance, disease resistance, etc. (Daneill et al. 1998).

With protoplast fusion technology, it is even possible to obtain inter specific (between species), and inter generic (between genera) hybrids (Figure 2). In mulberry and other plant species, through protoplast fusion, it is possible to produce novel hybrids to overcome the sexual incompatible barriers between parents due to floral asynchrony, incompatibility of pollen, unreceptive nature of stigma, and inter specific and inter generic incompatibility among parents (Eeckhaut et al. 2013). Hence, somatic hybridization can be used as a potential tool for the genetic improvement of mulberry.

Protoplasts are considered as the naked plant cells with plasma membrane as their outer boundary protecting the inner nuclear and cytoplasmic contents. Thus protoplasts have been used as tools for the plant genetic improvement by introduction of genes through plasma membrane using direct gene delivery methods such as electroporation, microinjection, particle bombardment, liposome and calcium phosphate mediated gene transfers (Yusuff et al. 2016). Direct uptake of exogenous DNA by protoplasts can be further stimulated by chemical fusogens like dextran sulphate and PEG (polyethylene glycol) (Shabir et al. 2013).

Transgenosis approach is easy way of DNA transfer into protoplasts and coupled with totipotent nature protoplasts provide a viable option for the genetic improvement of mulberry (Shabir et al. 2013). Thus, both the single protoplasts and fused protoplasts can be cultured under *in vitro* conditions for the regeneration of control and somatic hybrid plantlets of mulberry (Pavan et al. 2000b, Vijayan 2010a).

Several genomic tools have been used for the genetic improvement of mulberry (*Morus* spp.) (Vijayan 2010a). This book chapter reviews the research work specifically related to protoplasts and their role in mulberry improvement.

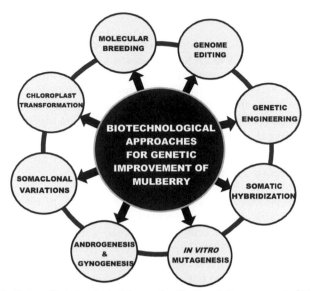

Figure 1. Various Biotechnological Approaches for Genetic Improvement of Mulberry.

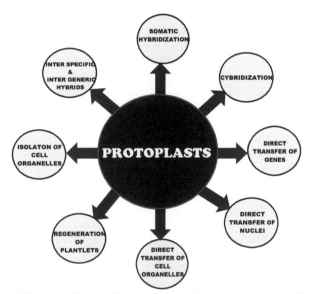

Figure 2. Utilization of Isolated Protoplasts for the Genetic Improvement of Mulberry.

Protoplast Isolation in Mulberry

Protoplasts can be isolated from any part of mulberry, i.e., leaf, stem, root, fruit, pollen grains, *in vitro* culture derived callus or suspension cultures (Pavan et al. 2000b, Salak et al. 2001). For successful isolation of protoplasts, pectin material found in intercellular spaces of tissues has to be removed and the cell wall should also to be removed (Eeckhaut et al. 2013). Protoplasts can be isolated by clearing pectin and cellulosic material present around the plant cells either through mechanical shearing by using sharp blades or through cellulase, macerozyme, pectinase and driselase mediated enzymatic method. Enzymatic method is preferred over mechanical method since there is less damage to the cells coupled with high yields of protoplasts and less osmotic damages.

Plant parts used for Isolation of Protoplasts

Most researchers isolated the protoplasts from mesophyll cells of leaves of the axenic cultures since they are healthy, soft and free from the microbial load (Table 1). Another advantage of mesophyll cells in protoplast isolation is spongy parenchyma cells which are loosely arranged with the packing of pectin material in the intercellular spaces (Salak et al. 2001, Kapur et al. 2003, Pavan 2010, Chakravarthy et al. 2011, Rohela et al. 2018a). Interestingly, some researchers were able to isolate the mulberry protoplasts successfully from cotyledonary cells (Tohjima et al. 1996), leaf derived callus cells (Tewary et al. 1995, Pavan et al. 2000b), root derived callus cells (Yukio et al. 1999), and suspension cells (Wang et al. 1994).

Enzymatic Method of Protoplast Isolation

Among the various reports of protoplast isolation in mulberry, majority of researchers have used enzymatic method. Using this method protoplasts were successfully isolated with high yields from mesophyll cells of leaf, callus cells, and suspension cells (Toshio et al. 1989, Tewary and Lakshmisita 1992, Wei et al. 1994, Wang et al. 1994, Chand et al. 1996, Tewary et al. 1995, Tohjima et al. 1996, Yukio et al. 1999, Salak et al. 2001, Kapur et al. 2003, Pavan 2010, Chakravarthy et al. 2011, Rohela et al. 2018a).

For isolation of mulberry protoplasts the plant parts (leaf, cotyledons and callus) were cut into small sized (0.5–2.0 mm) pieces and incubated in enzymatic solution with varied combinations and concentrations of enzymes having specific pH, temperature and incubation time. Most reports suggested that a combination of cellulase, macerozyme and pectinase to be most suitable medium for the maximum yield of protoplasts (Wei et al. 1994, Wang et al. 1994, Tohjima et al. 1996, Kapur et al. 2003, Rohela et al. 2018a). Yukio et al. (1999) isolated the protoplasts of mulberry by using only driselase enzyme instead of combination of enzymes.

During isolation of protoplasts, it is essential to have an osmoticum (mannitiol or sorbitol) in enzymatic solution, because the isolated protoplasts bound with only plasma membrane has to be protected from osmotic damages such as shrinkage or plasmolysis after the degradation of cell wall by cellulase and pectinase enzymes. Most of the researchers succeeded in isolating maximum number of viable protoplasts by using either 0.4–0.6M or 9–13% of Mannitol as osmoticum. The pH of enzymatic solution for the maximum yield of mulberry protoplasts was found in the range 5.5–5.8 (Chakravarthy et al. 2011, Toshio et al. 1989), temperature in the range 25–28°C (Rohela et al. 2018a, Wang et al. 1994) and duration of enzymatic treatment between 4–16 hrs (Pavan et al. 2000a, Kapur et al. 2003, Chakravarthy et al. 2011).

As optimum conditions and enzymatic combinations were standardized by various researchers, it became possible to isolate protoplasts in viable state for their further utilization in mulberry improvement programmes.

Regeneration from Isolated Protoplasts

The isolated protoplasts of mulberry are regenerated in culture by inducing divisions to form colony, callus, and complete plantlets making explicit the totipotency of protoplasts, i.e., regeneration of protoplasts into complete plantlets. Initially success was achieved in inducing the mulberry protoplasts to divide to form the colony (Katagiri 1989, Kapur et al. 2003). Subsequently, the culture media with desired hormonal combinations and additional supplements were identified for regeneration of complete plantlets from protoplasts (Oka and Ohyama 1985, Wei et al. 1994, Chen et al. 1995, Salak et al. 2001, Pavan et al. 2005). Protoplasts were also utilized for the efficient gene transfer and the expression pattern of transferred genes (Yukio et al. 1999, Kyozuka and Shimamoto 1991). A software system known as Mulberry Protoplast Information System (MPIS) was developed for an easier and better analysis of isolated protoplasts of mulberry to determine the accurate yield of protoplasts (Chakravarthy et al. 2011).

Table 1. Protoplast Isolation from Mulberry (*Morus* spp.) by Enzymatic Method.

#	*Morus* spp.	Plant part used	Enzymatic combination	Osmoticum	Other chemicals/ Buffer	pH of Enzymatic solution	Temperature and Incubation time	Reference
1	*Broussonetia papyrifera*	Leaf (Mesophyll Cells)	4.0% Cellulase & 2.0% Macerozyme	0.6 M Mannitol	2.0% Dithiothreitol, 20.0Mm MES buffer & 6.0 Mm CaCl2	pH of 5.8	--	Toshio et al. 1989
2	*Morus alba*	Leaf (Mesophyll Cells)	1.0% Cellulase, 1.0% Macerozyme & 0.05% Pectinase	9% Mannitol	18% Sucrose	pH of 5.7	4 Hours	Wei et al. 1994
3	*Morus* spp.	Leaf (mesophyll cells), Callus Cells & Suspension Cells	6 U/mL Macerozyme, 100 U/mL Cellulase and 150 U/mL Pectinase,	0.6 M Mannitol	-	pH of 5.8	28°C/6 Hours	Wang et al. 1994
4	*Morus* spp.	Cotyledons &Leaf (Mesophyll Cells)	3.0% Cellulase, 1.0% Macerozyme & 0.1% Pectinase	0.6 M Mannitol	-	-	27°C/12 Hours	Tohjima et al. 1996
5	*Morus alba* cv. Ichinose	Root derived callus	2.2% Driselase	0.6 M Mannitol	-	pH of 5.6	-	Yukio et al. 1999
6	*Broussonetia papyrifera*	Leaf (Mesophyll Cells)	2.0% Cellulase & 0.2% Pectinase	-	18% Sucrose	pH of 5.7	-	Salak et al. 2001
7	*Morus indica* var.K2	Leaf (Mesophyll Cells from axenic shoot cultures)	2.0% Cellulase, 0.1% Macerozyme & 0.1% Pectinase	0.6 M Mannitol	-	-	14–16 Hours	Kapur et al. 2003
8	*Morus indica.* L. cv. K2, S13 & S36	Leaf (Mesophyll Cells from axenic cultures)	2.0% Cellulase & 1% Macerozyme	0.5 M Mannitol	20% Sucrose	pH of 5.5	27°C/9 Hours	Pavan et al. 2000a
10	*Morus indica.* L. cv. Mysore local, S54, M5 and S34.	Leaf (Mesophyll Cells from axenic cultures)	2.0% Cellulase & 1.0% Macerozyme	0.6 M Mannitol	-	pH of 5.5	27°C/4–5 Hours	Chakravarthy et al. 2011
11	*Morus alba*	Callus	2.0% Cellulase & 1.0% Macerozyme	0.6 M Mannitol	-	pH of 5.8	12–13 Hours	Tewary et al. 1995
12	*Morus alba* cv. Ichinose, Chinese white	Leaf (Mesophyll Cells from axenic cultures)	2.0% Cellulase, 0.5% Macerozyme & 0.2% Pectinase	13% Mannitol	20% Sucrose	pH of 5.6	26°C/8 Hours	Rohela et al. 2018a

Cell Division and Culturing of Mulberry Protoplasts

Katagiri (1989) attempted culture of mesophyll derived protoplasts and observed the formation of colony from the divided protoplasts, but could not succeed in regeneration of plantlets. Kapur et al. (2003) tried to culture both free protoplasts and calcium alginate based immobilized protoplasts. They cultured protoplasts on solid as well as on liquid medium and succeeded in cell wall formation in isolated protoplasts only on solid media. Calcium alginate based immobilized protoplasts divided on solid media after 10 days of culture on Murashige and Skoog (MS) medium (Murashige and Skoog 1962) supplemented with 15% of coconut milk in combination with individual cytokinin or auxin either of 1 mg/L of thidiazuron (TDZ), or 6-benzyl aminopurine (BAP), or 2,4-dichlorophenoxy acetic acid (2,4-D). Initially four celled clumps formed from immobilized protoplasts and later colony formation was observed after one month of culture (Kapur et al. 2003).

Complete Plantlet Regeneration from Mulberry Protoplasts

First report about complete regeneration of mulberry plants from mesophyll derived protoplasts were reported by Oka and Ohyama (1985) in paper mulberry (*Broussonetia kazinoki* Sieb.). Initial cell division was observed on MS medium after 20 days of culture and callus formation after 4 weeks of culture. Later from these protoplast derived callus cultures, complete plantlets were regenerated on modified MS medium supplemented with combination of 0.1 mg/L of α-naphthalene acetic acid (NAA) and 1 mg/L of 6-benzyl adenine (BA). Callus induction and regeneration of complete plants of *Morus alba* were obtained from the cultured protoplasts by Wei et al. (1994). Initially callus was induced from protoplasts cultured on modified KM8p liquid medium amended with 1.0 mg/L of NAA, 0.2 mg/L of 2,4-D and 0.5 mg/L of BA after 2 weeks of culture, later regeneration of shoots occurred when callus was transferred on MS media supplemented with 0.5 mg/L of NAA and 0.5 mg/L of BA. Rooting of shoots was carried on 0.5 mg/L of indole-3-butyric acid (IBA) and 0.5 mg/L of BA (Wei et al. 1994).

Chen et al. (1995) cultured mulberry protoplasts and succeeded in regenerating the complete plantlets (Table 2). They observed cell divisions, colony formation and callus formation after 4, 10 and 30 days of culture. Chen et al. (1995) used KM8p media (Kao and Michayluk 1975) for callus induction and MS media (Murashige and Skoog 1962) for regeneration of plantlets, respectively. Regeneration of plantlets from callus was achieved on MS medium supplemented with phytohormones of BA and NAA, and rooting on media supplemented with indole-3-butyric acid (IBA). Similarly combination of KM8p media and MS media was used by Salak et al. (2001) for protoplasts division and regeneration of plantlets. They used MS media supplemented with 0.5–1.0 mg/L concentration of BA or Zeatin for the regeneration of paper mulberry (*Broussonetia papyrifera*) plants from protoplast derived callus.

Regeneration of complete plants of *Morus indica* cv. S36 from the isolated protoplasts was also reported by Pavan et al. (2005). The divisions of protoplasts was observed after 4 days of culture, among the various hormonal combinations tested combination of Zeatin either with 2,4-D or NAA has given maximum number of cell divisions. On these hormonal combinations, division of protoplasts was observed for short period only as cell divisions could not continue for prolonged durations. On media supplemented with 13.5 μM dicamba the cell divisions continued even for prolonged durations resulting in colony formation followed by callus formation. Pavan et al. (2005) achieved success in regeneration of shoots from protoplast derived callus on media supplemented with combination of TDZ and indole-3-acetic acid (IAA); rooting of shoots occurred on IBA supplemented medium. The plant growth regulators (TDZ, BAP, Zeatin, BA, 2,4-D, NAA, IBA & IAA) were also tried in regeneration studies of mulberry from leaf, shoot tip and nodal explants (Vijayan et al. 1998, 2000b, 2011a, Rohela et al. 2018c,d,e).

Table 2. Culturing of Protoplasts and Regeneration of Complete Plantlets of Mulberry

#	Type of Protoplasts	Species & Cultivar Type	Division of Protoplasts	Colony Formation	Callus Formation	Media for Division/colony and Callus formation	Regeneration of Plantlets	Media combination for Regeneration of Shootlets	Media combination for Rooting of Shootlets	References
1	Mesophyll derived protoplasts	*Broussonetia kazinoki* Sieb	√	√	√	MS media + 2,4-D or NAA	√	MS media + 0.1 mg/L NAA + 1.0 mg/L BA	MS media + IBA	Oka and Ohyama 1985
2	Mesophyll derived protoplasts	*Morus alba*	√	√	√	K8P media + 1.0 mg/L of NAA, + 0.2 mg/L of 2,4-D or 0.5 mg/L of BA	√	MS media + 0.5 mg/L NAA + 0.5 mg/L BA	MS media + 0.5 mg/L IBA + 0.5 mg/L BA	Wei et al. 1994
3	Mesophyll derived protoplasts	*Morus alba*	√	√	√	K8P liquid media	√	MS media + NAA + BA	MS media + IBA	Chen et al. 1995
4	Mesophyll derived protoplasts	*Broussonetia papyrifera*	√	√	√	K8P liquid media	√	MS media + 0.5–1.0 mg/L of BA or Zeatin	MS media + IBA	Salak et al. 2001
5	Mesophyll derived protoplasts	*Morus indica* var.K2	√	√	√	MS media + 15% of coconut milk + 1.0 mg/L TDZ or BAP or 2,4-D	×	×	×	Kapur et al. 2003
6	Mesophyll derived protoplasts	*Broussonetia papyrifera*	√	√	√	K8P liquid media	√	MS media + 0.5–1.0 mg/L of BA or Zeatin	MS media + IBA	Salak et al. 2001
7	Mesophyll derived protoplasts	*Morus indica* cv. S36	√	√	√	MS media + Zeatin + 2,4-D orNAA	√	MS media + 4.5 µM TDZ + 1.7 µMIAA	MS media + IBA	Pavan et al. 2005

√ = Achieved; × = Not achieved; K8 media (Kao and Michayluk 1975); MS media (Murashige and Skoog 1962); Thidiazuron (TDZ), 6-benzyl aminopurine (BAP), 2,4-dichlorophenoxy acetic acid (2,4-D),α-naphthalene acetic acid (NAA); 6-benzyl adenine (BA); Indole-3-butyric acid (IBA)and Indole-3-acetic acid (IAA)

Mulberry Protoplast Information System

Chakravarthy et al. (2011) has constructed a comprehensive software system for the analysis of isolated protoplasts known as Mulberry Protoplast Information System (MPIS). One can give the inputs easily regarding the results related to isolated protoplasts and it is user friendly with special graphical user interface (GUI). MPIS provides the information related to yield of protoplasts in a well formatted and convenient manner at a faster rate. Integrating the results of isolated protoplasts with MPIS has an advantage as it gives accurate assessment regarding the yield of protoplasts from the most suitable cultivar/variety, or plant part grown either *in vivo* or *in vitro*.

Gene Transfer in Mulberry Protoplasts

For genetic improvement desired alien genes were transferred into the isolated protoplasts of mulberry. Yukio et al. (1999) optimized the conditions for the successful transfer of genes into mulberry protoplasts by electroporation. During electroporation, an electric current of 330 uFd with pulse length of 3.3–3.7 m/sec was used to transfer β-glucuronidase (GUS) gene into protoplasts. Earlier Kyozuka and Shimamoto (1991) studied the expression pattern of transferred genes and found that transgene expression was affected by factors such as type of mulberry cultivar, explant type used for isolation of protoplasts, nature of gene transferred, promoter in gene constructs, physiological and viability status of protoplasts used in electroporation studies.

Chloroplast Isolation and Plastid Transformation

Compared to intact leaf tissues, protoplasts were regarded as ideal raw materials for chloroplasts isolation. Chloroplast transformation gained an importance over the nuclear transformation since later could lead to gene contamination through pollen grains (Khan and Maliga 1999). Further, chloroplast DNA encodes for important proteins of photosynthesis and it possess genes for antibiotic and insecticide resistance. Chloroplast transformation by overexpression of normal proteins or synthesis of modified proteins could lead to high yield of leaf biomass with increase in quality parameters.

Plastid transformation in mulberry was comprehensively described by Pavan (2010). Chloroplast-transformed mulberry protoplasts carrying *aadA* gene coding for aminoglycoside-3'-adenyltransferase conferred resistance to spectinomycin/streptomycin (Svab and Maliga 1993). The *aadA* gene expresses strongly with 16S rDNA promoter of *Nicotiana tabacum* and rbcL terminator of *Chlamydomonas reinhardtii* (Svab and Maliga 1993). During the selection, the plastid transformed protoplasts could only survive on selection media supplemented with streptomycin/spectinomycin (Umate et al. 2007). Hence, transformed protoplasts only could form callus and later regenerate transgenic plantlets. This approach of plastid transformation using isolated protoplasts will greatly contribute to utilize mulberry plants in '*molecular farming*'. Over expression of pharmaceutically/industrially important phytochemicals identified in mulberry leaf, like 1-deoxynojirimycin (Katsube et al. 2006, Park et al. 2013, Kiran et al. 2019), 2-arylbenzofuran (Chen et al. 2016), γ-amino butyric acid, melatonin (Wang et al. 2016) and cyanidin-3-O-beta-D glucopyranoside (Kang et al. 2005) will be of use in generation of economy.

Other approaches in utilizing mulberry protoplasts for the development of superior mulberry varieties are through somatic hybridization, cybridization and *in vitro* mutagenesis using protoplasts. To combat grasserie disease, which is a serious threat to sericulture industry (Gani et al. 2017), mulberry protoplasts could be used for the uptake of transgene and for expression of antiBmNPV proteins in mulberry plants.

Protoplast Fusion in Mulberry

For successful protoplast fusion and regeneration of somatic hybrids, it is a prerequisite to have good protocol both for fusion and regeneration. Protoplasts are commonly fused either by chemical fusion or by electrofusion. Chemical induced fusion of plant protoplasts is performed by using PEG (polyethylene glycol) or Ca^{+2} ions at higher pH values, whereas electrofusion is carried by using the altered pulses of AC and DC currents for specific durations with the help of an instrument known as electro cell manipulator. In mulberry, both chemical based fusion (Chand et al. 1996) and electric fusion (Pavan 2010) were reported successfully. Between the two types, electro fusion was regarded as more advantageous for mulberry as it is easy to fuse protoplasts, causes less damage to protoplasts upon fusion with highest percentage with 20 fold increase in fusion products (Cheng and Saunder 1995, Duquenne et al. 2007). Electrofusion can be finely tuned by automated regulation of electro cell manipulator through the inputs of specific voltage of currents and duration of electric treatment. After electrofusion, fused protoplasts divide more efficiently and regenerate easily under *in vitro* conditions (Assani et al. 2005).

Chemical Mediated Fusion of Mulberry Protoplasts

Protoplast membrane consists of phospholipids, due to which they have strong negative charges on outer surface. This makes protoplasts to repel from each other in protoplast medium. Under these circumstances chemical fusogens with positive charge such as Ca^{+2} ions or polyethylene glycol (PEG) are mostly used for protoplast fusion (Reinert and Yeoman 1982, Boss et al. 1984). In the present study, fusion of protoplasts of temperate mulberry species was carried out using polyethylene glycol as a fusogen in different concentrations (10–40%). Initially, protoplasts were diluted with osmoticum solution of 13% mannitol in such a way that every 10 µl of solution possesses single protoplast. For fusion, 10µl of protoplast solution of different cultivar types were added independently as distinct drops on a sterile slide and microscopic examination made to confirm the presence of single protoplast in every drop. This was followed by using a sterile micro tip to transfer two drops (one drop each of different mulberry species or cultivar) into sterile eppendorf or falcon tube of lesser volume. To this 1 ml of PEG (10–40%) solution was added and initial time was recorded. The eppendorf tube was shaken well and allowed to settle down for few minutes for microscopic observation. After the duration of 15–18 minutes in PEG (20%) solution, the protoplasts fusion will be observed. Later PEG was removed and fused protoplasts were cultured to induce divisions to form callus and subsequently to regenerate somatic hybrids.

Electrofusion of Mulberry Protoplasts

Pavan (2010) successfully carried out fusion of mulberry protoplasts isolated from *Morus indica* cv. S13 with *Morus indica* cv. S36 by using electric fusion process. Mesophyll derived protoplasts of S13 and S36 were pipetted into electrofusion chamber in 1:1 ratio. Differentiation of protoplasts of two cultivar types were made based on the size of protoplasts, hence, during fusion the small sized protoplasts of S36 type got fused with large sized protoplasts of S13 (Pavan 2010).

During the electric fusion of protoplasts, initially protoplasts got aligned in a straight line, instead of clumps, by using an electric AC current field of 50V/cm for 15 seconds. Passage of AC current was followed by DC current of about 400–500V/cm for short duration of 70 micro seconds which enabled membrane contact leading to fusion of protoplasts. After fusion, an AC field of 10V/cm was given for 05 seconds duration and the fused protoplasts (heterokaryons) identified according to developed protocol under microscopic observation (Sowers 1993, Reichert and Liu 1996).

The inputs regarding altered AC and DC currents for specific time limit were given in electro cell manipulator instrument as a prerequisite before starting the protoplast fusion process. Regeneration of cell walls in fused protoplasts was observed after 2–3 days of culture under dark conditions (Pavan 2010).

Applications of Protoplast Fusion in Mulberry Improvement

Production of Superior Mulberry Varieties via Somatic Hybridization

To produce superior mulberry varieties with desired traits, usually conventional breeding is carried out by cross pollinating two varieties of mostly the same species. But in certain varieties of same species cross pollination is not possible due to several barriers such as timing of flowering, protoandrous or protogynous conditions, pollen incompatibility, unreceptive nature of stigma and harsh environmental conditions (Vijayan et al. 1997, Tikader et al. 2014). To overcome these constraints somatic hybridization will probably be the best answer (Rohela et al. 2018a). Somatic hybridization through protoplast fusion followed by regeneration of somatic hybrids creates new mulberry genotypes (Narayan et al. 1992). Through somatic hybridization, entire genomes from the sexually incompatible parents can be combined and resultant hybrids are expected to be superior in characters (Melchers et al. 1978). Through somatic hybridization, the high biomass production ability of tropical mulberry genotypes can be combined with better leaf quality of temperate mulberry genotypes (Narayan et.al. 1992). Hence, it is possible to transfer the characters of both the parents into a somatic hybrid.

Overcoming the Sexual Incompatibility through Inter Specific Somatic Hybrids

To the best of knowledge and according to reviewed literature there is scanty information about the interspecific hybrids in mulberry. This is mainly due to sexual incompatibility among different species, i.e., *Morus alba*, *M. indica*, *M. lavaegata*, *M. multicaulis*, *M. rubra*, *M. serrata*, *M. nigra*, etc. The incompatibility may be due to failure of pollen grain germination of one species on unreceptive stigma of the other species. Sometimes, the fertilization is possible between different species, but it may not result in production of viable embryos and seeds. With protoplast fusion technology and totipotent nature of fused protoplasts, it is possible to have inter specific hybrids of mulberry. Thus fused protoplasts can be used to overcome the sexual incompatibility barrier of different species of *Morus* for production of inter specific hybrids (Figure 3).

Production of Inter Generic Somatic Hybrids

Through inter generic hybridization, some important characters present in different genera can be brought together in a single hybrid plant, which may be a distant possibility in conventional breeding due to incompatibility barriers, but through protoplast fusion protoplasts of one genus can be fused to protoplasts of another genus resulting in inter generic hybrids. To substantiate this inter generic hybrids of *Brassica* and *Arabidopsis* (Yamagishi et al. 2002), *Triticum* and *Zea* (Xu et al. 2003), *Triticum* and *Avena* (Xiang et al. 2003), *Lathyrus* and *Pisum* (Durieu and Ochatt 2000), *Hyoscyamus* and *Nicotiana* (Zubko et al. 2002), *Atropa* and *Nicotiana* (Yemets et al. 2000), *Citrus and Fortunella* (Takami et al. 2004), *Helianthus* and *Cichorium* (Varotto et al. 2001) and many more combinations have been successfully produced. Moreover, the characteristics like disease resistance (Hu et al. 2002, Du et al. 2009, Xiao et al. 2009), herbicide resistance (Yemets et al. 2000, Han et al. 2009) and salinity tolerance (Yue et al. 2001, Xia et al. 2003) could be transferred into other intergeneric hybrids through protoplast fusion. In mulberry, however, considerable efforts are required to produce intergeneric somatic hybrid plants having improved genetic traits particularly for sustainable growth of various *Morus* species.

Conclusion and Future Prospectives

Most of the research studies related to mulberry protoplasts during 1985–2009 demonstrated the success achieved in protoplast isolation and subsequent regeneration of plantlets. However, while protoplast fusion can be carried out successfully but there has been little success with regard to production of somatic hybrids of mulberry. However, with advancements in application of biotechnological methods in research, there is a huge scope for use of protoplasts in the genetic improvement of mulberry particularly with regard to yield and sustainable development under changing environment.

Protoplasts of mulberry thus have an important role in the genetic improvement of *Morus* spp. by producing new genotypes/varieties with important undermentioned characteristic features:-

- Developing the abiotic and biotic stress resistant mulberry plants through inter specific somatic hybridization.
- Evolving superior mulberry varieties through *in vitro* mutagenesis using protoplast as raw materials.
- Raising of mulberry cybrids with improved characters.
- Utilization of mulberry in production of industrially important compounds, i.e., *molecular farming*.
- Chloroplast transformation for high yield of leaf biomass and growth under lower inputs of fertilizers.

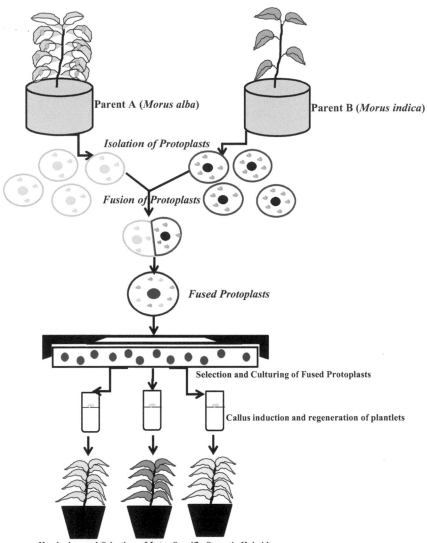

Figure 3. Schematic Representation of Production of Inter Specific Somatic Hybrids in Mulberry.

- Bio fortification of mulberry leaf for quality silk production.
- Development of mulberry varieties with resistance to harsh environmental conditions.
- Expression of antiBmNPV proteins in mulberry leaves to combat grasserie disease in silkworms.
- Application of mulberry for nanoparticle/nanofertilizer synthesis.

References

Assani, A., D. Chabane, R. Haicour, F. Bakry, W. Gerhard and F.W. Barbel. 2005. Protoplast fusion in banana (*Musa* spp.): Comparison of chemical (PEG: Polyethylene glycol) and electrical procedure. Plant Cell, Tissue and Organ Culture 83: 145–151.

Bao, T., Y. Xu, V. Gowd, J. Zhao, J. Xie, W. Liang and W. Chen. 2016. Systematic study on phytochemicals and antioxidant activity of some new and common mulberry cultivars in China. J. Funct. Foods. 25: 537–547. doi: 10.1016/j.jff.2016.07.001.

Boss, W.F., H.D. Grimes and A. Brightman. 1984. Calcium-induced fusion of fusogenic wild carrot protoplasts. Protoplasma 120: 209. https://doi.org/10.1007/BF01282601.

Chakravarthy, R., M. Ashish and M. Raja. 2011.Protoplast information system in *Morus* sps.: An interface for research data on mesophyll protoplasts isolated from mulberry cultivars S34, S54, M5 and mysore-local. In. J. Sft. Com. & Bioi. 2: 21–30.

Chand, P.K., Y. Sahoo, S.K. Patnaik and S.N. Patnaik. 1996. Evaluation of Procedures for Inter-Specific Protoplast Fusion in Mulberry. Ind. J. Seric. 35: 46–49.

Chauhan, S.S., P. Saini, G.K. Rohela, P. Shukla, A.A. Shabnam and M.K. Ghosh. 2018. Present postion and furture strategies of mulberry breeding for North West India: An Overview. Seri Breeders Meet, 20–21 Feb, 2018, pp. 22–33.

Chen, A., Y. Wang and G. Ni. 1995. Culture of mulberry protoplasts and regeneration of plantlets. Acta Sericologica Sinica 21(3): 154–157.

Chen, H., X. He, Y. Liu, J. Li, Q. He, C. Zhang, W. Benjun, Z. Ye and W. Jie. 2016. Extraction, purification and anti-fatigue activity of γ-aminobutyric acid from mulberry (*Morus alba* L.) leaves. Sci. Rep. 6: 18933.

Chen, K., W. Yanpeng, Z. Rui, Z. Huawei and G. Caixia. 2019. CRISPR/Cas genome editing and precision plant breeding in agriculture. Annual Review of Plant Biology 70: 667–697.

Cheng, J. and J.A. Saunders. 1995. Methods in molecular biology. Plant Cell Electroporation and Electrofusion Protocols 55: 181–186.

Daniell, H., R. Datta, S. Varma, S. Gray and S.B. Lee. 1998. Containment of herbicide resistance through genetic engineering of the chloroplast genome. Nature Biotechnology 16(4): 345–348. doi:10.1038/nbt0498-345.

Du, X.Z., X.H. Ge, X.C. Yao, Z.G. Zhao and Z.Y. Li. 2009. Production and cytogenetic characterization of intertribal somatic hybrids between *Brassica napus* and *Isatis indigotica* and backcross progenies. Plant Cell Reports 28(7): 1105–1113.

Du, Z.H., J.F. Liu and G. Liu. 2001. Study on mulberry trees as both water and soil conservation and economy trees. Journal of Guangxi Sericulture 38: 10–12.

Duquenne, B., T. Eeckhaut, S. Werbrouck and J. Van Huylenbroeck. 2007. Effect of enzyme concentrations on protoplast isolation and protoplast culture of *Spathiphyllum* and *Anthurium*. Plant Cell, Tissue and Organ Culture 91(2): 165–173.

Durieu, P. and S.J. Ochatt. 2000. Efficient inter generic fusion of pea (*Pisum sativum* L.) and grass pea (*Lathyrus sativus* L.) protoplasts. Journal of Experimental Botany 51: 1237–1242.

Eeckhaut, T., P.S. Lakshmanan, D. Deryckere, D.B. Erik Van and H. Johan Van. Planta. 2013. Progress in plant protoplast research. Plants. 238: 991. https://doi.org/10.1007/s00425-013-1936-7.

Ercisli, S. and E. Orhan. 2007. Chemical composition of white (*Morus alba*), red (*Morus rubra*) and black (*Morus nigra*) mulberry fruits. Food Chem. 103:1380–1384. doi:10.1016/j.foodchem.2006.10.054.

Gani, M., S. Chouhan, R.K. Gupta, G. Khan, N. Bharath Kumar, P. Saini and M.K. Ghosh. 2017. *Bombyx mori* nucleopolyhedrovirus (BmBPV): its impact on silkworm rearing and management strategies. J. Biol. Control. 31: 189–193. https://doi.org/10.18311/jbc/2017/16269.

Ghosh, A., G. Debnirmalya and C. Tanmay. 2017. Economical and environmental importance of mulberry: A review. International Journal of Plant and Environment 3(2): 51–58.

Hammond Kosack, K.E. and J.D. Jones. 1997. Plant disease resistance genes. Annu Rev. Plant Physiol. Plant Mol. Biol. 48: 575–607.

Han, L., C. Zhou, J. Shi, D. Zhi and G. Xia. 2009. Ginsenoside Rb$_1$ in asymmetric somatic hybrid calli of *Daucus carota* with *Panax quinquefolius*. Plant Cell Reports 28(4): 627–638.

Han, S.Y. 2007. The ecological value of mulberry and its ecological cultivation models for planting mulberry from eastern to western areas in Guizhou. Journal of Guizou Agrical Sciences 35: 140–142.

Hu, Q., S.B. Anderson, C. Dixelius and L.N. Hansen. 2002a. Production of fertile inter generic somatic hybrids between *Brassica napus* and *Sinapis arvensis* for the enrichment of the rapeseed gene pool. Plant Cell Reports 21: 147–152.

Jian, Q., H. Ningjia, W. Yong and X. Zhonghuai. 2012. Ecological issues of mulberry and sustainable development. Journal of Resources and Ecology 3: 330–339.

Jothimani, P., S. Ponmani and R. Sangeetha. 2013. Phytoremediation of heavy metals—A review. International Journal of Research: Studies in Bioscience 1: 17–23.

Kang, T.H. and H.R. Oh, S.M. Jung, J.H. Ryu, M.W. Park, Y.K. Park and S.Y. Kim. 2005. Enhancement of neuroprotection of mulberry leaves (*Morus alba* L.) prepared by the anaerobic treatment against ischemic damage. Biological and Pharmaceutical Bulletin 29 (2):270-274.

Kao, K.N. and M.R. Michayluk. 1975. Nutritional requirement for growth of *Vicia hajaastana* cells and protoplasts at a very low population density in liquid media. Planta 126: 105–110.

Kapur, A., S. Bhatnagar and P. Khurana. 2003. Isolation and culture of leaf protoplasts of Indian mulberry (*Morus indica* var. K-2). Sericologia 43(1): 95–107.

Katagiri, K. 1989. Colony formation in culture of mulberry mesophyll protoplasts. J. Seric. Sci. Jpn. 58: 267–26.

Katsube, T., N. Imawaka, Y. Kawano, Y. Yamazaki, K. Shiwaku and Y. Yamane. 2006 Antioxidant flavonol glycosides in mulberry (*Morus alba* L.) leaves isolated based on LDL antioxidant activity. Food Chemistry 97: 25–31.

Khan, M.S. and P. Maliga. 1999. Fluorescent antibiotic resistance marker for tracking plastid transformation in higher plants. Nat. Biotechnol. 17: 910–915.

Kiran, T., Y.Z. Yuan, M. Andrei, Z. Fang, G.Z. Jian and J.W. Zhao. 2019. 1-Deoxynojirimycin, its potential for management of non-communicable metabolic disorders. Trends in Food Science & Technology 89: 88–99.

Kyozuka, J. and K. Shimamoto. 1991. Electroporation. Kagaku to Seibutsu 29: 734–739.

Lu, M., R.Q. Wang and X.S. Qi. 2004. Reaction of planting tree species on chlorine pollution in the atmosphere. Journal of Shandong University 39: 98–101.

Melchers, G., D.S. Maria and A.H. Anthony. 1978. Somatic hybrid plants of potato and tomato regenerated from fused protoplasts. Carlsberg Research Communications 43(4): 203–218. DOI: 10.1007/BF02906548.

Murashige, T. and F. Skoog. 1962. A revised medium for rapid growth and bio assays with tobacco tissue cultures. Physiologia Plantarum. 15(3): 473–497.

Narayana, P., S.P. Chakraborti, M.K. Raghunath and K. Sengupta. 1992. A new method to improve mulberry, Indian silk 19–20.

Nicky, J.K. and E.U. Peter. 2012. The interaction of plant biotic and abiotic stresses: from genes to the field. Journal of Experimental Botany 63(10): 3523–3543.

Oka, S. and K. Ohyama. 1985. Plant regeneration from leaf mesophyll protoplasts of *Broussonetia kazinoki* Sieb (Paper mulberry). J. Plant Physiol. 119: 455–460.

Park, E., S.M. Lee, J.E. Lee and J.H. Kim. 2013. Anti-inflammatory activity of mulberry leaf extract through inhibition of nf-kb. Journal of Functional Foods 5: 178v186.

Parmar, N., K.H. Singh, D. Sharma, L. Singh, P. Kumar, J. Nanjundan and A.K. Thakur. 2017. Genetic engineering strategies for biotic and abiotic stress tolerance and quality enhancement in horticultural crops: a comprehensive review. 3 Biotech 7(4): 239.

Pavan, U., A.V. Rao, V. Yashodhara, N. Ramaswamy and A. Sadanandam. 2000a. Evaluation of specific parameters in the isolation of protoplasts from mesophyll cells of three mulberry cultivars. Sericologia 40(3): 469–474.

Pavan, U., A.V. Rao, V. Yashodhara, N. Rama Swamy and A. Sadanandam. 2000b. A simple protocol for rapid and efficient isolation of protoplast from callus cultures of mulberry (*Morus indica* L.) cv. S13. Sericologia 40: 647–651.

Pavan, U., R. Venugopal, K. Kiranmayee, T. Jaya Sree and A. Sadanandam. 2005. Plant regeneration of mulberry (*Morus indica*) from mesophyll-derived protoplasts. Plant Cell Tiss Org Cult. 82: 289–293.

Pavan, U. 2010. Mulberry improvements via plastid transformation and tissue culture engineering. P. S. Behav. 5(7): 785–87.

Perez-Gregorio, M.R., J. Regueiro, E. Alonso-Gonzalez, L.M. Pastrana-Castro and J. Simal-Gandara. 2011. Influence of alcoholic fermentation process on antioxidant activity and phenolic levels from mulberries (*Morus nigra* L.). LWT-Food Science and Technology 44: 1793–1801.

Reichert, N.A. and D. Liu. 1996. Protoplast isolation, culture and fusion protocols for kenaf (*Hibiscus cannabinus*) Plant Cell Tiss. Org. Cult. 44: 201–210.

Reinert, J. and M.M. Yeoman. 1982. Protoplast Fusion Induced by Polyethylene Glycol (PEG). pp. 44–45: Plant Cell and Tissue Culture. Springer, Berlin, Heidelberg.

Rohela, G.K., S. Pawan, M.A. Ravindra, G. Mudasir, A.S. Aftab, Y. Srinivasulu and S.P. Sharma 2016. Somatic hybridization as a potential tool for mulberry improvement: A review. Indian Hortic. J. 6(Special): 46–49.

Rohela, G.K., A.A. Shabnam, P. Shukla, A.N. Kamili and M.K.Ghosh. 2018a. Effect of various factors in protoplast isolation from temperate mulberry. International Journal of Advanced Research in Science and Engineering 7 (Special Issue 04): 3056–3065.

Rohela, G.K., P. Jogam, A. A. Shabnam, P. Shukla, S. Abbagani and M.K. Ghosh. 2018c. *In vitro* regeneration and assessment of genetic fidelity of acclimated plantlets by using ISSR markers in PPR-1 (*Morus* sp.): An economically important plant. Scientia Horticulturae 241: 313–321.

Rohela, G.K., A. A.Shabnam, P. Shukla, R. Aurade, M. Gani, S. Yelugu and S.P. Sharma. 2018d. *In vitro* clonal propagation of PPR-1, a superior temperate mulberry variety. Indian Journal of Biotechnology 17: 619–625.

Rohela, G.K., A.A. Shabnam, P. Shukla, A.N. Kamili and M.K. Ghosh. 2018e. Rapid one step protocol for the *in vitro* micro propagation of *Morus multicaulis* var. Goshoerami, an elite mulberry variety of temperate region. Journal of Experimental Biology and Agricultural Sciences 6(6): 936–946.

Salak, P., C. Rommanee, P. Ploenchit and T. Takeshi. 2001. Plant Regeneration from Protoplasts of Paper Mulberry, Thai Variety. Pro of Mulberry and Handmade paper for Rural Dev. 19–24 Mar 2001, Rama Gardens Hotel, Bangkok.

Shabir, A.W., M. Ashraf Bhat, A.S. Kamilli, G.N. Malik, M.R. Mir, Z. Iqba, M.A. Mir, F. Hassan and A. Hamid. 2013. Transgenesis: An efficient tool in mulberry breeding. African Journal of Biotechnology 12(48): 6672–6681. DOI: 10.5897/AJB2013.12090.

Shabnam, A.A., S.S. Chauhan, G.K. Rohela, P. Shukla, P. Saini and M.K. Ghosh. 2018. Mulberry breeding strategies for North and North West India 7(4): 2124–2133.

Shivakumar, G.R., K.V. Anantha Raman, S.B. Magadum and R.K. Datta. 1995. Medicinal values of mulberry. Indian Silk 34: 15v16.

Shukla, P., G.K. Rohela, A.A. Shabnam and S.P. Sharma. 2016. Prospect of cold tolerant genes and its utilization in mulberry improvement. Indian Horticulture Journal 6(Special): 127–129.

Shukla, P., A.R. Ramesha, K.M. Ponnuvel, G.K. Rohela, A.A. Shabnam, M.K. Ghosh and R.K. Mishra. 2019. Selection of suitable reference genes for quantitative real-time PCR gene expression analysis in Mulberry (*Morus alba* L.) under different abiotic stresses. Molecular Biology Reports 6(2): 1809–1817. doi: 10.1007/s11033-019-04631-y.

Sowers, A.E. 1993. Membrane electro fusion: a paradigm for study of membrane fusion mechanisms. Methods Enzymol. 220: 196–211.

Svab, Z. and P. Maliga. 1993. High frequency plastid transformation in tobacco by selection for a chimeric *aadA* gene. Proc. Natl. Acad. Sci. USA. 90: 913–917.

Takami, K., A. Matsumara, M. Yahata, T. Imayama, H. Kunitake and H. Komatsu. 2004. Production of intergeneric somatic hybrids between round kumquat (*Fortunella japonica* Swingle) and Morita navel orange (*Citrus sinensis* Osbeck). Plant Cell Reports 23: 39–45.

Tewary, P.K. and G. Lakshmisita. 1992. Protoplast isolation, purification and culture in Mulberry. Sericologia 32: 651–657.

Tewary, P.K., H.Y. Mohan Ram and N.S. Rangaswamy. 1995. Isolation and culture of mulberry (*Morus alba* L.) protoplasts from callus tissues. Indian Journal of Plant Physiology 38: 114–117.

Tikader, A., B. Saratchandra, K. Vijayan and S. Ravindranath. 2014. A text book on Mulberry germplasm management and utilization. A.P.H. Publishing Corporation, Ansari Road, Darya Ganj, New Delhi.

Tohjima, F.Y., H. Amanouchi, A. Koyama and H. Machii. 1996. Effects of various Enzymes on the efficiency of protoplast isolation from mulberry mesophylls. J. Seric. Sci. Japan. 65: 485–489.

Toshio, O., S. Keizo and T. Hiroshi. 1989. On the protoplast fusion of mulberry and paper-mulberry in polyethylene glycol. J. Seric. Sci. Jpn. 58(2): 145–149.

Umate, P., S. Schwenkart, I. Karbat, C. Dal Bosco, L. Mlcochova, S. Volz, H. Zer, R.G Hermann and O.I. Meurer. 2007. Deletion of PsbM in tobacco alters the QB site properties and the electron flow within photosystem II. J. Biol. Chem. 282: 9758–9767.

Varotto, S., E. Nenz, M. Lucchin and P. Parrini. 2001. Production of asymmetric somatic hybrid plants between *Cichorium intybus* L. and *Helianthus annuus* L. Theoretical and Applied Genetics 102: 950–956.

Vijayan, K., A. Tikader, S.P. Chakraborti, B.N. Roy, S.M.H. Qadri and T. Pavan Kumar. 1997. Studies on stigma receptivity in mulberry (*Morus* spp.) Sericologia 37(2): 34–346.

Vijayan, K., S.P. Chakraborti and B.N. Roy. 1998. Regeneration of plantlets through callus culture in mulberry. Indian Journal of Plant Physiology 3: 310–313.

Vijayan, K., S.P. Chakraborti and B.N. Roy. 2000b. Plant regeneration from leaf explants of mulberry: influence of sugar, genotype and 6-benzyladenine. Ind. J. Exp. Biol. 38: 504–508. www.ncbi.nlm.nih.gov/pubmed/11272418.

Vijayan, K., S.P. Chakraborti and P.D. Ghosh. 2003. *In vitro* screening of mulberry for salinity tolerance. Plant Cell Reports 22: 350–357.

Vijayan, K. 2010a. The emerging role of genomic tools in mulberry (*Morus*) genetic improvement. *Tree Genetics & Genomes,* 6: 613–625.

Vijayan, K. and C.H. Tsou. 2010b. DNA barcoding in plants: taxonomy in a new perspective. Current Science 99(11): 1530–1541.

Vijayan, K., A. Tikader and J.A.T. da Silva. 2011a. Application of tissue culture techniques propagation and crop improvement in mulberry (*Morus* spp.). Tree and Forestry Science and Biotechnology 5(1): 1–13.

Wang, C., L.Y. Yin, X.Y. Shi, H. Xiao, K. Kang, X.Y. Liu, J.C. Zhan and W.D. Huang. 2016. Effect of cultivar, temperature, and environmental conditions on the dynamic change of melatonin in mulberry fruit development and wine fermentation. J. Food Sci. 81: M958–M967.

Wang, Y., C. Peng, Y. Wei-Zheng, Z.W. Chuan and I. Hong. 1994. Optimization of Conditions for Mulberry Protoplast Isolation.

Wei, Z., C. Zhihong, H. Jianqiu, X. Nong and H. Minren. 1994. Plants regenerated from mesophyll protoplasts of white mulberry. Cell Research 4: 183–189.

Wolter, F., S. Patrick and P. Holger. 2019. Plant breeding at the speed of light: the power of CRISPR/Cas to generate directed genetic diversity at multiple sites. BMC Plant Biology 19: Article number: 176 (2019).

Xia, G.M., F.N. Xiang, A.F. Zhou, H. Wang, S.X. He and H.M. Chen. 2003. Asymmetric somatic hybridization between wheat (*Triticum aestivum* L.) and *Agropyrone longatum* (Host) Nevski. Theoretical and Applied Genetics 107: 299–305.

Xiang, F., G. Xia and H. Chen. 2003. Effect of UV dosage on somatic hybridization between common wheat (*Triticum aestivum* L.) and *Avena sativa* L. Plant Science 164: 697–707.

Xiao H.X., Z. Lingli and G Xueling. 2017. Phenolic compounds content and antioxidant activity of mulberry wine during fermentation and aging. Am. J. Food Technology 12: 367–373.

Xiao, W., X. Huang, Q. Gong, X.M. Dai, J.T. Zhao, Y.R. Wei and X.L. Huang. 2009. Somatic hybrids obtained by asymmetric protoplast fusion between *Musa* Silk cv. Guoshanxiang (AAB) and *Musa acuminate* cv. Mas (AA). Plant Cell, Tissue and Organ Culture 97(3): 313–321.

Xu, C.H., G.M. Xia, D.Y. Zhi, F.N. Xiang and H.M. Chen. 2003. Integration of maize nuclear and mitochondrial DNA into the wheat genome through somatic hybridization. Plant Science 165: 1001–1008.

Yamagishi, H., M. Landgren, J. Forsberg and K. Glimelius. 2002. Production of asymmetric hybrids between *Arabidopsis thaliana* and *Brassica napus* utilizing an efficient protoplast culture system. Theoretical and Applied Genetics 104: 959–964.

Yemets, A.I., O.P. Kundel'chuk, A.P. Smertenko, V.G. Solodushko, V.A. Rudas, Y.Y. Gleba and Y.B. Blume. 2000. Transfer of amiprophosmethyl resistance from a *Nicotiana plumbaginifolia* mutant by somatic hybridization. Theoretical and Applied Genetics 100: 847–857.

Yue, W., G.M. Xia, D.Y. Zhi and H.M. Chen. 2001. Transfer of salt tolerance from *Aeleuropus littorulis* Sinensis to wheat (*Triticum aestivum*) via asymmetric somatic hybridization. Plant Science 161: 259–263.

Yukio, S., M. Junji, Y. Kazunori, K. Eiji and F. Toshiharu. 1999. Gene transfer by electroporation into protoplasts isolated from mulberry calli. J. Seric. Sci. Jpn. 68(1): 49–53.

Yusuff, O., M.Y. Rafii, A. Norhani, H. Ghazali, R. Asfaliza, A.R. Harun, M. Gous and U. Magaji. 2016. Principle and application of plant mutagenesis in crop improvement: a review, Biotechnology & Biotechnological Equipment 30: 1–16, DOI: 10.1080/13102818.2015.1087333.

Zhou, M., Q.Q. Chen, J.F. Bi, Y.X. Wang and X.Y. Wu. 2017. Degradation kinetics of cyanidin 3-O-glucoside and cyanidin 3-Orutinoside during hot air and vacuum drying in mulberry (*Morus alba* L.) fruit: A comparative study based on solid food system. Food Chem. 229: 574–579.

Zubko, M.K., E.I. Zubko and Y.Y. Gleba. 2002. Self-fertile cybrids *Nicotiana tabacum-Hyoscyamus aureus* with a nucleo–plastome incompatibility. Theoretical Applied Genetics 105: 822–828.

CHAPTER 11

Applications of Genetic Engineering in Mulberry

Zhao Weiguo,[1,2,]*** *Fang Rong Jun,*[1,2] *Liu Li*[1,2] *and Li Long*[1,2]

INTRODUCTION

Mulberry (*Morus* spp.) is one of the essential tree plants cultivated in Asian countries as it is a vital component of the sericulture industry because the silk-producing insect *Bombyx mori* L. feeds only on the leaves of mulberry. Good mulberry plantations are essential for the success of sericulture as the cost of mulberry leaf covers more than 60% of the cocoon production cost (Das and Krishnaswami 1965). Therefore, much attention has been made on the genetic improvement of mulberry (Pan 2000). Traditional breeding, with its limitations in mulberry, has developed many new mulberry varieties suitable for a wide range of agronomic conditions and cultural practices in China, India, Japan, Russia, and many other Asian countries (Vijayan 2009). Because of the tremendous economic importance attached to mulberry leaf, significant efforts have been made in different countries to develop new varieties with high leaf yield potential, resistance to biotic and abiotic stress, and greater response to fertilizer. With the advancement of mulberry breeding technology since 1980s, the technology of sexual hybridization, mutagenic breeding, polyploid breeding were developed to select and breed a series of good mulberry varieties, which can adapt to different environmental conditions, cultivation technology, rearing technology, and other usages (Zhao et al. 2006). In China, a number of triploid mulberry varieties were also produced and bred, for instance, Shansang 305, Dazhonghua, Jialing 16, and Jialing 20; their yield increased 15% than control, particularly, of a triploid mulberry raised from hybrid seed "Yusang 2" (Pan 2003, Pan 2000).

Although conventional breeding plays an essential role in crop improvement, but it usually entails growing and examining large populations of crops over multiple generations, which is a lengthy and labor-intensive process (Wang et al. 2017). In addition, mulberry production is severely affected by many abiotic and biotic stresses (Kashyap and Sharma 2006, Li et al. 2017). Therefore, developing mulberry varieties that can better adapt to biotic and abiotic stress is a key issue for the plant scientist to increase mulberry yield in many parts of the world today (Parmar et al. 2017). In the past years, some progress was made through classical breeding with regard to these problems (Vijayan 2010). Accumulation of favorable alleles that contribute to stress tolerance/resistance and other agronomic and economic traits in a plant genome is one of the main objectives of plant breeding (Wang et al. 2017). Genes that confer stress tolerance/resistance or other traits can be sourced from local germplasm resources, or through introduced landraces, breeding lines obtained from other breeding programs, wild species, or genera

[1] School of Biology and Technology, Jiangsu University of Science and Technology, Sibaidu, Zhenjiang 212018, Jiangsu Province, China, Emails: frj1971@163.com; touchliu@163.com; seri68@hotmail.com

[2] Sericultural Research Institute, Chinese Academy of Agricultural Sciences, Sibaidu, Zhenjiang 212018, Jiangsu Province, China.

* Corresponding author: wgzsri@126.com

(Varshney et al. 2011). Conventional plant breeding has had little success in improving mulberry plants with these traits due to constraints such as (1) long juvenile period, breeding programs for evolving such plants can involve the professional lifetime of several generations of scientists (2) erosion of naturally occurring genetic variability (3) transfer of undesirable genes along with desirable traits and (4) reproductive obstacles that limit the transfer of favorable alleles from diverse genetic resources (Gómez-Lim et al. 2004, Varshney et al. 2011). Tissue culture technologies, for *in vitro* selection have also been applied to some mulberry plants for crop improvement (Rai et al. 2011).

Genetic engineering, which refers to the direct alteration of an organism's genetic material using biotechnology, possesses several advantages compared with conventional breeding. First, it enables the introduction, removal, modification, or fine-tuning of specific genes of interest with minimal undesired changes to the rest of the crop genome. As a result, crops exhibiting desired agronomic traits can be obtained in fewer generations compared with conventional breeding. Second, genetic engineering allows the interchange of genetic material across species. Thus, the genetic materials that can be exploited for this process are not restricted to the genes available within the species. Third, plant transformation during genetic engineering allows the introduction of new genes into vegetatively propagated crops such as banana (*Musa* sp.) and cassava (Wang et al. 2017). Genetic engineering methods based on the introduction of transgenes and the development of transgenic plants have been successfully adopted to improve crops. Genetic transformation (changing the genetic characteristics of an organism by introducing a specific piece of DNA from another source) began with research on bacteria (Avery et al., 1944; Griffith 1928). Genetic transformation has emerged as a significant tool for genetic improvement of perennial tree species, including mulberry plants. Development of new cultivars is often constrained by their long generation time, high heterozygosity, nucellar embryony and other reproductive barriers (Gómez-Lim et al. 2004, Vijayan 2010, Gambino and Gribaudo 2012, Rai et al. 2014). In the last 20 years, genetic transformation of mulberry crops focused mainly on enhancing disease resistance to viruses, fungi, and bacteria (Fang et al. 2019), increasing tolerance to abiotic stresses, e.g., drought and salt (Lal et al. 2008, Das et al. 2011, Checker et al. 2012, Li et al. 2018), modifying plant growth habit and mulberry quality. There are few reports on field evaluation and commercial application of these transgenic plants (Litz and Padilla 2012). Mulberry being the backbone of the sericulture industry, global climatic changes and scarcity of land as well as water in the near future make it mandatory to develop mulberry varieties suitable for different agro-climatic conditions for sustainable development of the sericulture industry (Khurana and Checker 2011). The present article provides an overview of the progress achieved in recent years on the development of genetically engineered mulberry plants tolerant to biotic/abiotic stress and in improvement of mulberry quality.

Important Ecological and Economic Roles of Mulberry

The global environmental problems of common concern are closely linked up with forests. Mulberry is widely distributed geographically, easily adapted to different agro-climatic conditions, quick in regrowth, amenable to various trimming and pruning methods, propagated both asexually and sexually, and easily hybridized naturally and artificially (Zhao et al. 2005), thus providing an excellent opportunity for the development of mulberry eco-industry. Because of the long-term cultivation of corresponding ecotypes, these mulberry varieties posses different characters that adapt to different ecological conditions, such as nature of branching, leaf size, budding time, resistance to biotic stress from diseases and pests, and abiotic stresses like drought and cold (Zhao et al. 2016).

The economic importance of mulberry is primarily due to its leaf, which is used to feed the silk-producing insect Bombyx mori. In China, India, Japan, Korea, Pakistan, Bangladesh, and in many other Asian countries, sericulture (rearing of silkworms for the production of silk fibers) is one of the significant rural industries that employ a large number of people (Vijayan 2009). Of recent the value of mulberry as multipurpose resource other than silkworm feed has been realized and explored. Mulberry leaf is also used as animal fodder because it is highly nutritious, palatable, and digestible (70–90%) for herbivorous animals like cow, sheep goat, and buffalo (Narayana and Setty 1977, Islam et al. 2014). The protein content in leaves and young stems varies from 15 to 28%, depending on the variety. The mineral content in a mulberry leaf is also reported to be high, and no antinutritional factors or toxic compounds have been identified. Experiments in countries like China, Cuba, Tanzania, and Guatemala showed that mulberry is an excellent forage for livestock, especially during the dry seasons. In addition to leaves, mulberry fruits are rich in anthocyanins, which hold potential use for health benefits and as natural food colorants (Zhang et al. 2016). As the use of synthetic pigments is being discouraged, the demand for natural food colorants like anthocyanins has increased significantly in recent years. Mulberry is also used for landscaping in Asia, Europe, and America (Tipton 1994).

Genetic Transformation Methods for Mulberry: Mode of Transformation and Regeneration

Transgenic plants have been developed through a number of gene delivery methods. The original method devised for the production of the first GM plants in 1983 depended on the use of the natural bacterial vector *Agrobacterium tumefaciens*. Agrobacterium and particle bombardment mediated gene transfer is the most popular methods used for the development of transgenic plants (Rai et al. 2014). Genetic improvement through gene transfer has not been very successful in mulberry due to the recalcitrant nature of the plant. The success of gene transfer is generally decided by the availability of techniques for plant regeneration and the delivery of the desired gene into the genome of the host plant. Successful gene delivery through particle bombardment, electroporation, and *Agrobacterium tumefaciens* cocultivation has been achieved in many woody plants. The initial attempts to deliver genes using *Agrobacterium tumefaciens* and particle bombardment failed to some extent because they only transformed the calli but failed to develop transgenic plants (Machii 1990, Machii et al. 1996, Nouze et al. 2000, Oka and Tewari 2000). Later Bhatnagar and coworkers succeeded in the regeneration of transgenic plants with stable incorporation of GUS genes by particle bombardment and *Agrobacterium tumefaciens* method (Bhatnagar et al. 2002, Bhatnagar et al. 2003). Oka and Tewary (2000) induced hairy roots in *in vitro* grown mulberry hypocotyls using Japanese wild *Agrobacterium rhizogenes* strains. Specific amplification of DNA fragment by PCR showed that portions of the rol genes in the T-DNA core region of the Ri plasmid were integrated into the hairy roots. Yukio et al. (1999) transferred the GUS gene by electroporation into protoplasts isolated from mulberry calli. Histochemical observation showed that successful transient expression of GUS gene was accomplished in 20–30% of protoplasts at the specified pulse. The OC (oryzacystatin) gene was introduced into callus tissues of mulberry by using biolistic method (Wang et al. 2003). Transgenic mulberry plants were also obtained by infection of Agrobacterium tumefaciens T-DNA directly on the meristematic tissues of the axiallary buds of the plants after their apical buds were removed (Lu et al. 2005). The differences in morphological characters of the T_1 and T_0 were identified. Subsequently, the first genetically modified plants overexpressing a barley *HVA1* gene were developed in mulberry using *Agrobacterium*-mediated transformation (Lal et al. 2008). Molecular analysis of the transgenic plants revealed stable integration and expression of the transgene in the transformants. Li et al. (2018) constructed a miR166f binary overexpression vector, pCAMBIA-35S-*GUS*-miR166f, and established a transient transformation system in mulberry. Histochemical β-glucuronidase (GUS) staining of miR166f-overexpression in transient transgenic mulberry leaves showed the best effect and highest *GUS* gene expression compared to the wild-type control plants at day four post-transformation using an optimized concentration of transformation liquid (OD600 = 0.7).

Virus-induced gene silencing (VIGS) is a post-transcriptional gene silencing (PTGS) phenomenon discovered in recent years, and it has been developed to be a new technology of reverse genetics for rapid identification of plant gene function. VIGS has many advantages compared to commonly used biotechnological tools such as gene knockout, gene transformation, and mutant screening (Gould and Kramer 2007). VIGS does not need to build the transgenic plants and has many merits such as a short cycle, simple operation, quick phenotypic expression, and low cost (Pflieger et al. 2013). VIGS has been widely applied in the field of functional gene research in plants. Li et al. (2018) established mulberry VIGS transformation system and provided technical support for the analysis and research of the mulberry functional genes.

In mulberry plants, both plant regeneration under *in vitro* conditions and genetic transformation were found slow to develop, therefore, they are usually considered as 'recalcitrant' for *in vitro* culture and genetic transformation. Lack of an efficient regeneration system in mulberry crops, mainly perennial woody plants, is one of the bottlenecks for applying gene transfer technologies to these plants. In addition, the difficulty in regeneration from elite or mature phase selections is another major problem of mulberry trees. Usually, the success of genetic transformation highly depends on the regeneration pathway adopted by individual species, which is influenced by several factors, e.g., genotype or cultivar, source of the explant, and the degree of determination in the tissue (Litz and Padilla 2012). Therefore, tissue culture conditions for targeted species must be optimized to achieve efficiency in rate of transformation. Organogenesis and somatic embryogenesis are usually used for the regeneration of transgenic plants. Organogenesis is the process by which plant regeneration occurs by organ formation on explants or explant derived callus. Somatic embryogenesis is the formation of bipolar embryos from cells other than gametes or the products of gametic fusion (Peña and Séguin 2001, Rai et al. 2010). Somatic embryogenesis appears to have many advantages over organogenesis that include its potentially high multiplication rates, potential for scale-up via bioreactor, and delivery through synthetic seeds. Therefore, somatic embryogenesis has been emphasized as a suitable target for gene transfer (Merkle and Dean 2000). In mulberry, plantlets were successfully regenerated from callus, leaves, cotyledons, and isolated protoplast (Narayan et al. 1989, Bhatnagar et al. 2002).

Common Tools for DNA Transfer and Detection

Reporter Genes

A reporter gene confers a readily detectable phenotype on a recipient organism and is often attached to a regulatory sequence or a gene of interest to monitor transgenic events or gene expression. Although more than 50 reporter genes have been described, only a few of them, such as GUS, and green fluorescent protein (GPF) have been used extensively for plant research and crop development (Wang et al. 2017).

Selectable marker genes are pivotal to plant genetic transformation and are present in the vector along with the target gene. The commonly used selectable markers are antibiotic- or herbicide-resistance genes, which confer resistance to the toxicity of antibiotic or herbicide. The most widely used antibiotic selectable marker genes are neomycin phosphotransferase II (nptII) and hygromycin phosphotransferase (hpt), both from *Escherichia coli*. The nptII gene, encoding a neomycin phosphotransferase inactivates aminoglycoside antibiotics such as kanamycin, neomycin, and geneticin by phosphorylation. Among these aminoglycoside antibiotics, kanamycin and geneticin are mostly used as selective agents. Although widely used in a diverse range of plant species, kanamycin is ineffective for selecting several gramineae and legumes. For example, the hpt gene is more suitable than nptII for the selection of *Setaria italica* and *Brassica napus* plants. The product of the hpt gene inactivates hygromycin B, an aminocyclitol antibiotic interfering with protein synthesis (Merkle and Dean 2000). The bialaphos resistance (bar) gene is a classic herbicide selectable marker, which inactivates the herbicide phosphinothricin by converting it into the acetylated form (De Block et al. 1987, Akama et al. 1995). To date, the bar gene has been used successfully in many plant genetic transformation systems, including *Arabidopsis thaliana*, rice (Wang et al. 2017).

Gene Promoters

The promoters used in plant genetic transformation largely determine the expression profile of the added gene. Gene promoters are traditionally divided into three categories: constitutive, induced and tissue-specific promoters. The importance of the cauliflower mosaic virus (CaMV) 35S promoter was first highlighted in 1981 (Covey et al. 1981) . This promoter controls the synthesis of the 35S major transcript of the CaMV virus and is the most commonly used constitutive promoter in mulberry genetic transformation.

Target Traits

Biotic and Abiotic stresses due to insect, disease, virus, cold, drought, and salinity are the significant environmental constraints limiting the production and productivity of mulberry since the vegetative as well as reproductive stages of growth and development are impacted. These stresses generally trigger a series of physiological, biochemical and molecular changes in plants, which often result in damage to the cellular machinery (Rai et al. 2011). Conventional plant breeding has not proved that much successful in addressing stress mitigation till date. The reason might be that the traits are controlled by some genes present at a quantitative trait locus (QTL). The genetic engineering technique offers myriads of applications in the improvement of mulberry for target traits such as biotic and abiotic stress tolerance, better adaptability to climate change and the quality enhancement.

Drought Tolerance

Drought is one of the major natural disasters, and with the intensification of the greenhouse effect, dry weather is occurring with greater frequency. Drought results in costly losses to industrial and agricultural production and also causes severe damage to the ecological environment. With the completion of mulberry high-throughput genome sequencing and the construction of small RNA libraries, relevant data are now available for the identification of miRNAs and molecular biology studies in mulberry (He et al. 2012, 2013, Jia et al. 2014, Wu et al. 2015). The dehydration responsive element binding (DREB) transcription factors have been reported to be involved in stress responses. Most studies have focused on *DREB* genes in subgroups A-1 and A-2 in herbaceous plants, but there have been few reports on the functions of DREBs from the A-3–A-6 subgroups and in woody plants. Mulberry transcription factor MnDREB4A protein is localized to the nucleus where it activates transcription. The promoter of *MnDREB4A* can direct prominent expression downstream of the β-glucuronidase (GUS) gene under heat, cold, drought, and salt stress. The *MnDREB4A*-overexpression transgenic tobacco showed the improved phenotypic growth such as greener leaves, longer roots, lower water loss, and senescence rates. Overexpression of *MnDREB4A* in tobacco can significantly enhance tolerance to heat, cold, drought, and salt stresses in transgenic plants (Liu et al. 2015). Transcriptome profiling revealed that miR166f is induced by drought treatment, and its target genes encode homeobox-leucine zipper (HD-Zip) transcription factors and histone arginine demethylase, which are evolved in responses to heat, cold, drought, and salt stresses (Li et al. 2017). The target genes of miR166f - two HD-Zips and one histone arginine demethylase gene (*JMJD6*)-showed markedly lower expression levels in miR166f-overexpression transient transgenic mulberry leaves (Li et al. 2018). Investigation shows that transient transgenic mulberry trees have higher relative water content, free proline content, soluble protein content, superoxide dismutase, peroxidase activities, and lower malondialdehyde content compared to the wild-type control plants, which suggested that overexpression of miR166f in mulberry could enhance tolerance to drought stresses in transient transgenic mulberry. Shaggy-like protein kinase (SK) plays essential roles in plant growth development, signal transduction, abiotic stress, and biotic stress and substance metabolism regulation. Li et al. (2018) successfully silenced *MmSK* gene by VIGS, and after *MmSK* was silenced, the expression of *MmSK* in pTRV2-*MmSK*-VIGS plant (transgenic mulberry) dropped to 34.02% when compared with the negative control. Under drought stress, the soluble protein content, proline content, superoxide dismutase (SOD) and peroxidase (POD) activities in transgenic mulberry decreased in different degree compared to the SK.

In contrast, the accumulation of malondialdehyde (MDA) increased significantly in transgenic mulberry. With the extension of drought stress treatment time, the soluble protein content, proline content, and MDA content gradually increased. The SOD activity and POD activity under drought stress steadily rose to the maximum on the fifth day. It then decreased, which was consistent with the changing trend of *MmSK* gene expression. The results suggested that *MmSK* gene could function as a positive regulator of drought stress in mulberry.

Salinity Tolerance

Salt is the primary factor limiting crop productivity in saline soils and is controlled by various genes. Mulberry (*Morus*) is a deciduous woody tree with moderate tolerance to salinity. The accumulation of salt in the soil causes harmful effects and leads to a reduction in production. Improving the salt tolerance of mulberry is one of the most important objectives of mulberry breeding programs, especially in coastal areas (Vijayan 2009). HNX1 encodes the vacuolar membrane protein and plays the function of vacuolar-type Na/H antiporter. Wang et al. (2006) cloned the mulberry mNHX1 gene, the homologous gene of *Arabidopsis thaliana*. They created *Arabidopsis* overexpression mutant with the agrobacterium-mediated mulberry gene HNX1 transformation. The germination rates of transgenic seeds were significantly higher than that of wild type plants when seeds were planted on MS medium plates containing 100 ~ 200 Mm/L NaCl. Furthermore, the transgenic plants demonstrated stronger salt resistance when these plants were subjected to salt stress. Lal et al. (2008) overexpressed a barley *HVA1* gene in mulberry by using *Agrobacterium*-mediated transformation. Molecular analysis of the transgenic plants revealed stable integration and expression of the transgene in the transformants. Upon testing for the increased stress tolerance ability, the transgenic plants showed better cellular membrane stability (CMS), photosynthetic yield, less photo-oxidative damage, and better water use efficiency as compared to the non-transgenic plants under both salinity and drought stress. Under salinity stress, transgenic plants show a many-fold increase in proline concentration than the non-transgenic plants, and under water deficit conditions proline was accumulated only in the non-transgenic plants. Osmotin and osmotin-like proteins are stress proteins belonging to PR-5 group of proteins induced in several plant species in response to various types of biotic and abiotic stresses. Das et al. (2011) overexpressed the tobacco osmotin in transgenic mulberry plants under the control of a constitutive promoter (CaMV 35S) as well as a stress-inducible rd29A promoter. When subjected to simulated salinity and drought stress conditions transgenic plants showed better cellular membrane stability (CMS) and photosynthetic yield than non-transgenic plants under conditions of both salinity and drought stress. Proline levels appeared very high in transgenic plants with the constitutive promoter related to those with the stress-inducible promoter. An encoding late embryogenesis abundant gene from barley (HVA1) was also introduced into mulberry plants by Agrobacterium-mediated transformation (Checker et al. 2012). Transgenic mulberry with barley Hva1 under a constitutive promoter actin1 was shown to enhance drought and salinity tolerance. Overexpression of barley Hva1 also confers cold tolerance in transgenic mulberry. Further, the barley Hva1 gene under control of a stress-inducible promoter rd29A can effectively negate growth retardation under non-stress conditions and confer stress tolerance in transgenic mulberry.

Insect and Disease Tolerance

Biotic stresses are major constraints limiting the leaf quality and productivity of mulberry. Wang et al. (1998) obtained 14 transgenic plants when mulberry cotyledons were treated with agrobacteria of which plasmids carried cecropin gene. Inoculation with bacteria *Pseudomonas solanacearum* Dowson showed that 14 transgenic plants of 5 lines had a strong resistance to the infection. The OC (oryzacystatin) gene was introduced into callus tissues of mulberry by using a biolistic method (Wang et al. 2003) and the resultant regenerative transgene mulberry plants were gotten through resistant selection in tissue culture. Three Fungal species known to cause severe losses to mulberry cultivation are *Fusarium pallidoroseum*, *Colletotrichum gloeosporioides*, and *Colletotrichum dematium*. Experiment revealed that transgenic plants with osmotin under control of the constitutive promoter had a better resistance to the fungi than those with osmotin under the control of the stress-inducible promoter (Das et al. 2011). Gai et al. (2017) cloned Latex protein MLX56 from mulberry when the HMLX56 gene was ectopically expressed in *Arabidopsis*, the transgenic plants showed enhanced resistance to aphids, the fungal pathogen *Botrytis cinerea* and the bacterial pathogen *Pseudomonas syringae pv. tomato DC3000*. To gain insight into the response of mulberry to phytoplasma-infection, the expression profiles of mRNAs and proteins in mulberry phloem sap were examined (Gai et al. 2018b). A total of 955 unigenes and 136 proteins were found to be differentially expressed between the healthy and infected phloem sap. These differentially expressed mRNAs and proteins are involved in signaling, hormone metabolism, stress responses, etc. Interestingly, it was found that both the mRNA and protein levels of the major latex protein-like 329 (*MuMLPL329*) gene increased in the infected phloem saps. Ectopic expression of *MuMLPL329* in *Arabidopsis* enhances transgenic plant resistance to *Botrytis cinerea*, *Pseudomonas syringae* pv *tomato DC3000* (*Pst.* DC3000) and phytoplasma. LncRNAs have emerged as important regulators in response to biotic and abiotic stresses in plants. A novel lncRNA designated as MuLnc1 was found to be cleaved by mul-miR3954 and produce secondary siRNAs in a 21nt phase in mulberry (Gai et al. 2018a). It was demonstrated that one of the siRNAs produced, si161579, which can silence the expression of the calmodulin-like protein gene CML27 of mulberry (MuCML27). When MuCML27 was heterologously expressed in *Arabidopsis*, the transgenic plants exhibited enhanced resistance to *Botrytis cinerea* and *Pseudomonas syringae pv tomato DC3000*. In addition, the transgenic MuCML27-overexpressing *Arabidopsis* plants are more tolerant to salt and drought stresses. The plant *NPR1* and its homologous genes are important for plant systemic acquired resistance. Xu et al. (2019) cloned the mulberry *NPR1* and *NPR4* genes and developed *MuNPR1* or *MuNPR4* transgenic Arabidopsis producing an early flowering phenotype with the expression of the pathogenesis-related 1a gene as promoter in *MuNPR1* transgenic *Arabidopsis*. The *MuNPR1* transgenic plants showed more resistance to *Pseudomonas syringae* pv. tomato DC3000 (*Pst*. DC3000) than did the wild-type *Arabidopsis*. Moreover, the ectopic expression of *MuNPR1* might lead to enhanced scavenging ability and suppress collase accumulation. In contrast, the *MuNPR4* transgenic *Arabidopsis* were hypersensitive to *Pst*. DC3000 infection. Additionally, transgenic *Arabidopsis* with the ectopic expression of either *MuNPR1* or *MuNPR4* showed sensitivity to salt and drought stresses.

Other Traits

Low-temperature resistance of plants originates from their intrinsic genetic make up as a result of long-term genetic variation and natural selection. During the response process to low temperature, plants can adjust their physiological and biochemical processes like levels of carbohydrates, soluble proteins, free amino acids, and endogenous hormones to alleviate the adverse effects of low-temperature stress. Lu et al. (2008) cloned low-temperature response gene (*Wap25*) from Mongol mulberry, a wild type species of genus *Morus* in cold regions. Then *Wap25* gene was genetically transformed into leaf discs of *Petunia hybrida Vilm* mediated by *Agrobacterium* which then expressed in the plant. The transgenic plant improved cold resistance of *Petunia hybrida Vilm* as a result of genetic transformation.

In order to investigate the possibility of improving mulberry quality by transgenic techniques, expression vector carrying glycinin subunit Alablb gene was constructed and transferred into *Agrobacterium tumefaciens*, and subsequently to leaf disc as well as stem apex of mulberry as receptor materials for obtaining transgenic mulberry plants (Tan et al. 2001).

Engineering Multiple Traits through Gene Stacking in Mulberry

The rapid advances in genome sequencing, bioinformatics, and understanding of metabolic pathways have led to more and more candidate genes becoming available for trait modification or enhancement. Consequently, the focus is shifting from introducing traditional single traits, such as herbicide tolerance or insect resistance, to a combination of multiple traits or complicated metabolic pathway engineering in plants, especially in main crop species. Co-transformation has been used extensively in gene testing and product development in various species (Wang et al. 2017). Both direct (such as biolistics) and indirect (such as *Agrobacterium*-mediated) co-transformation methods have been used. Early co-transformation studies showed the feasibility of stacking genes in one transformation without the need to put all genes on a single construct. The ability to modify multiple genes in complex metabolic pathway through a single transformation process is being strongly pursued. For plant species such as trees, which usually take a longer time to go through the regeneration cycle, co-transformation offers the opportunity to study multiple genes in one transformation event and can significantly accelerate the gene evaluation process (Li et al. 2003). Co-transformation tends to produce unlinked genes which would be useful to remove undesired components such as selectable marker genes by subsequent breeding.

Marker-free Transgenic Technology Application

Generally, the methods of genetic transformation employ selection markers such as antibiotic resistance genes or herbicide tolerance genes for the selection of desirable transgene expression in transformed cells (Akama et al. 1995, Bevan et al. 1983). However, except for the role as a selectable marker, these genes do not have any relevant function inside the plant cell and, thus, they exert an extra burden on the plant genome. Also, the constitutive expression of these genes encoded proteins affects the plant metabolism in a negative way. Further, the use of marker genes, particularly those coding for antibiotic resistance, has been facing a strong criticism and opposition, particularly in edible crops. Developing marker-free plants or finding out suitable alternatives for antibiotic or herbicide tolerance genes has been proposed with the hope of increasing consumers' acceptance for genetically modified crops. A set of new technologies has been developed which involve the elimination of marker genes during transgenic plant development, which under the recent molecular advent are called as 'marker-free transgenic technology' or 'Clean-gene technology'. de Vetten et al. (2003) suggested the use of marker-free gene construct for genetic transformation of potato followed by PCR-based selection of transformed cells for the identification of transformants.

Genome Editing Technology for Genetic Modification in Mulberry Improvement

Genome editing is recent technology that relies on specific engineered endonucleases (EEN) that cleave DNA in a sequence-specific manner due to the presence of a sequence-specific DNA-binding domain. These endonucleases recognize specific DNA sequence and, thus, efficiently and precisely cleave the target genes. The double-strands breaks (DSBs) of DNA result in cellular DNA repair mechanisms, including homology-directed repair (HDR) and non-homologous end-joining breaks (NHEJ), leading to gene modification at the target sites in the genome of plants. Generally, this technology employs three types of engineered endonucleases viz. Zinc finger nucleases (ZFNs), TALENs, and CRISPR/Cas9 for site-specific cleavage. Genome editing in plants has been revolutionized with the development of CRISPR/Cas9 technology. The clustered regularly interspersed short palindromic repeats (CRISPR)-associated protein 9 (CRISPR/Cas9) technology has revolutionized genome editing by overcoming the disadvantages of ZFNs and TALENs due to a high efficiency, low cost involvement, simplicity and versatility (Cardi and Stewart 2016). Recently, this technology has found application in developing resistance against many viruses (Ali et al. 2015). Chandrasekaran et al. (2016) successfully developed virus resistance in cucumber using Cas9/subgenomicRNA (sgRNA technology). Tian et al. (2017) demonstrated the usefulness of genome editing technology, CRIPSR/Cas9, as a powerful tool to effectively create knockout mutations in watermelon. So CRISPR-Cas guides is the future of genetic engineering (Knott and Doudna 2018).

More Application of miRNA and ncRNAs in Mulberry

Previous research on molecular mechanisms underlying plant stress tolerance is largely focused on the functions of protein-coding genes. MicroRNAs (miRNAs) are evolutionarily highly conserved and endogenous single-stranded non-coding small RNAs containing 21–24 nucleotides (nts). miRNAs are distributed widely in plants, animals, and microorganisms, and participate in a series of important biological processes. miRNAs regulate gene expression accurately and effectively at the post-transcriptional level by repressing translation or directly degrading target mRNAs (Bartel 2009). Jia et al. (2014) identified 85 conserved miRNAs belonging to 31 miRNA families and 262 novel miRNAs at 371 loci in mulberry by high-throughput sequencing. Wu et al. (2015) identified 48 conserved miRNAs, and 162 novel miRNAs under drought and saline treatment in mulberry. Gai et al. (2018b) found that miRNA-seq-based profiles of miRNAs in mulberry phloem sap provide insight into the pathogenic mechanisms of mulberry yellow dwarf disease. A large number of non-coding RNAs (ncRNAs) have been identified and found to play important roles in various biological processes, including plant responses to environmental stresses (Zhao et al. 2016). Gai et al. (2017) found that novel lncRNA, MuLnc1, was found to be cleaved by mulmiR3954 and produce secondary siRNAs in a 21-nt "phased" pattern in mulberry. One of the siRNAs produced, si161579, can silence the expression of the MuCML27 gene. The transgenic *Arabidopsis* plants ectopically expressing MuCML27 gene showed more resistant to biotic and abiotic stresses. However, to date, little is known about mulberry miRNAs and their target genes.

Enhancing Mulberry Leaves Quality by Genetic Engineering

The primary purpose of growing mulberry is to get the leaf to feed the silk-producing insect *Bombyx mori* L. as this monophagous insect feeds only on mulberry leaf. The quality of the mulberry leaf greatly influences the growth and development of the silkworm. It is estimated that the fifth instar (mature) silkworm uses nearly 70% of the leaf protein for making the cocoon. Thus, it is essential to feed the silkworm with best quality leaf, i.e., disease-free leaf with adequate moisture, protein, and carbohydrates. In the future, it is necessary to focus on increasing the quality of mulberry leaves by genetic engineering.

References

Akama, K., H. Puchta and B. Hohn. 1995. Efficient Agrobacterium-mediated transformation of Arabidopsis thaliana using the bar gene as selectable marker. Plant Cell Rep. 14: 450–454.

Ali, Z., A. Abulfaraj, A. Idris, S. Ali, M. Tashkandi and M.M Mahfouz. 2015. CRISPR/Cas9-ediated viral interference in plants. Genome Biol. 16: 238.

Avery, O.T., C.M. Macleod and M. McCarty. 1944. Studies on the chemical nature of the substance inducing transformation of pneumococcal types: induction of transformation by a desoxyribonucleic acid fraction isolated from pneumococcus type III. J. Exp. Med. 79(2): 137–158.

Bartel, D.P. 2009. MicroRNAs: Target recognition and regulatory functions. Cell 136(2): 215–33.

Bevan, M.W., R.B. Flavell and M.D. Chilton. 1983. A chimeric antibiotic resistance gene as a selectable marker for plant cell transformation. Nature 304: 184–187.

Bhatnagar, S., A. Kapur and P. Khurana. 2002. Evaluation of parameters for high efficiency gene transfer via particle bombardment in Indian mulberry, *Morus indica* cv. K2. Indian J. Exp. Biol. 40: 1387–1392.

Cardi, T. and C.N. Stewart. 2016. Progress of targeted genome modification approaches in higher plants. Plant Cell Rep. 35: 1401–1416.

Chandrasekaran, J., M. Brumin, D. Wolf, D. Leibman, C. Klap, M. Pearlsman, A. Sherman, T. Arazi and A. Galon. 2016. Development of broad virus resistance in non-transgenic cucumber using CRISPR/Cas9 technology. Mol. Plant Pathol. 17(7): 1140–1153.

Checker, V.G., K.A. Chhibbar and P. Khurana. 2012. Stress-inducible expression of barley *Hva1* gene in transgenic mulberry displays enhanced tolerance against drought, salinity and cold stress. Transgenic Res. 21: 939–957.

Covey, S.N., G.P. Lomonossoff and R. Hull. 1981. Characterization of cauliflower mosaic virus DNA sequences which encode major polyadenylated transcripts. Nucleic Acids Res. 9(24): 6735–6748.

Das, B.C. and S. Krishnaswami. 1965. Some observations on Interspecific hybridization in mulberry. Indian J. Sericulture 4: 1–8.

Das, M., H. Chauhan, A. Chhibbar, H.Q.M. Rizwanul and P. Khurana.2011. High-efficiency transformation and selective tolerance against biotic and abiotic stress in mulberry, *Morus indica* cv. K2, by constitutive and inducible expression of tobacco osmotin. Transgenic Res. 20: 231–246.

De Vetten, N., A.M. Wolters, K. Raemakers, M.I. Van Der, R. Stege and E.A. Heeres. 2003. Transformation method for obtaining marker-free plants of a cross-pollinating and vegetatively propagated crop. Nat. Biotechnol. 21: 439–442.

Fang, L.J., R.L. Qin, Z. Liu, C.R. Liu, Y.P. Gai and X.L. Ji. 2019. Expression and functional analysis of a *PR-1* Gene, *MuPR1*, involved in disease resistance response in mulberry (*Morus multicaulis*). J. Plant Interact. 14(1): 376–385.

Gai, Y.P., Y.N. Zhao, H.N. Zhao, C.Z. Yuan, S.S. Yuan, S. Li, B.S. Zhu and X.L. Ji. 2017. The latex protein MLX56 from mulberry (*Morus multicaulis*) protects plants against insect pests and pathogens. Front. Plant Sci. 8: 1475.

Gai, Y.P., H.N. Zhao, Y.N. Zhao, B.S. Zhu, S.S. Yuan, S. Li, F.Y. Guo and X.L. Ji. 2018a. MiRNA-seq-based profiles of miRNAs in mulberry phloem sap provide insight into the pathogenic mechanisms of mulberry yellow dwarf disease. Scientific Reportse 8: 812.

Gai, Y.P., S.S. Yuan, Y.N. Zhao, H.N. Zhao, H.L. Zhang and X.L. Ji. 2018b. A novel LncRNA, *MuLnc1*, associated with environmental stress in mulberry (*Morus multicaulis*). Front. Plant Sci. 9: 669.

Gambino, G. and I. Gribaudo. 2012. Genetic transformation of fruit trees: current status and remaining challenges. Transgenic Res. 21: 1163–1181.

Gomez-Lim, M.A. and R.E. Litz. 2004. Genetic transformation of perennial tropical fruits. *In Vitro* Cell Dev. Biol. Plant 40: 442–449.

Gould, B. and E.M. Kramer. 2007. Virus-induced gene silencing as a tool for functional analyses in the emerging model plant *Aquilegia* (columbine, Ranunculaceae). Plant Methods 3(1): 6–12.

He, N.J., A.C. Zhao, J. Qin, Q.W. Zeng and Z.H. Xiang. 2012. Mulberry genome project and mulberry industry. Zhongguo Zhong yao za zhi Sci. 38(1): 140–145.

He, N.J., C. Zhang, X.W. Qi et al. 2013. Draft genome sequence of the mulberry tree *Morus notabilis*. Nat. Commun. 4: 2445.

Islam, M.R., M.N. Siddiqui, A. Khatun, M.N.A. Siddiky, M.Z. Rahman, A.B.M.R. Bostami and A.S.M. Selim. 2014. Dietary effect of mulberry leaf (*Morus alba*) meal on growth performance and serum cholesterol level of broiler chickens. SAARC J. Agri. 12: 79–89.

Jia, L., O. Zhang, X. Qi, B. Ma, Z. Xiang and N.J. He. 2014. Identification of the conserved and novel MiRNAs in mulberry by high-throughput sequencing. PloS One 9(8): e104409.

Kashyap, S. and S. Sharma. 2006. *In vitro* selection of salt tolerant *Morus alba* and its field performance with bioinoculants. Hort. Sci. (Prague) 33(2): 77–86.

Khurana, P. and V.G. Checker. 2011. The advent of genomics in mulberry and perspectives for productivity enhancement. Plant Cell Rep. 30: 825–838.

Knott Gavin, J. and A. Doudna Jennifer. 2018. CRISPR-Cas guides the future of genetic engineering. Science 361: 866–869.

Lal, S., V. Gulyani and P. Khurana. 2008. Over expression of *HVA1* gene from barley generates tolerance to salinity and water stress in transgenic mulberry (*Morus indica*). Transgenic Res. 17: 651–663.

Li, L., Y. Zhou, X. Cheng, J. Sun, J.M. Marita, J. Ralph and V.L Chiang. 2003.Combinatorial modification of multiple lignin traits in trees through multigene co-transformation. PNAS USA 10(8): 4939–4944.

Li, R., D. Chen, T. Wang, R. Li, Y. Wan, L. Liu, R. Fang, Y. Wang, F. Hu, L. Li and W. Zhao. 2017. High throughput deep degradome sequencing reveals microRNAs and their targets in response to drought stress in mulberry (*Morus alba*). PLoS One 12(2): e0172883.

Li, R., L. Liu , K. Dominic, T. Wang, T. Fan, F. Hu, Y. Wang, L. Zhang, L. Li and W. Zhao. 2018. Mulberry (*Morus alba*) MmSK gene enhances tolerance to drought stress in transgenic mulberry. Plant Physiol. Biochem. 132: 603–611.

Li, R., T. Fan, T. Wang, K. Dominic, F. Hu, L. Liu, L. Zhang, R. Fang, G. Pan, L. Li and W. Zhao. 2018. Characterization and functional analysis of miR166f in drought stress tolerance in mulberry (*Morus multicaulis*). Mol. Breeding 38: 132.

Litz, R.E. and G. Padilla. 2012. Genetic transformation of fruit trees. pp. 117–153. *In*: Priyadarshan, P.M. and R.J. Schnell (eds.). Genomics of tree crops. Springer, Berlin.

Liu, X.Q., C.Y. Liu, Q. Guo, M. Zhang, B.N. Cao, Z.H. Xiang and A.C. Zhao. 2015. Mulberry transcription factor MnDREB4A confers tolerance to multiple abiotic stresses in transgenic tobacco. PloS One 10(12): e0145619.

Lu, X., C. Lou, F. Shen and M. Kojim. 2005. Molecular Identification and phenotype analysis of transgenic mulberry plant by agrobacterium tumefaciens T-DNA introduction. J. Agri. Biotechnol. 13(2): 157–161.

Lu, X., B. Sun, F. Shen, G. Pan, J. Wang and C. Lou. 2008. Cloning of low-temperature induced gene from *Morus mongolica* C.K. Schn and its transformation into *Petunia hybrida* Vilm. Afr. J. Biotechnol. 7(5): 579–586.

Machii, H. 1990. Leaf disc transformation of mulberry plants (*Morus alba* L.) by agrobacterium Ti plasmid. J. Seric. Sci. Jpn. 59: 105–110.

Machii, M., G.B. Sung, H. Yamnuchi et al. 1996. Transient expression of GUS gene introduced into mulberry plant by particle bombardment. J. Seric. Sci. Jpn. 65(6): 503–506.

Merkle, S.A. and J.F. Dean. 2000. Forest tree biotechnology. Curr. Opin. Biotechnol. 11: 298–302.

Narayan, P., S. Chakraborty and G.S. Rao. 1989. Regeneration of plantlets from the callus of stem segments of mature plants of *Morus alba* L. P. Natl. A Sci. India 55: 469–472.

Narayana, H. and S.V.S. Setty. 1977. Studies on the incorporation of mulberry (*Morus indica*) leaves in layers mash on health, production and egg quality. Indian J. Animal Sci. 47: 212–215.

Nozue, M., W. Cai, L. Li, W. Xu, H. Shioiri, M. Kojima and H. Saito. 2000. Development of a reliable method for Agrobacterium tumefaciens-mediated transformation of mulberry callus. J. Seric. Sci. Jpn. 69: 345–352.

Oka, S. and P.K. Tewary. 2000. Induction of hairy roots from hypocotyls of mulberry (*Morus indica* L.) by Japanese wild strains of Agrobacterium rhizogenes, J. Seri. Sci. Jpn. 69: 13–19.

Pan, Y.L. 2000. Progress and prospect of germplasm resources and breeding of mulberry. Acta Sericologic Sinica 2000, 26(supplement): 1–8.

Pan, Y.L. 2003. Popularization of good mulberry varieties and sericultural development. Acta Sericologic Sinica 1: 1–6.

Parmar, N., H.S. Kunwar, D. Sharma, S. Lal, P. Kumar, J. Nanjundan, Y.J. Khan, D.K. Chauhan and A.K. Thakur. 2017. Genetic engineering strategies for biotic and abiotic stress tolerance and quality enhancement in horticultural crops: a comprehensive review. 3 Biotech 7: 239.

Pena, L. and A. Seguin. 2001. Recent advances in the genetic transformation of trees. Trends Biotechnol. 19: 500–506.

Pflieger, S., R. Mms, S. Blanchet, C. Meziadi and V. Geffroy. 2013. VIGS technology: an attractive tool for functional genomics studies in legumes. Funct. Plant Biol. 40(12): 2–7.

Rai, M.K., P. Asthana, V.S. Jaiswal and U. Jaiswal. 2010. Biotechnological advances in guava (*Psidium guajava* L.): recent developments and prospects for further research. Trees-Struct. Funct. 24: 1–12.

Rai, M.K., R.K. Kalia, R. Singh, M.P. Gangola and A.K. Dhawan. 2011. Developing stress tolerant plants through *in vitro* selection-an overview of the recent progress. Environ. Exp. Bot. 71: 89–98.

Tan, J., C. Lou, H. Wang and M. Zhong. 2001. Transgenic plants via transformation of glycinin gene to mulberry. J. Agri. Biotechnol. 9(4): 400–402.

Tian, S., L. Jiang, Q. Gao, J. Zhang, M. Zong, H. Zhang, Y. Ren, S. Guo, G. Gong, F. Liu and Y. Xu. 2017. Efficient CRISPR/Cas9-based gene knockout in watermelon. Plant Cell Rep. 36: 399–406.

Tipton, J. 1994. Relative drought resistance among selected southwestern landscape plants. J. Arboriculture 20(3): 151–155.

Varshney, R.K., K.C. Bansal, P.K Aggarwal, S.K. Datta and P.G. Craufurd. 2011. Agricultural biotechnology for crop improvement in a variable climate: hope or hype? Trends Plant Sci. 16: 363–371.

Vijayan, K. 2009. Approaches for enhancing salt tolerance in mulberry (*Morus* L.)—A review. Plant Omics J. 2(1): 41–59.

Vijayan, K. 2010. The emerging role of genomic tools in mulberry (*Morus*) genetic improvement. Tree Genet. Genomes 6: 613–625.

Wang, H., C. Lou, Y. Zhang, J. Tan and F. Jiao. 2003. Primarily report on Oryzacystatin gene transferring into mulberry and production of transgenic plants. Acta Sericologica Sinica 29(3): 291–294.

Wang, X., S. Chang, J. Lu, R. Fray, D. Grierson and Y. Han. 2017. Plant genetic engineering and genetically modified crop breeding: history and current status. Front. Agr. Sci. Eng. 4(1): 5–27.

Wang, Y., S. Jia, A. Chen, Y. Tang, Z. Xia and X. Wang. 1998. Transform from mulberry cotyledons by cecropin gene and regeneration of transgeneic plants resistance to *Pseudomonas solanacearum* Dowson. Acta Sericologica Sinica 24(3): 136–140.

Wu, P., S.H. Han, W.G. Zhao, T. Chen, J.C. Zhou and L. Li. 2015. Genome-wide identification of abiotic stress-regulated and novel microRNAs in mulberry leaf. Plant Physiol. Biochem. 95: 75–82.

Xu Yu-Qi, H. Wang, Qin Rong-Li, Fang Li-Jing, Liu Zhuang, Yuan Shuo-Shuo, GaiYing-Ping and Jia Xian-Ling. 2019. Characterization of NPR1 and NPR4 genes from mulberry (*Morus multicaulis*) and their roles in development and stress resistance. Physiologia Plantarum 167(3): 302–316.

Yukio, S., M. Junji, Y. Kazunori, K. Eiji and F. Toshiharu. 1999. Gene transfer by electroporation into protoplasts isolated from mulberry calli. J. Seri. Sci. Jpn. 68: 49–53.

Zhao, J., Q. He, G. Chen, L. Wang and B. Jin. 2016. Regulation of noncoding RNAs in heat stress responses of plants. Front. Plant Sci. 7: 1213.

Zhao, W., X. Miao, Y. Pan and Y. Huang. 2005. Isolation and characterization of microsatellite loci from the mulberry, *Morus* L. Plant Sci. 168(2): 519–525.

Zhao, W., X. Miao, B. Zang, L. Zhang, Y. Pan Yi-Le and Y. Huang. 2006. Construction of fingerprinting and genetic diversity of mulberry cultivars in China by ISSR markers. Acta Genetica Sinica 33(9): 851–860.

Zhang, J., T. Yang, L. Li, Y. Zhou, Y. Pang, L. Liu, J. Fang, Q. Zhao, L. Li and W. Zhao. 2016. Association analysis of fruit traits in mulberry species (*Morus* L.). J. Hort. Sci. Biotechnol. 91(6): 645–655.

Section C

Sustainable Growth of Mulberry in Context of Climate Change

CHAPTER 12

Impact of Climate Change on the Sustainable Growth of *Morus alba*

Tsvetelina Nikolova

INTRODUCTION

Cocoon is related to the establishment of an effective forage base for the production of mulberry leaves. Of great importance is the variety of mulberry and its good adaptability to climatic conditions meat.

The climate change, has immense impact on the growth and development of plants including mulberry (*Morus*) species. As a result plant species adapt to new environmental conditions for survival and sustainability.

The increased amount of carbon dioxide in the atmosphere, heat stress, longer droughts and more intense rainfalls associated with global warming continue to affect not only our daily lives, but they also affect the quantity and quality of the plant yield considering the fact that mulberry leaves serve as food to silk worm (*Bombyx mori*) which contributes to raw silk production in sericulture industry.

An obligatory condition for solving issues related to agrotechnology of mulberry and its commercial exploitation is the knowledge of the mulberry tree requirements to various environmental factors since the external environment plays an important role in the individual plant development.

Under optimal environment conditions, the genetic capabilities for high productivity are naturally developed, but adverse environmental factors can inhibit the development productivity of *Morus alba* L. As a tree species, it prefers culture conditions best available in tropical and subtropical conditions. In these regions there is enough heat, light, air humidity, and soil rich in humus. As a consequence the growing period can even get extended. Climate change poses a great risk as sometimes yield losses due to unfavorable weather conditions may reach 45–50% even 70%, or more, when combined with several adverse factors (late frost, sharp frost, drought).

In countries around the world researchers are trying to develop technologies that can adapt agriculture to climate change. A wide range of adaptive action plans such as farm technology, improved farming management, and policy tools are followed in order to appropriate genetic improvement of plant cultivations. Agro-forestry could be a way to reduce the climate risks in rural plant development and of other areas.

The changes in the climate have affected the mulberry tree in recent years as plant stress caused by unfavorable conditions is viewed as a major physiological issue in *Morus* sps. Further determining the conditions for development of leaves in mulberry varieties under different climatic conditions is required.

The studies of Munns (2002) and Ohashi (2009) are mainly aimed at revealing mechanisms for acclimatization and adaptation of plants to various adverse environmental conditions. For the proper cultivation of *Morus alba* var. Chemista L., soil, light, air, water, temperature and humidity are the most important external ecological factors. The

University of Forestry-Sofia, Faculty of Agronomy, Department of "Perennials and gardening" 10, Kliment Ochridsky Blvd. 1756 Sofia, Bulgaria.
Email: c.alipieva@abv.bg

requirements for these factors are closely linked to the age of the mulberry trees and the applied agrotechnology. Young mulberry plants have higher requirements for optimal rearing and respond to deviations mostly to humidity and temperature. The role of these ecological factors in cultivation of *Morus alba* plants/trees is examined in this chapter.

Requirements to Temperature

Morus alba L. has different requirements of this factor dependant on different ages and phenological phases. The air temperature determines the distribution range of the species. *Morus alba* L. is referred to as warm-climatic plant species, but is also found in the more northern areas. Temperature is extremely important for the proper growth and development of these mulberry plants. At about 25°C seed germination begins in order to develop young plants. After the winter rest the cold temperature at 6 to 8°C starts the soft proliferation. When the air temperature reaches 10–12°C the buds burst for appearance of the first leaves.

Cultivated varieties of mulberries are more demanding to heat than the related wild forms. During the growing season the mulberry is, however, very sensitive to the sharp drops in temperature. Even at optimum temperature, it stops developing if temperature drops to form low frosts. In such cases, the plants mobilize their stocks and develop their spare buds which give rise to new shoots after about 14 days. It is important to note that in case the plants are properly quenched, their sensitivity to winter frosts is not so high.

The duration of the vegetation period is also influenced by the air temperature. At temperature of 30–32°C, the most intense growth occurs in mulberry plants. Lowering the temperature to 12°C or increasing to 40°C slows the growth of trees.

Over the last decades, climate change notably global warming has had different impacts on the natural systems of all continents. Climate change often subjects plants to ecological stress by limiting their growth and development, yield formation, and production quality. The lack of sufficient moisture during the growing season of crops is an increasingly challenging task and requires the creation of stress-tolerant varieties which experience achieving high enough quality yields.

The climate change over recent years has adversely affected the mulberry tree. Stress caused by unfavorable conditions is a major issue in mulberry plant physiology. As a warm-climatic plant, *Morus alba* is better tolerant to low temperatures, but it also has a negative impact. Stress factors like high temperature and resulting droughts cause the quality of the mulberry leaves being impaired. At high temperatures there are disturbances in the process of assimilation and thus wilting of leaves takes place. High temperature combined with low air and low soil humidity is particularly damaging to mulberry plants, and with longer exposure it causes drying and dying of plants.

Climate impacts causing temperature variations affect the development and productivity of different species and varieties of mulberries in a different way. A number of authors have identified species and varietal differences in production biology of mulberry through the phenological development phases and changes in the growing period (Ho et al. 1985, Kumar 1990, Bari et al. 1990).

During the winter, plants are in deep tranquility. In this state, all parts of the plant have the best cold resistance. Trees tolerate lower temperatures better when their wood is richer in organic matter. The cold resistance of the different mulberry varieties of *M. alba* is not the same as apart from the organic matter they also depend on their hardening before they come to rest. The degree of damage is dependant on the rate of lowering of the air temperature. Larger tissue damage is caused by a rapid drop in temperature. Wind speed also affects the freezing of trees. During the winter mowing of the mulberry plants is required as the low temperatures can partially or totally damage the annual branches and even the entire plant. The cold-resistance of the mulberry species depends on agro-practices, genetic make up, and nature of variety.

Spring frosts, an another climatic aspect, greatly affects the development and productivity of the mulberry. Frost can occur on different parts of the plant depending on the sensitivity of their tissues. This being the reason for damaging the young leaves and mulberry blossoms. The level of resistance to frosts varies with different varieties (Sakai and Larcher 1987, Wisniewski et al. 2003). Late spring frosts equally lead to the death of buds, young leaves, and shoots. Sleeping or dormant buds do not develop at all, and more advanced the phase of development of tree's generative organs, the greater is the damage. If cold resistance reduces in the middle of winter, plant growth is still lost until the spring.

Brassinosteroids are the group of polyhydroxy steroids that regulate a wide range of physiological responses in plants. In addition to regulating growth, they also play a significant role in protecting plants from stress. Brassinosteroids have been successfully used to increase the resistance of plants to drought (Nilovskaya et al. 2001, Kagale et al. 2007, Vayner et al. 2014). This proves that salicylate and succinic acid are able to induce plant resistance to abiotic stress factors along with the effect of these acids on plant antioxidant protection systems (Kolupaev et al. 2011).

Moisture Requirements

Water plays a major role in the transport of mineral substances, synthesis of organic substances, and especially in the cooling of leaves through the process of transpiration under the heat of the sun. For the proper development of the mulberry trees and the flow of the physiological processes in them it is of great importance that the sufficient quantity of water and the air is retained in the soil to maintain normal turgor, succulence and leanness of the leaves for the production of quality leaf, and overall growth of the mulberry tree. Need for an optimal amount of water particularly required in the spring and during the summer period of development. For proper plant growth, the soil moisture content should be over 70%. Water scarcity leads to wilting with premature fall of the leaves.

The second major problem is the temperature stress to plants that are increasingly exposed if there is the lack of sufficient moisture. Lack of moisture forces plants to limit water losses by reducing leaf surface or to develop their root system deep into soil which is a disadvantage for mulberry. Water deficiency causes breakdown in cell structure, biologically active substances, and completely inhibits physiological processes. Plant protection mechanisms at different levels such as osmotic self-regulation and expression of proteins with protective and regulatory functions are equally hampered (Vasilev et al. 2010).

For the quality and nutritional value of mulberry leaves, the amount of water is very important especially for trees growing during the summer period, because during drought the water content decreases. It has been found that for the synthesis of 1 kg of dry matter it consumes 700–800 kg of water. Usually the leaves consume 70–75%, shoots 60–63%, and the roots 55–56% water. On the water content of the leaves depends the extent of their eating by the silkworm larvae. That is why maintenance of the juicy leaves is necessary for which supply of sufficient water should be available to trees.

Both soil and air humidity influence the growth and development of the mulberry tree. Optimal soil moisture content in soil within the range of 75–80% is referred to as marginal soil moisture content (MFN). Accordingly, the lower limit for normal growth of mulberry trees in gray forest soils is 70% of MFN, and for meadow soils 75% of MFN, respectively.

Water and nutrients in the soil-plant system show many interactions that have multiple mechanisms of self-regulation. A balanced fertilizer system increases the efficiency of water use and helps crops achieve optimal productivity under limited humidity conditions. The efficient management of nutrients for soil enrichment is one major measure that has proven to increase water efficiency by 10–25%. Studies have shown that soil water content in agrochemical plant nutrition management measures is an important factor in strengthening plant adaptation to extreme weather conditions during the growing season (Miroshnychenko et al. 2017).

A combination of low soil nutrients, air moisture content (humidity) and a low temperature in the spring adversely affect the mulberry tree growth. Under such conditions there is abundant flowering that quickly exhausts the plant, the foliage develops slowly thereby having low nutritional value for silkworm larvae.

Requirements of Light

Mulberry refers to relatively light-loving species. Crucial to the formation of organic matter and the quality of the leaf mass under the light regime.

In excess light, photosynthesis processes are disturbed, resulting in damage to leaves and branches; under intense light the leaves are thicker and fleshy but slightly smaller. Shaded leaves at the bottom of the crown have higher water and ash content but are significantly poorer in proteins and soluble sugars. When feeding the silkworm larvae with leaves from well-lit parts of the crown, a higher vigor in larvae is observed and heavier and swollen cocoons obtained. In the temperate belt, under normal sunlight, 8 to 10 thousand shoots are produced per one decare of plantation. Under insufficient light the quality of the leaves and particularly the seeds deteriorates. Shrubs of the mulberry tree have smaller amounts of carbohydrates and proteins, and leaf growth is significantly slower.

There is a greater requirement for light during the period of fruit growing and flowering season of mulberry plants. All types of mulberry trees need to be grown in such a manner that individual trees and parts of them have sufficient access to light.

Planting density is of utmost importance and it has to be optimal. Equally of great importance is the type of crown formation that allows good sunlight which is essential for proper ripening of the shoots and there should be 5-fold leaf surface in the area of the plantation. For the normal course of photosynthesis, the spectrum of the sun's rays is also affected by plantation density. Particularly advantageous are the yellow-red rays, which are best absorbed by the chlorophyll and contribute to the production of more asymylates in photosynthesis. The lighting regime is thus influenced by a number of factors, such as site exposure, planting density, layouts, and tree formation in plantations.

Requirements for the Soil

Morus alba trees have a high adaptability to the soil. The choice of planting site is directly related to quality of soil as it affects the yield of mulberry leaf. The development of the root system depends on the physico-mechanical properties and the chemical composition of the soil. On soil depends nutrient and water requirements for growth and sustainability through entire life of trees.

Morus alba trees are best grown in soil-rich humus, with good aeration. In these conditions the trees develop a powerful root system, which is important to provide the plants sufficient water and nutrients. Lighter black soils are suitable for planting mulberry. In the valley of the rivers, the mulberry tree can be grown successfully on alluvial-meadow soils. To obtain high yields of mulberry leaves it is very important to add organic and mineral fertilizers to soil. Young mulberry trees, however, absorb relatively small amounts of nutrients from the soil.

Mulberry trees used for silkworm rearing develop a large amount of plant biomass since they extract significant amounts of nutrients from the soil.

Moisture and soil temperature are key determinants of nutrient availability and root growth, but climate change would affect plant nutrition. According to Jungk (2002), the availability of nutrients depends on soil properties and presence of nutrients around the root surface. So the absorption of nutrients by the plant reflects a number of physiological processes that determine the transport of nutrients within the roots and aspects such as quantum of chemicals and nutrients in the soil. The effects of drought on plant growth and soil water content have been extensively studied in recent years (Munns 2002, Valliyodan and Nguyen 2006, Ashraf and Foolad 2007, Silva et al. 2009, Molden et al. 2010, Kano et al. 2011, Stikić et al. 2015). However, studies related to nutrient intake and its consequences on physiological processes are scarce. Warming up and reduction in rainfall during the spring/summer season affects C, N and P balance in soils and the presence of basic microelements for plants.

Studies have shown that new methods for assessing changes in mode and application of fertilizer under changed climate conditions, and increased impact of external stressors (drought, temperature rise, etc.), can lead to innovative crop management solutions under stressful conditions. Thus, adequate fertilizer can simultaneously improve soil quality and resilience against the negative impacts of climate change in order to sustain agricultural productivity. Balancing and optimal use of minerals in fertilizers is, therefore, of utmost importance in adapting plants to abiotic stress (International Food Policy Research Institute, USA 1995).

Nutrients as a Factor for Crop Resistance under Extreme Weather Conditions

Agrochemical plant nutrition measures comprise an important factor for adaptation to extreme weather conditions during the growth period of mulberry trees. Properly nutrient soils support the rapid expansion of the leaf area, thus increasing the efficiency of evapotranspiration. Therefore, high levels of nutrients in the soil also have an impact on maintenance of water efficiency even under adverse weather conditions (Schmidhalter and Studer 1998); so water and nutrients interact to provide high yields (Prihar et al. 1985, Aggarwal 2000). Adequate nutrients of fertilizer may also increase plant tolerance to drought (Lahiri 1980, Wang et al. 2011).

Role of Nutrients

Availability of nutrients ensure that plant roots can absorb enough of each nutrient throughout the growing season. Thus each nutrient has important role in growth and development of plants including the trees of *Morus alba*.

Nitrogen - (N) is important component of amino acids that are building block of proteins and other biologically active substances that take part in many important plant processes. Without nitrogen the formation and growth of roots, shoots and leaves is impossible. In the absence or lack of assimilative nitrogen, growth is suppressed, leaf yield sharply decreases, the color of the leaves changes to pale green. These leaves are poor in protein.

Phosphorus - (P) essentially contributes to the sustainable development of plants by regulating energy in cells while accelerating the overall growth of plants with increasing yields. It influences cold-resistance and root system development. Also impacts processes of photosynthesis and respiration.

Potassium - (K) plays a vital role in the physiological and biochemical functions of plants. It improves the resistance of stem and leaf diseases, activates various enzymes in plants, contributes to the development of a waxy layer of the epidermis (which prevents disease and loss of water), regulates cell pressure to prevent dilation, increases the size and improves the taste of the fruit, helps to synthesize amino acids (building blocks for proteins), chlorophyll, and transfer of starch and sugars from the leaves to the roots. Potassium also affects the cold response of mulberry trees. Enriching potassium in fertilizer increases not only the leaf surface but also affects metabolism and regulation of transpiration.

Further to above essential nutrients the trace elements like the manganese, zinc, copper, boron, magnesium, and iron are equally important for the normal development of the mulberry tree.

Manganese (Mn) plays a role in many of the vital processes of growth. It usually works on the enzyme systems of plants involved in carbohydrate break down, nitrogen metabolism and many other processes.

Zinc - (Zn) is important for the transformation of carbohydrates into other organic compounds and in regulation of sugar consumption in plants. It forms part of the enzyme systems that regulate plant growth and the synthesis of chlorophyll.

Copper - (Cu) is very important in the reproductive stage of plant growth and plays an indirect role in the production of chlorophyll. It is essential in maintenance of normal metabolism, yields increase, and resistance to unfavorable climatic conditions.

Boron - (B) required for enzyme synthesis associated with increased cellular activity on maturity that provide greater number of colors, fruits, richer crops and better quality. It also affects the nitrogen and carbohydrate metabolism, and osmosis in the cells. Its rapid absorption favors the onset of fruit buds, seeds, and flowering.

Magnesium - (Mg) is involved in the synthesis of chlorophyll and through it the process of carbon dioxide assimilation. As an activator of enzymes, magnesium plays an important role in the carbohydrate metabolism thereby facilitating the formation as well as decomposition of glucose and phosphorus compounds. It also plays a role in the synthesis of nucleoproteins, fats and ascorbic acid. Magnesium exhibits a general regulating effect in the absorption of nutrients and water through the roots, thus affecting the condition of the root cell colloid systems. A particular role therefore is attributed to magnesium in the absorption and transport of phosphates in plants.

Zinc - (Zn) is important for the transformation of carbohydrates and the regulation of sugar consumption in plants. It forms part of the enzyme systems that regulate plant growth. Zinc is also a necessary element in the synthesis of chlorophyll. It is apparent that each nutrient element is effective if the other elements are in sufficient quantity. This requires the presence of proportionate fertilizer and nutritional equilibrium in the soil.

Requirements for Air

All parts of the mulberry tree are in close interaction with the air. From the air, trees receive the oxygen and carbon dioxide needed in photosynthesis and respiration. Humidity in air affects temperature and light. Of all aerodisperse systems, dust contributes to air pollution. Strong air dust hinders respiratory and photosynthesis processes making mulberry leaves unsuitable and, in some cases, even harmful to feeding the silkworm larvae.

The effects of sulfur dioxide in air on plants mostly depend on the impact of the mechanism associated with the functioning of the stomata located on the surface of the leaves. This mechanism regulates the processes related to transpiration, exchange of carbon dioxide and oxygen during respiration and photosynthesis. Impacts that change the functioning of this mechanism subsequently result in damage to plants. The stomata of the leaf responds to low concentrations of sulfur dioxide in the air, which can be used as an early warning signal ("signal") before damage to the plant itself. Concentrations of sulfur dioxide in the range of 30–280 $\mu g/m^3$ usually cause stomatal opening in young plants (especially in combination with high relative air humidity). Concentrations of sulfur dioxide higher than 280 $\mu g/m^3$ alone or in combination at low concentrations of SO_2, NO_2, and O_3 cause stomata closure.

Sulfur dioxide directly and indirectly affects photosynthesis. The direct effect is expressed by the shift in the production of adenosine triphosphate and nicotinamide-dinucleotide phosphate and their transport on photosynthetic membrane. On the other hand, Sulfur dioxide suppresses important respiratory enzymes involved in the Calvin cycle, such as ribulose-1,5 diphosphate carboxylase (the CO_2-digesting enzyme), phosphoenol-pyruvate carboxylase and maltodehydrogenase, thus indirectly effecting photosynthesis. It has also been established that sulfur dioxide retains "cellular charge" by affecting the transport of leaf sugars to plant roots at sulfur dioxide concentrations lower than photosynthesis inhibitors. Minor concentrations also affect the vital activity of the microorganisms interacting with the plants.

All the internal damage to the plants caused by sulfur dioxide, after a different period of time, is manifested by external visible lesions such as: leaf chlorosis, growth retardation and changes in the root system (Stancheva 2000).

Mulberry is an anemophilous plant and in this connection the air movement has a positive influence on its pollination. It is also useful for removing the excess water after precipitation from the surface of the leaves with a view to their use for feeding the silkworm larvae. Further the strong wind tilts the rooted trees, especially the young plantations. It may also cause mechanical damage to leaves, especially in large leaf size varieties. Dry winds inculate major damage leading to wilting and drying of leaves.

Concluding Remarks

The present study focusses on role of various ecological factors in sustainable growth of mulberry (*Morus alba*) under changed climatic conditions. It is apparent that impact of these factors is synergistic and with proper manipulation of these factors and effective agro-chemical management *Morus alba* can sustain *in situ* in context of climate change.

References

Aggarwal, P.K. 2000. Application of system simulation for understanding and increasing yield potential of wheat and rice. Ph.D. thesis, Wageningen, The Netherlands.

Ashraf, M. and M.R. Foolad. 2007. Roles of glycine betaine and proline in improving plant abiotic stress resistance. Environmental and Experimental Botany 59: 206–216.

Bari, M., M. Quyym and S. Ahmed. 1990. Stability of lield in some selected genotypes of mulberry Indian J. Seric 29(1): 88–92.

Borsh, T. and S. Kuznetsova. 2008. Effective use of herbicides in maize inbred lines. Corn and Sorghum, N2, 7–9 (Ru).

Grimalovski, A.M. 1995. Herbicide reaction. Corn and Sorgum, N2, p. 14–15 (Ru).

Ho, D., M. Ono, T. Ichihashi and S. Kobayashi. 1985. Study on productivity of mulberry varieties. Bll. Seric. Exp. Station 127: 71–86.

Jungk, A.O. 2002. Dynamics of nutrient movement at the soil-root interface. pp. 587–616. *In*: Waisel, Y., A. Eshel and U. Kafkafi (eds.). Plant Roots: The Hidden Half, 3rd Edn. Marcel Dekker, New York.

Kagale, S., U.K. Divi, J.E. Krochko, W.A. Keller and P. Krishna. 2007. Brassinosteroids confers tolerance in *Arabidopsis thaliana* and *Brassica napus* to a range of abiotic stresses. Planta. 225(2): 353–364.

Kano, M., Y. Inukai, H. Kitano and A. Yamauchi. 2011. Root plasticity as the key root trait for adaptation to various intensities of drought stress in rice. Plant and Soil 342: 117–128.

Krasteva, L., K. Uzundzhalieva and R. Ruseva. 2012. Plant Genetic Resources as a part of the biodiversity.—Agroznanje 13(1): 5–14.

Kumar, N. 1990. Studies on productive biology of mulberry. Sericologia 30(4): 477–487.

Lahiri, M. 1980. Increased drought tolerance of mycorrhizal onion plants caused by improved phosphorus nutrition. Vol. 154, No. 5 (1982), pp. 407–413 (7 pages).

Masilamani, S., A. Reddy, A. Sarkar, B. Srinivas and C. Kamble. 2000. Heritability and genetic advance of quantitative traits in mulberry (*Morus* spp.) Indian J. Seric. 39: 16–20.

Malakanova, V.P. et al. 2013. Effectiveness of chemical measures to protect parental forms of corn from weeds in Krasnodar. Corn and sorgum, N1, 25–28 (Ru).

Miroshnychenko, M., Y. Hladkikh, I. Pachev, B. Nosko, A. Khristenko and A. Revt. 2017. Plants nutrient management as a factor enhancing the resilience agricultural production in conditions of extreme weather events. Journal of Mountain Agriculture on the Balkans 20(4): 1–21.

Molden, D., T. Oweis, P. Steduto, P. Bindraban, M.A. Hanjra and J. Kijne. 2010. Improving agricultural water productivity: between optimism and caution. Agricultural Water Management, Comprehensive Assessment of Water Management in Agriculture 97(4): 528–535.

Munns, R. 2002. Comparative physiology of salt and water stress. Plant, Cell and Environ. 25: 239–250.

Nilovskaya, N.T., N.V. Ostapenko and I.I. Seregina. 2001. Effect of epibrassinolide on the productivity and drought resistance of spring wheat. Agrokhimiya 2: 46–50 (Ru).

Ohashi, Y., N. Nakyama, H. Saneoka, P.K. Mohapata and K. Fujita. 2009. Differences in the responses of stem diameter and pod thickness to drought stress during the grain filling stage in soybean plants. Acta Physiol. Plant 1: 271–277.

Petkov, Z. 2000. About new methods for mulberry vegetative resources investigation, Bulgarian Journal of Agricultural Science 6: 481–482.

Prihar et al. 1985. Nitrogen fertilization of wheat under limited water supplies. Fertilizer research 8: 1–8.

Rahman, M., S. Doss, S. Debnath, S. Roy Chowdhari, P. Ghosh and A. Sarkar. 2006. Genetic variability and correlation studies of leaf characters in some mulberry (*Morus* spp.) germplasm accessions. Indian J. Genet. 66(4): 359–360.

Sakai, A. and W. Larcher. 1987. Frost Survival of Plants. Springer, Berlin, 321.

Schmidhalter and Studer. 1998. Water use efficiency as influenced by plant mineral nutrition. 1st Sino-German Workshop Impact of Plant Nutrition on Sustainable Agricultural Production.

Silva, E.C., R.M. Nogueira, F.H.A. Vale, N.F. Melo and F.P. Araujo. 2009. Water relations and organic solutes production in four umbu tree (*Spondias tuberosa*) genotypes under intermittent drought. Brazilian Journal of Plant Physiology 21: 43–53.

Stancheva. 2000. Ecological bases of agriculture. p. 346.

Stefanovic, L. 1980. Directions in researching to establish the impact of herbicides on corn inbred lines. Status and problems of corn production in prbreport of the Scientific and Technical Conference(Bg).

Stikić, R., Z. Jovanović, M. Marjanović and S. Đorđević. 2015. The effect of drought on water regime and growth of Quinoa (Chenopodium quinoa Willd.). Ratar. Povrt. 52: 2.

Stoimenova, I. et al. 2008. *Sorghum halepensis* L./Pers. and the fight against it. Soil Science, Agrochemistry and Ecology, 42(1): 38–41 (Bg).

Valliyodan, B. and H.T. Nguyen. 2006. Understanding regulatory networks and engineering for enhanced drought tolerance in plants.

Vasilev, A., Zl. Zlatev, M. Berova and N. Stoeva. 2010. Tolerantnost na rasteniiata kam zasuchavane i visoki temperature – fiziologchni mehanizmi i podhodi zapodbor na tolerantni genotipove. – Agrarni nauki, g. II, vol 4: 59–68.

Vayner, A.A., M.M. Miroshnychenko, Yu.E. Kolupaev, T.O. Yastreb, 21 V.A. Khripach and Yu.A. Sotnikov. 2014. The influence of 24-epibrassinolide on heat resistance and productivity of millet (*Panicum miliaceum*) plants. Bull. Kharkiv Nat. Agr. Univ. Ser. Biol. Is. 3(33): 35–42.

Wang et al. 2011. Plant responses to drought, salinity and extreme temperatures: Towards genetic engineering for stress tolerance. Planta 218(1): 1–14.

Wisniewski, M., C. Bassett and L. Gusta. 2003. An overview of cold hardiness in woody plants: seeing the forest through the trees. Hort Science 38: 952–959.

CHAPTER 13

Stress Tolerant Traits in Mulberry (*Morus* spp.) Resilient to Climate Change

An Update on Its Genetic Improvement

Tanmoy Sarkar,[1,]* *S Gandhi Doss,*[1] *Vankadara Sivaprasad*[2] *and Ravindra Singh Teotia*[1]

INTRODUCTION

Mulberry (*Morus* spp.; family *Moraceae*) is a cross-pollinated, highly heterozygous, woody perennial tree or shrub with deep root system, rapid growth pattern and high biomass production ability (Dhanyalakshmi and Nataraja 2018). Mulberry grows worldwide under diverse climatic conditions between latitudes of 50° N and 10° S and from a sea level to as high as 4000 m which includes China, India, Bangladesh, Pakistan, Thailand, Brazil, Uzbekistan and other countries (Sarkar et al. 2017). Mulberry leaves are the sole food source for the monophagous mulberry silkworm (*Bombyx mori* L.) (Vijayan et al. 2018). Further, mulberry leaf is rich in proteins, carbohydrates and moisture content, which are the primary criteria for silkworm rearing. The sustainability of silk industry is very much dependent on production and continuous supply of high-quality and sufficient quantity of mulberry leaves for silkworm rearing. In most of the mulberry growing countries especially in China and India, focuses have been paid towards its genetic improvement for enhancing the quantum of foliage yield and quality. In terms of mulberry silk production, China and India stand first and second, respectively; while the later occupies the first place for highest consumption in the world (Sarkar et al. 2017, 2018). However, there is ever increasing demand for high-quality mulberry silk not only in India, but also across the world. Hence, there is need to bridge the gap between the supply and demand for high quality mulberry silk. Further, one of the important determining factors for enhancing the production of quality silk is to produce sufficient quantity and high-quality mulberry leaves for silkworm rearing by cultivation of high-yielding mulberry varieties that can adapt to different climatic conditions and drastically changing environment. Further, developing mulberry cultivars for better nitrogen use efficiency (NUE) could be one of the important thrust areas in sericultural research to decrease nitrogen (N) fertilizer application under climate change scenario (Nguyen et al. 2016).

In India, mostly mulberry plants are cultivated under irrigation system for commercial silkworm rearing. Mulberry is grown as low bush plantation with a short cultivation period of 70 d (Rukmangada et al. 2018a). Mulberry leaf is repeatedly collected (4–5 times in a yr) by leaf plucking or shoot pruning for silkworm rearing (Rukmangada et al. 2018b). However, high bush mulberry plantation with drip irrigation system could minimize the water budget under water scarcity. Various abiotic stresses can cause 50–60 percentage of its foliage yield

[1] Central Sericultural Research & Training Institute (CSRTI), Mysuru, Karnataka-570008, India.
[2] Central Sericultural Research & Training Institute (CSRTI), Berhampore, West Bengal-742 101, India.
* Corresponding author: tanmoy.dgr@gmail.com

reduction (Sajeevan et al. 2017). Climate change scenario raises the CO_2 level and atmospheric temperature across the globe. It is estimated that CO_2 level has steadily increased during the last decade at the rate of 1.9 ppm per yr. Further, the global average surface temperature is likely to rise by 1.8–4.0°C in this Century; and by up to 6.4°C in the worst case situation. Climate change scenario has catastrophic effects, by causing frequent drought and salinity stress, not only on agricultural productivity but also on sericulture (Senapati et al. 2013, Kumar and Gautam 2014). Further it is predicted that the rise in sea level due to climate change will increase the risk of permanent or seasonal saline water intrusion from the sea into the ground water, and rivers, leading to enhancing of salinity affected areas across the world (Senapati et al. 2013, Kumar and Gautam 2014). Most importantly, climate change also lead to erratic and scanty rainfall resulting in frequent droughts. However, under climate change scenario, increased concentrations of CO_2 may boost up crop productivity by stimulating photosynthesis, only where soil moisture is not a constraint.

Further, ozone layer depletion has turned into a burning environmental issue across the globe and the thinning of ozone layer results in increased UV-B radiation (more than 10 percentage higher than ambient) at the Earth's surface. Enhanced UV-B radiation has profound negative effects on morphology, physiology, biomass allocation and leaf structure of female mulberry plants than do males (Chen et al. 2016a).

In view of these stress related facts, it is imperative to put concerted efforts to develop mulberry variety which can provide sustained high leaf yields under imminent stress conditions. This review discusses the physio-biochemical traits that have been introgressed in mulberry for enhancing their adaptive capacity to climate change scenario with a reasonable improvement in quality and quantity of foliage yields. The application of various genetic improvement strategies, such as traditional and molecular breeding methods, have also been discussed to develop climate resilient mulberry varieties. Further this review sheds a light on prospects of genetic engineering and genomic tools in accelerating mulberry trait improvement programmes.

Diversity in Mulberry (*Morus* spp.) and the Rationale behind its Trait Improvement

Approximately 68 mulberry species have been reported across the world, and majority of them are available in Asian Countries like China (24 spp.), Japan (19 spp.) and India (4 spp.) (Tikader and Vijayan 2017, Sarkar et al. 2018).* Only a few mulberry species such as *M. alba, M. indica, M. bombycis, M. latifolia* and *M. multicaulis* are widely cultivated for foliage yield to feed silkworms, while M. *laevigata, M. rubra, M. alba* and *M. nigra* are grown for their edible fruits. Further, *M. serrata* is grown for its timber (Sarkar et al. 2018). Approx 4800 mulberry germplasm resources are being maintained in China, Japan, South Korea, France, Italy and Bulgaria, while 1291 accessions are being conserved at Central Sericultural Germplasm Resources Centre, India (Pinto et al. 2018, Sarkar et al. 2018). *Morus* species also shows extensive variation at ploidy level, ranging from diploid ($2n = 2x = 28$) to decosoploid ($2n = 22x = 308$) including a naturally available haploid mulberry species *M. notabilis* (Yile and Oshigane 1998).

Diploid ($2n = 2x = 28$) and triploid ($2n = 3x = 42$) mulberry is cultivated in 2.20 lakhs ha of land in India and the sustainability of Indian silk Industry is dependent on production and continuous supply of high-quality mulberry leaves for silkworm rearing (Sarkar et al. 2018). As stated earlier, to meet ever-increasing demand for quality silk at domestic and international markets, horizontal expansion of mulberry cultivation in traditional agricultural land may not be possible due to competition with other food and cash crops. Hence, it is imperative to utilize marginal, problematic soils and non-vegetational areas, affected by various abiotic stresses such as alkalinity, acidity, water-logging, low-temperature, salinity and soil moisture deficit for mulberry cultivation. Abiotic stress tolerance in plant system is a polygenic trait and results from expression and interaction among several inducible genes through signal transduction pathways (Sarkar 2014, Liu et al. 2015, Sarkar et al. 2014, 2016). Further, in mulberry plant system, there are cross-talks and overlapping mechanisms for abiotic stress responses. To combat environmental constraints/abiotic stress, a range of stress adaptation traits such as efficient water conservation, wider and deeper

---

* See also Chapter 2.

root system, improved photosynthetic yield, water use efficiency (WUE), NUE, maintenance of macromolecule and ionic homeostasis, protection of biomolecules and cellular tolerance need to be introgressed in mulberry.

Improvement of Climate Resilient Traits in Mulberry through Traditional Breeding

The traditional breeding methods, relying on agronomic and morpho-physiological based phenotyping, have been successfully utilized (Table 1) in developing drought, alkalinity, salinity, water logging and frost-tolerant productive mulberry varieties (Susheelamma et al. 1992, Mogili et al. 2008, Vijayan et al. 2009, Doss et al. 2012, Ghosh et al. 2012).

Mulberry germplasm collections including exotic and indigenous genotypes, and wild spp. have been characterized and conserved. They show extensive sexual polymorphism (dioecious, monoecious, bisexual), genotypic and phenotypic variations for important agronomic traits like rapid growth and biomass production, plant insect/microbe interaction, biotic/abiotic stress tolerance and the traits associated with nutritional and medicinal properties (Sarkar et al. 2017, Tikader and Vijayan 2017, Dhanyalakshmi and Nataraja 2018). A portion of the germplasm collections has been utilized in developing stress tolerant mulberry varieties. Physiological trait (Carbon isotope discrimination-Δ^{13}C) based breeding strategies helped in development of introgression lines of mulberry with high WUE and root traits from a cross Dudia white x MS3 (Mishra 2014). These introgression lines could be used for further foliage yield evaluation trials, as a pre-breeding resource in breeding programmes, and validation of quantitative trait loci (QTLs) associated with drought adaptive traits in mulberry (Thinnaluri 2015, Sarkar et al. 2017).

Although a significant number of mulberry varieties used for commercial cultivation are diploid in nature, high yielding triploid mulberry varieties are being used for silkworm rearing in various parts of India. Triploid mulberry genotype is developed through hybridization between selected parents (natural/induced tetraploid x diploid) or it is selected from natural triploid populations. Majority of the triploid mulberry plants are superior

Table 1. List of climate resilient mulberry varieties developed through conventional breeding methods in India.

Variety name	Trait/Type of cultivation area/soil type	Pedigree	Method of breeding	Ploidy	Institute/Country
AR12	Alkaline soil	S41(4x) x C776 (2x)	Polyploid breeding	Triploid	CSRTI, Mysuru, India
MSG2	Soil moisture stress prone area	BR4 x S13	Controlled hybridization	Diploid	CSRTI, Mysuru, India
AGB8	Soil moisture stress prone area	(Sujanpur5 x Phillipines) x (Kanva2 x Black cherry)	Controlled hybridization	Diploid	CSRTI, Mysuru, India
S13	High temperature/Soil moisture stress prone area	Selection from OPH of Kanva2 (variety)	Selection	Diploid	CSRTI, Mysuru, India
S34	Soil moisture stress prone area	S30 x Ber. C776	Controlled hybridization	Diploid	CSRTI, Mysuru, India
PPR1	Temperate region (frost tolerance)	Goshoerami x Chinese white	Controlled hybridization	Diploid	CSRTI, Pampore, India
C776	Saline soils	*M. multicaulis* x Black cherry	Controlled hybridization	Diploid	CSRTI, Berhampore, India
C2028	Water logged condition	China white x S1532	Controlled hybridization	Diploid	CSRTI, Berhampore, India
BC259	Temperate region	Back crossing of F_1 hybrid (Matigara x Kosen) with Kosen twice	Backcross breeding	Diploid	CSRTI, Berhampore, India

Source: Rohela et al. 2018, Vijayan et al. 2018

to the diploids, especially in terms of growth pattern, leaf yield and nutritive qualities. Desired traits of triploid mulberry could be maintained through clonal propagation without any loss of foliage yield potential due to their sterile nature and non-ability to set seeds (Vijayan et al. 2011; Chapter 2). Further, physiological traits exhibited in mulberry germplasm were C_4 like photosynthetic 'trait' system (an energy efficient mechanism that contributes to improved CO_2 assimilation rate), enhanced water, nitrogen and radiation use efficiency; increased tolerance to high temperature and drought stresses. Although, mulberry is C_3 perennial tree, an earlier report showed that there is C_4 photosynthetic 'syndrome' in triploids; these cultivars showed greater amount of CO_2 fixation in the form of four-carbon compound than do the diploid high-yielding cultivars. Furthermore, triploid cultivars showed more phosphoenolpyruvate carboxylase (PEPC) activity than ribulose 1, 5-bisphosphate carboxylase-oxygenase (Rubisco) (Das et al. 1997).

Hybridization-based traditional breeding strategies that are utilized for genetic improvement of mulberry, include a well-defined set of activities such as collection of exotic/indigenous germplasm within the country or across the globe, evaluation of germplasm collections for a trait of interest, selection of suitable parents for use in hybridization based on traits pyramiding, crossing selected parents (diploid x diploid, tetraploid x diploid) and backcrossing F_1 hybrids with recurrent parent. This follows collection of F_1 seeds from a cross/backcross, raising of seedlings in the nursery beds, planting of F_1 seedlings from a single, multiple cross, backcross populations to

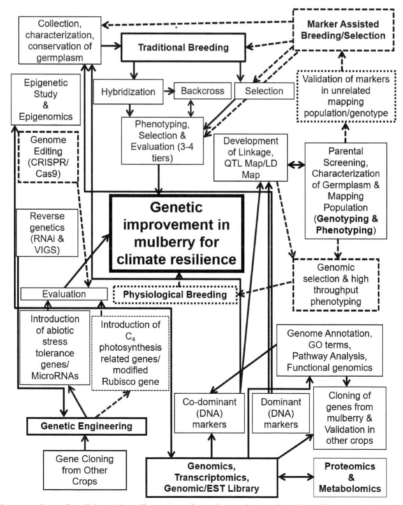

Figure 1. Integrated approaches of traditional breeding, genomic tools, marker assisted breeding, genetic engineering for improving climate resilient traits in mulberry.

progeny-row-trail (PRT) plots for initial screening, and selection of F_1 seedlings/backcross progenies which are subjected to evaluation based on traits of agronomic interest and foliage-yield (Tikader and Kamble 2008; Table 1, Figure 1). Other important breeding strategies utilized in genetic improvement of mulberry include natural selection methods viz. clonal selection, selection from natural variability, selection of hybrids from open pollinated cross/polycross, etc. (Vijayan 2010, Vijayan et al. 2018). Breeding, selection and evaluation of mulberry hybrids, conservation and characterization of germplasm collections follows two experimental designs (Tikader and Vijayan 2017) such as (i) randomized block design (RBD) and (ii) augmented randomized complete block design (ARCBD).

Mulberry tree is highly heterozygous and characterized by long juvenile period. Traditional breeding methods mostly rely on the production of F_1 hybrids and subsequent propagation of hybrids by vegetative means (e.g., stem/shoot cuttings and bud grafting) in order to raise saplings for field-cultivation. Mulberry F_1 hybrids with desirable agronomic traits, once identified through PRT, are taken to advanced evaluation process like primary-yield-trials (PYT) for analyzing their potential agronomic traits, foliage yield, biochemical indices and silkworm feeding qualities (Vijayan 2010, Vijayan et al. 2018). In this trial, 20–40, or even more hybrids are evaluated with 12–16 plants per hybrid or more number of plants per hybrid per replication. The 5–10 percentage of hybrids which performed better than others in PYT are selected for detailed assessment in final-yield-trials (FYT) using 3–5 replications and 25–49 plants per hybrid per replication. In this trial, the plants of each hybrid are subjected to thorough assessment for tolerance to abiotic stress, leaf yield, leaf quality, susceptibility to pest and diseases, rooting percentage and sprouting ability, response to agronomic practices, and silkworm feeding qualities, etc. Thereafter, the best performing hybrid progeny from FYT is selected and multiplied through vegetative propagation for further testing at different regions/hotspots through multi-location-trials (MLT). Usually 8–9 hybrids are used for MLTs and the hybrid(s) which shows consistently better performance than the check varieties in terms of improved leaf yield, quality and stress tolerance across the seasons and locations are selected. These selected mulberry hybrids/triploids are further tested in All India Coordinated Experimental Trials for Mulberry (AICEM) to evaluate their performance in different agro-climatic conditions in India for a period of four years (Vijayan 2010, Vijayan et al. 2018). Then the best performing hybrid progeny is recommended and released as productive and stress tolerant/climate resilient variety for commercial cultivation for silkworm rearing. It takes 15 yr to develop a mulberry variety for commercial cultivation (Vijayan et al. 2018, Pinto et al. 2018). Lately, to reduce the duration of breeding and evaluation programmes, FYT has been clubbed with MLT. Thus FYT is conducted at various locations/hotspots, across the seasons and years. The mulberry varieties which are developed through selection (clone/open-pollinated hybrids/natural triploids) and backcross breeding strategies are also passed through the stated three-tiers/four-tiers rigorous evaluation process.

Further, physiological breeding approaches, through high throughput phenotyping platforms and capacity building, in combination with genomic selection in mulberry, could open up a new dimension to introgress the traits associated with climate resilience, or stress tolerance, and foliage yield (Reynolds and Langridge 2016).

Genome, Transcriptome, Degradome Analysis for Identification of Novel genes and Marker Resources for Improvement of Climate Resilient Traits in Mulberry

Realizing the economic importance of mulberry and the needs for its trait improvement under climate change scenario, the draft genome (357 Mb) of mulberry species (*M. notabilis*) has been generated, the genome annotation revealed 128 Mb repetitive sequences with 27,085 high confidence protein coding loci in tandem with complete gene structure (He et al. 2013). The RNA-seq data (21 Gb) from various tissues of *M. notabilis* generated 5833 unique ESTs (expressed sequence tag) and five miRNAs (He et al. 2013). The genomic sequences, transcriptomic data, predicted genes and unigenes, ESTs, and transposable elements are available on Morus Genome Database (MorusDB). A dedicated database on transposable elements (TEs), namely, MnTEdb is available on public domain and this database consists of 5925 TEs belonging to 13 superfamilies and 1062 families (Ma et al. 2015). SSR markers, mined from whole genome sequencing data and ESTs of *Morus notabilis*, were deposited on a database known as MulSatDB for further use in mulberry breeding programmes. Subsequently, chloroplast genomes of various *Morus* spp. have been sequenced which helped in identification of protein-coding genes, rRNA genes, tRNA genes and SSR marker resources (Ravi et al. 2006, Chen et al. 2016b, Kong and Yang 2016). Further, wealth of information on transcriptomes, proteomes and metabolomes has been generated in mulberry and the applicability of such genomic resources in trait improvement programmes have been extensively reviewed (Sarkar et al. 2017, Dhanyalakshmi and Nataraja 2018, Khurana and Checker 2011). Transcriptomic analysis showed that 609 (2.87 percentage) genes out of 21, 229 genes were differentially expressed between diploid and autotetraploid mulberry plants. Among them, 30 genes were associated with the biosynthesis and signal transduction of plant hormones, while 41 genes were associated with photosynthesis, respectively (Dai et al. 2015a). Drought and salt stress specific transcriptomic and proteomic data of mulberry helped in identification of differentially expressed genes, proteins and molecular markers resources which could be used in climate resilient trait improvement programmes (Wang et al. 2014, Liu et al. 2017, Liu et al. 2019). Comparative transcriptomic analysis of *M. laevigata* and *M. serrata* lead to identification of 24,049 SSRs, 1,201,326 SNPs and 67,875 Insertion-Deletions (InDels) (Saeed et al. 2016). Transcriptome data, GO (Gene Ontology) terms, pathway analysis helped in identification of key functional genes responsible for genotype-specific and tissue-specific functions (Liu et al. 2017), early flower development (Shang et al. 2017), mulberry growth and development mechanisms (Dai et al. 2015b), alkaloid biosynthetic pathways (Wang et al. 2018) and abiotic stress responses (Dhanyalakshmi et al. 2016). An integrated approach of metabolomics, proteomics and transcriptomics (Figure 1) could help in identification of novel secondary metabolites in mulberry and their production at commercial scale through cell suspension culture (Mena et al. 2016, Sarkar et al. 2017, 2018, Liu et al. 2019).

MicroRNAs (miRNAs) play important role in regulation of gene expression by targeting mRNAs for cleavage or translational repression during environmental stress responses, and plant's growth and development (Li et al. 2018a). MiRNA-seq data analysis leads to identification of miRNAs and their target genes, thus unraveling MiRNA-target based molecular mechanism of biotic and environmental stress responses in mulberry (Gai et al. 2018, Li et al. 2017). More details on these aspects are mentioned in Chapter 11.

Applications of Molecular Breeding in Genetic Improvement of Mulberry for Resilience to Climate Change

Dominant and co-dominant molecular markers have found numerous applications in genetic improvement programmes such as analysis of genetic diversity, molecular characterization of germplasm collections and varieties, DNA finger printing, conservation of germplasm collections, determining genetic fidelity of tissue culture based regenerated plants, development of core collections, development of linkage and QTL map, association mapping, parental selection schemes and marker-assisted selection (MAS) (Khurana and Checker 2011, Vijayan et al. 2014, Figure 1). Sometimes, a combination of multiple marker types such as two or more types of dominant markers (RAPD and ISSR), or these dominant markers with and co-dominant markers (SSR) have been used in germplasm characterization, genetic diversity analysis and development of linkage maps for better coverage of mulberry genome as compared to using single marker type (Vijayan et al. 2004, Venkateswarlu et al. 2006, Zhao et al. 2007).

With the advancement of molecular biology and genomics, the genomic- and genic-SSR (microsatellites) markers have been isolated from genomic clones or identified from genomic library, EST (expressed sequence tag) sequences, transcriptome data of mulberry and these markers show transferability to other closely related species (Thumilan et al. 2013, 2016, Dhanyalakshmi et al. 2016). A path breaking study which lead to development of first genetic linkage map in mulberry was possible by adapting two-way pseudo-testcross mapping strategy. In this study, 50 F_1 full-sib progenies of S36 x V1 cross were used for developing linkage map using combination of dominant and co-dominant markers such as RAPD, ISSR and genomic-SSR (Venkateswarlu et al. 2006). Subsequently, efforts have been made to develop QTL maps using dominant markers such as RAPD and ISSR; for specific important agronomic traits such as WUE, root traits, and yield contributing traits in mulberry, but few markers showed linkage to the QTLs governing the trait of interest (Naik et al. 2014, Mishra 2014). These QTLs specific markers have, however, not been used for further validation in unrelated mapping populations for marker assisted selection in mulberry. This may be due to their dominant nature and not being tightly linked to QTLs. Hence, the genetic linkage maps need further investigation for identification of tightly linked co-dominant markers to the QTLs governing trait of interest.

Table 2. QTLs maps for biometric traits developed in mulberry using co-dominant (SSR) markers.

Trait	QTL NAME	Linkage Group	Name of Flanking Markers	LOD	Phenotypic Variance-R^2 (%)	Country
Root Dry Weight to Total Leaf Area (R/LA)	q R_LA 2-1	2	MUL3SSR191-MUL3SSR318	6.48	22.35	India
	q R_LA 4-1	4	MUL3SSR193-MUL3SSR183	2.68	14.33	
Leaf Dry Weight (L Wt)	q L Wt 4-1	4	MUL3SSR143-MUL2SSR97	7.34	12.69	
	q L Wt 4-2	4	MUL3SSR124-MUL3SSR143	7.56	11.58	
Total Shoot Dry Weight (S Wt)	q S Wt 11-1	11	MUL2SSR20-MUL2SSR58	3.80	5.44	
	q S Wt 2-1	2	MUL3SSR59-MUL3SSR50	6.10	5.35	
Specific Leaf Area (SLA)	q SLA 7-1	7	MUL3SSR317-MUL3SSR58	7.18	10.36	
Length of Longest Shoot (LS)	q LS 4-1	4	MUL3SSR229-MUL3SSR300	2.71	16.73	
Total Leaf Area (TLA)	q TLA 4-1	4	MUL3SSR143-MUL2SSR97	2.74	7.69	

Source: Thinnaluri 2015

Drought stress tolerance trait in mulberry could be introgressed through traditional/molecular breeding methods for improving important agronomic traits such as water use efficiency (WUE), water mining (high root biomass and length) and water conservation. Genic and genomic SSR markers have been isolated/identified from genomic clones and transcriptome data of drought-stressed mulberry cv. Dudia white having high root traits (Biradar 2013, Thumilan et al. 2013, 2016, Thinnaluri 2015). Subsequently, a total of 134 SSR markers (111 genomic and 23 genic) segregated in 1:1 test ratio in F_1 population of Dudia white x UP105 cross (Biradar 2013, Thinnaluri 2015, Thumilan et al. 2016) and were used to construct genetic linkage map consisting of 14 linkage groups (LG). A total of nine QTLs associated with biometric traits have been mapped in LG 2, 4, 7 and 11 of mulberry genome based on 2 yr phenotypic data (Thinnaluri 2015). Six pairs of flanking markers (with logarithm of odd (LOD) > 3) linked to QTLs governing important biometric traits could be used for further confirmation studies (Table 2). Recently, chalcone synthase (CHS) based, candidate gene specific, cleaved amplified polymorphic sequence (CAPS) marker, putatively associated with progression of powdery mildew disease in mulberry, has been identified (Arora et al. 2017).

The identified QTL specific markers/flanking markers need to be validated in the other unrelated mapping populations to detect their accurate association with the particular phenotypic trait of interest (Table 2). Subsequently, the flaking SSR markers associated with QTLs could be used in recombinant selection of mulberry hybrids having biometric/drought adaptive traits. The QTL maps so far developed using SSR markers further needs to be saturated with most abundant co-dominant marker resources (SNPs) for identification of tightly linked markers to the QTLs responsible for variations in drought adaptive trait. Subsequently, the tightly linked co-dominant markers could be used in MAS in mulberry breeding programmes (Figure 1).

Improvement of Climate Resilient Traits in Mulberry through Genetic Transformation

Plant tissue culture is applied for growing, multiplying, and maintenance of plant cells, tissues or organs isolated from the mother plant, under nutritionally and environmentally supportive (*in vitro*) and sterile conditions (Thorpe

2007). Plant tissue culture has found many applications in mulberry genetic improvement, propagation and by-product utilization programmes over the past few decades. For example, micropropagation, regeneration (through organogenesis and embryogenesis), protoplast fusion, development of somaclonal variants, gynogenic haploid and triploid plants have been extensively attempted for developing improved desired traits in mulberry. One of the most important aspects of plant tissue culture viz. genetic engineering has been reported to develop mulberry genotypes resilient to climate change. Genetic engineering is technique which allows transfer of gene(s) from alien sources across taxa, when traditional or molecular breeding approaches may not work due to particularly sexual incompatibility of the parental partners, asynchronous flowering of exotic and indigenous germplasm collection, the unavailability of specific gene(s)/QTLs governing particular trait of interest in natural plant populations. Genetic engineering and *in vitro* regeneration of whole transgenic plant are essentially based on the principle of cellular totipotency/concept of the cell theory proposed by Matthias Jakob Schleiden and Theodor Schwann and the concept of breakthrough discovery of genetic transformation in bacteria by Frederick Griffith (Vasil 2008).

In vitro regeneration in mulberry is highly dependent on the genotype of explant (Raghunath et al. 2013). The success rate of genetic transformation experiments is also dependent on various factors such as regeneration potential of genotype, choice of explant (Sarkar et al. 2016), *Agrobacterium* strains and recombinant plasmids (Bhatnagar and Khurana 2003). Regeneration in mulberry might be influenced by season as also reported in various tree species. Direct and indirect (through callus phase) organogenesis routes have been reported in mulberry for whole plant regeneration (Vijayan et al. 2000, Raghunath et al. 2013, Rohela et al. 2018). Direct organogenesis has the advantages over the indirect one, as the former might retain genetic fidelity of the regenerated plants and the later might induce much genetic variations resulting in the development of somaclonal variants (Sarkar et al. 2018). Genetic transformation in mulberry with various transgenes has been attempted since last three decades *via* direct and indirect methods viz. particle bombardment, *Agrobacterium rhizogenes*-mediated, electroporation and *in planta,* and the relevance of these techniques in mulberry genetic engineering programmes has been reviewed (Sarkar et al. 2017, 2018).

Over the years, several attempts have been made to develop efficient regeneration and genetic transformation protocols by various research groups, not only in India, but also across the globe, using various type of explants, virulent starins of *Agrobacterium tumefaciens*, plant ransformation vectors, dosage of acetosyringone, and methods of genetic transformation and regeneration (Sarkar et al. 2017). Glycinin encoding gene (*AlaBlb*) has been introduced in mulberry through *Agrobacterium tuumefaciens* mediated genetic transformation (Jianzhong et al. 2001). *In planta* genetic transformation has been successfully applied in mulberry for development of transgenic mulberry (Ping et al. 2003). Subsequently, an efficient and reproducible protocol of genetic transformation with 6 percentage transformation frequencies has been developed using different explants such as hypocotyl, cotyledon, leaf and leaf induced callus of mulberry (*Morus indica*) cv. K2 through *Agrobacterium tumefaciens*-mediated genetic transformation. In this study, leaf derived callus was found to be the best explants for development of transgenic plants (Bhatnagar and Khurana 2003). Over the time, cotyledon/hypocotyl explants of K2/M5 genotypes became the choice of explants for development of transgenic mulberry with tolerance to various abiotic stresses, which might be due to their appreciable level of regeneration potential. Since last decade, various abiotic stresses-associated functional genes have been introduced in transgenic mulberry *M. indica* cv. K2 for ectopic expression of transgene and tolerance to various abiotic stresses (Table 3). Gain-of-function based genetic transformation approach helped in development of abiotic stress tolerant stable transgenic lines in mulberry (Lal et al. 2008, Sajeevan et al. 2017). Further heterologus expression of osmotin gene in transgenic mulberry showed tolerance to both biotic and abiotic stress. Thus, this study indicated that there might have osmotin regulated cross-talks between biotic and abiotic stress responses in transgenic mulberry. Ectopic expression of *bch1* gene from mulberry (*M. indica*) cv. K2 showed higher levels of carotenoids and tolerance to various oxidative stresses in transgenic mulberry lines than non-transgenic mulberry under non-stress and stress conditions (Saeed et al. 2015). Thus this study has paved the way to cisgenic approach in mulberry for improving its climate resilient traits (Table 3). The details on some of these aspects are also mentioned in some chapters of this book.

On the other hand, loss-of function based transformation approach through virus induced gene silencing (VIGS) in mulberry helped in development of transient *MmSK* transgenic lines with sensitivity to drought stress (Li et al. 2018b). These above-cited studies showed that the transformation efficiency was enhanced upto 20–60 percentage using cotyledon and hypocotyl explants. As stated earlier, MicroRNAs (miRNAs) play key

Table 3. Development of transgenic mulberry lines resilient to climate change through *Agrobacterium tumefaciens*-mediated transformation.

Transgene/ miRNAs	Source of transgene	Genotype	Gain-of-function/ loss-of-function approach	Explants	Selectable marker gene	Reporter gene	Promoter	Method of genetic transformation	Type of transformation/ transgene expression	Performance	Improved Physio-biochemical parameters	References
Hva1	Barley	*M. indica* (cv. K2)	gain-of-function	Hypocotyl, Cotyledon	*nptII*	*gus*	Act1	*A. tumefaciens*	Stable	Tolerance to drought and salinity stress	Higher cellular membrane stability, photosynthetic yield and water use efficiency, less photo-oxidative damage	Lal et al. 2008
Osmotin	Tobacco	*M. indica* (cv. K2)	gain-of-function	Leaf-derived callus, Hypocotyl, Cotyledon	*nptII*	-	CaMV35S, rd29A	*A. tumefaciens*	Stable	Tolerance to drought and salinity stress; resistance to fungi (*Fusarium pallidoroseum, Colletotrichum gloeosporioides and Colletotrichum dematium*)	Higher cellular membrane stability and photosynthetic yield	Das et al. 2011
Hva1	Barley	*M. indica* (cv. K2)	gain-of-function	Hypocotyl, Cotyledon	*nptII*	No	rd29A	*A. tumefaciens*	Stable	Tolerance to drought, salinity and cold stress	Higher free proline content, membrane stability index and photosynthetic yield	Checker et al. 2012
bchI	*M. indica*	*M. indica* (cv. K2)	gain-of-function	Leaf-derived callus, Hypocotyl, Cotyledon	*nptII*	*gus*	CaMV 35S	*A. tumefaciens*	Stable	Tolerance to UV, high temperature and irradiance stress	Higher membrane stability, carotenoid and chlorophyll content, lower ROS level	Saeed et al. 2015

Table 3 contd. ...

...Table 3 contd.

Transgene/ miRNAs	Source of transgene	Genotype	Gain-of-function/ loss-of- function approach	Explants	Selectable marker gene	Reporter gene	Promoter	Method of genetic transformation	Type of transformation/ transgene expression	Performance	Improved Physio-biochemical parameters	References
SHN1	*A. thaliana*	*M. indica* (cv. M5)	gain-of-function	Hypocotyl, Cotyledon	*nptII*	-	CaMV35S	*A. tumefaciens*	Stable	Enhanced leaf moisture retention capacity	Higher leaf surface wax content, leaf moisture retention capacity	Sajeevan et al. 2017
MmSK	*Morus alba*	*Morus alba*	loss-of-function	Cotyledon of seedling	-	-	CaMV35S	*A. tumefaciens*	Transient	Sensitivity to drought stress	Decrease in soluble protein content, proline content, superoxide dismutase (SOD) and peroxidase (POD) activities	Li et al. 2018b
pre-miR166f	*Morus multicaulis*	*Morus multicaulis*	Gain-of-function	Leaf	-	*gus*	CaMV35S	*A. tumefaciens*	Transient	Tolerance to drought stress	Higher relative water content, proline content, soluble protein; higher superoxide dismutase and peroxidase activities and lower malondialdehyde content	Li et al. 2018a

regulatory role on gene expression at the posttranscriptional and translational levels. More than 40 miRNA families genes have been identified in plant systems including woody plants. A pre-miR166f of mulberry has been introduced in mulberry leaves through *Agrobacterium* mediated transient transformation method. Overexpression of miR166f in transgenic mulberry lines downregulated the expression of two homeobox-leucine zipper (HD-Zip) genes and one histone arginine demethylase gene. However, overexpression of miR166f in transient transgenic mulberry lines enhanced tolerance to drought stress (Li et al. 2018a). This study demonstrated that MicroRNAs could be the potential candidates for genetic engineering in mulberry for enhancing its resilience to climate change (Figure 1). For recent updates refer to Chapter 11 on applications of genetic engineering in mulberry in this book.

Conclusion

Mulberry is an economically important perennial plant having a key role in the trade and tariff across the world, particularly in silk industry. Furthermore, mulberry has the intrinsic features to be considered as a potential perennial model tree system (Dhanyalakshmi and Nataraja 2018). Traditional breeding practices rely on selection of mulberry genotypes based on morphological traits, for foliage-yield and other yield contributing traits. Subsequently, various breeding programmes contributed immensely towards development of stress-tolerant, high leaf-yielding and high leaf-quality mulberry varieties. As mulberry fruit has considerable medicinal properties, breeding strategies need to be streamlined towards enhancement of its fruit productivity even under stressful environments. However, there seems lot of scope to utilize physiological trait based breeding approaches for development of climate resilient mulberry cultivars. The genomic and transcriptomic resources of *Morus* species generated genic- and genomic-SSRs, SNPs and InDel markers. The SSR markers are being utilized in identification of trait-specific QTLs through linkage mapping using bi-parental mapping population. At present, a wealth of transcriptomic and genomic data and comprehensive co-dominant marker resources are being generated due to advancement of cost effective next generation sequencing (NGS) platforms. So far dominant markers like AFLPs have been utilized to establish marker-trait association with root rot resistance and survival of mulberry shoot cutting, ascertained through association mapping (Pinto et al. 2018). As mulberry is a heterozygous and perennial tree plant, genome wide association study (GWAS)/Linkage disequilibrium (LD)/association mapping using co-dominant markers like SSRs, SNPs could be a viable option for identification of QTLs and candidate genes associated with climate resilient traits such as drought adaptation, nitrogen use efficiency, foliage as well as other yield-components, and disease resistance (Fodor et al. 2014, Sarkar et al. 2017, Arora et al. 2017, Du et al. 2018, Rukmangada et al. 2018a). The identification of tightly linked markers or candidate genes associated with trait specific QTLs will facilitate MAS in breeding programmes for genetic improvement of mulberry with climate resilience (Figure 1).

Transgenic mulberry lines with climate resilient traits so far developed need to be evaluated under real field conditions for commercial cultivation by farmers. Further, the abiotic stress tolerant transgenes from well-characterized transgenic mulberry lines could be introgressed into different genetic backgrounds of elite cultivars through traditional/marker assisted breeding strategies as attempted in mulberry and bread wheat (Tikader and Kamble 2008, Shavrukov et al. 2016). Considering that C_4 photosynthesis mechanism of plant is more efficient than C_3 photosynthesis system, key genes coding for enzymes or transporter proteins responsible for C_4 photosynthesis or modified form of gene coding for Rubisco, can be introduced in mulberry for engineering its CO_2 concentration mechanism to make it resilient to climate change (Lin et al. 2014, Chen et al. 2017). Genome editing tool such as clustered regulatory interspaced short palindromic repeats (CRISPR)/CRISPR-associated nuclease 9 (Cas9) systems (CRISPER/Cas9) can be explored to unravel molecular basis of abiotic stress tolerance and to improve climate resilient traits in mulberry (Sarkar et al. 2017). Integrated approaches of traditional breeding, VIGS and RNA interference (RNAi) based reverse genetics, genomic tools, integrated *in vitro* plant regeneration and genetic transformation protocols, marker assisted breeding techniques in mulberry along with adequate fund flow, capacity building, scientific networking, and concerted efforts on biosafety clearance could be the essential milestones towards sustained development and commercialization of climate resilient mulberry cultivars.

References

Arora, V., M.K. Ghosh, S. Pal and G. Gangopadhyay. 2017. Allele specific CAPS marker development and characterization of chalcone synthase gene in Indian mulberry (*Morus* spp., family Moraceae). PLoS ONE 12(6): e0179189. doi.org/10.1371/journal.pone.0179189.

Bhatnagar, S. and P. Khurana. 2003. *Agrobacterium tumefaciens*-mediated transformation of Indian mulberry, *Morus indica* cv. K-2: A time phased screening strategy. Plant Cell Rep. 21(7): 669–675.

Biradar, J. 2013. Molecular characterisation of root specific mapping population of mulberry by SSR markers and identification of QTLs Governing drought tolerance traits Ph.D. Thesis. University of Agricultural Sciences, Bengaluru, India.

Checker, V.G., A.K. Chibbar and P. Khurana. 2012 Stress-inducible expression of barley hva1 gene in transgenic mulberry displays enhanced tolerance against drought, salinity and cold stress. Transgen. Res. 21(5): 939–957.

Chen, C., W. Zhou, Y. Huang and Z.Z. Wang. 2016b. The complete chloroplast genome sequence of the mulberry *Morus notabilis* (*Moreae*). Mitochondrial DNA A DNA Mapp. Seq. Anal. 27(4): 2856–2867. doi:10.3109/19401736.2015.1053127.

Chen, M., Y. Huang, G. Liu, F. Qin , S. Yang and X. Xu. 2016a. Effects of enhanced UV-B radiation on morphology, physiology, biomass, leaf anatomy and ultrastructure in male and female mulberry (*Morus alba*) saplings. Environ. Exp. Bot. 129: 85–93.

Chen, P.Y., Y.T. Tsai, C.Y. Ng, M.S.B. Ku and K.Y. To. 2017. Transformation and characterization of transgenic rice and Cleome spinosa plants carrying the maize phosphoenolpyruvate carboxylase genomic DNA. Plant Cell Tissue Organ Cult. 128(3): 509–519.

Dai, F., Z. Wang, G. Luo and C. Tang. 2015a. Phenotypic and transcriptomic analyses of autotetraploid and diploid mulberry (*Morus alba* L.). Int. J. Mol. Sci. 22, 16(9): 22938–22956.

Dai, F., C. Tang, Z. Wang, G. Luo, L. He and L. Yao. 2015b. De novo assembly, gene annotation, and marker development of mulberry (*Morus atropurpurea*) transcriptome. Tree Genet. Genomes 11(2): 26.

Das, C., B.K. Das, S.K. Sen and T.P. Kumar. 1997. Studies on photosynthetic $^{14}CO_2$ fixation: occurrence of the C_4 photosnthetic syndrome in mulberry (*Morus* spp.). Sericologia 37(1): 89–97.

Das, M., H. Chauhan, A. Chhibbar, Q.M.R Haq and P. Khurana. 2011. High efficiency transformation and selective tolerance against biotic and abiotic stress in mulberry, *Morus indica* cv. K-2, by constitutive and inducible expression of tobacco Osmotin. Transgen. Res. 20(2): 231–246.

Dhanyalakshmi, K.H., M.B.N. Naika, R.S. Sajeevan, O.K. Mathew, K.M. Shafi, R. Sowdhamini and K.N. Nataraja. 2016. An approach to function annotation for proteins of unknown function (PUFs) in the transcriptome of Indian mulberry. PLoS One 11(3): e0151323. doi:10.1371/journal.pone.0151323.

Dhanyalakshmi, K.H. and K.N. Nataraja. 2018. Mulberry (*Morus* spp.) has the features to treat as a potential perennial model system. Plant. Signal. Behav. 13(8): p.e1491267.

Doss, S.G., S.P. Chakraborti, S. Roychowdhuri, N.K. Das, K. Vijayan and P.D. Ghosh. 2012. Development of mulberry varieties for sustainable growth and leaf yield in temperate and subtropical regions of India. Euphytica 185(2): 215–225.

Du, Q., W. Lu, M. Quan, L. Xiao, F. Song, P. Li, D. Zhou, J. Xie, L. Wang, and D. Zhang. 2018. Genome-Wide association studies to improve wood properties: Challenges and prospects. Front. Plant Sci. 9: 1912. doi: 10.3389/fpls.2018.01912.

Fodor, A., V. Segura, M. Denis, S. Neuenschwander, A. Fournier-Level, A. Chatelet, P. Homa, F.A.A., T. Lacombe, P. This and L. Le Cunff. 2014. Genome-Wide Prediction Methods in Highly Diverse and Heterozygous Species: Proof-of-Concept through Simulation in Grapevine. PLoS ONE 9(11): e110436. doi:10.1371/journal.pone.0110436.

Gai, Y.P., H.N. Zhao, Y.N. Zhao, B.S. Zhu, S.S. Yuan, S. Li, F.Y. Guo and X.L. Ji. 2018. MiRNA-seq-based profiles of miRNAs in mulberry phloem sap provide insight into the pathogenic mechanisms of mulberry yellow dwarf disease. Sci. Rep. 8(1): 812.

Ghosh, M.K., A.K. Misra, C. Shivnath, T. Sengupta, P.K. Ghosh, M.K. Singh and B.B. Bindroo. 2012. Morpho-physiological evaluation of some mulberry genotypes under high moisture stress condition. J. Crop Weed 8(1): 167–170.

He, N., C. Zhang, X. Qi, S. Zhao, Y. Tao, G . Yang, T.H. Lee, X. Wang, Q. Cai, D. Li, M. Lu, S. Liao, G. Luo, R. He, X. Tan, Y. Xu, T. Li, A. Zhao, L. Jia, Q. Fu, Q. Zeng, C. Gao, B. Ma, J. Liang, X. Wang, J. Shang, P. Song, H . Wu, L. Fan, Q. Wang, Q. Shuai, J. Zhu, C. Wei, K. Zhu-Salzman, D. Jin, J. Wang, T. Liu, M. Yu, C. Tang, Z. Wang, F. Dai, J. Chen, Y. Liu, S. Zhao, T. Lin, S. Zhang, J. Wang, J. Wang, H. Yang, G. Yang, J. Wang, A.H. Paterson, Q. Xia, D. Ji and Z. Xiang. 2013. Draft genome sequence of the mulberry tree *Morus notabilis*. Nature Commun. 4: 2445. doi:10.1038/ncomms3445.

Jianzhong, T., L. Chengfu, W. Hongli and C Mingqi. 2001. Transgenic plants via transformation of glycinin gene to mulberry. J. Agric. Biotechnol. 9(4): 400–402.

Khurana, P. and V.G. Checker. 2011. The advent of genomics in mulberry and perspectives for productivity enhancement. Plant Cell Rep. 30: 825–838.

Kong, W. and J. Yang. 2016. The complete chloroplast genome sequence of *Morus mongolica* and a comparative analysis within the Fabidae clade. Curr. Genet. 62(1): 165–172.

Kumar, R. and H.R. Gautam. 2014. Climate Change and its Impact on Agricultural Productivity in India. J. Climatol. Weather Forecasting 2: 1. doi:10.4172/2332-2594.1000109.

Lal, S., V. Gulyani and P. Khurana. 2008. Overexpression of hva1 gene from barley generates tolerance to salinity and water stress in transgenic mulberry (*Morus indica*). Transgen. Res. 17: 651–663.

Li, R., D. Chen, T. Wang, Y. Wan, R. Li, R. Fang, Y. Wang, F. Hu, H. Zhou, L. Li and W. Zhao. 2017. High throughput deep degradome sequencing reveals microRNAs and their targets in response to drought stress in mulberry (*Morus alba*). PLoS ONE 12(2): e0172883. doi: 10.1371/journal.pone.0172883.

Li, R., T. Fan, T. Wang, K. Dominic, F. Hu, L. Liu, L. Zhang, R. Fang, G. Pan, L. Li and W. Zhao. 2018a. Characterization and functional analysis of miR166f in drought stress tolerance in mulberry (*Morus multicaulis*). Mol. Breed 38(11): 132.

Li, R., L. Liu, K. Dominic, T. Wang, T. Fan, F. Hu, Y. Wang, L. Zhang, L. Li and W. Zhao. 2018b. Mulberry (*Morus alba*). MmSK gene enhances tolerance to drought stress in transgenic mulberry. Plant Physiol. Biochem. 132: 603–611.

Lin, M.T., A. Occhialini, P.J. Andralojc, M.A. Parry and M.R. Hanson. 2014. A faster Rubisco with potential to increase photosynthesis in crops. Nature 513(7519): 547.

Liu, C.Y., X.Q. Liu, D.P. Long, B.N. Cao, Z.H. Xiang and A.C. Zhao. 2017. *De novo* assembly of mulberry (*Morus alba* L.) transcriptome and identification of candidate unigenes related to salt stress responses. Russ. J. Plant Physiol. 64(5): 738–748.

Liu, X.Q., C.Y. Liu, Q. Guo, M. Zhang, B.N. Cao, Z.H. Xiang and A.C. Zhao. 2015. Mulberry transcription factor MnDREB4A confers tolerance to multiple abiotic stresses in transgenic tobacco. PLoS One 10(12): e0145619. doi:10.1371/journal.pone.0145619.

Liu, Y., D. Ji, R. Turgeon, J. Chen, T. Lin, J. Huang, J. Luo, Y. Zhu, C. Zhang and Z. Lv. 2019. Physiological and proteomic responses of mulberry trees (*Morus alba*. L.) to combined salt and drought stress. Int. J. Mol. Sci. 20: 2486. doi.org/10.3390/ijms20102486.

Ma, B., T. Li, Z. Xiang and N. He. 2015. MnTEdb, a collective resource for mulberry transposable elements. Database, 2015. doi:10.1093/database/bav004.

Mena, P., E.M. Sanchez-Salcedo, M. Tassotti, J.J. Martinez, F. Hernandez and D.D. Rio. 2016. Phytochemical evaluation of eight white (*Morus alba* L.) and black (*Morus nigra* L.) mulberry clones grown in Spain based on UHPLC-ESI-MSn metabolomics profiles. Food Res. Int. 89: 1116–1122.

Mishra, S. 2014 Genetic analysis of traits controlling water use efficiency and rooting in mulberry (*Morus* spp.) by molecular markers. Ph.D. Thesis, University of Mysore, Mysuru, India.

Mogili, T., K. Rajashekar, P.M. Tripathi, K. Sathyanarayana, R. Balakrishna and M.M. Reddy. 2008. Screening mulberry genotypes for tolerance to alkalinity stress. Adv. Plant. Sci. 2: 621–629.

Naik, V.G., B. Thumilan, A. Sarkar, S.B. Dandin, M.V. Pinto and V. Sivaprasad. 2014. Development of genetic linkage map of mulberry using molecular markers and identification of QTLs linked to yield and yield contributing traits. Sericologia 54(4): 221–229.

Nguyen, H.T.T., D.T. Dang, C. Van Pham and P. Bertin. 2016. QTL mapping for nitrogen use efficiency and related physiological and agronomical traits during the vegetative phase in rice under hydroponics. Euphytica 212(3): 473–500.

Ping, L.X., M. Nogawa, H. Shioiri, M, Nozue, N. Makita, M. Takeda, L. Bao and M. Kojima. 2003. *In planta* transformation of mulberry trees (*Morus alba* L.) by *Agrobacterium tumefaciens*. J. Insect Biotechnol. Sericol. 72(3): 177–184.

Pinto, M.V., H.S. Poornima, M.S. Rukmangada, R. Triveni and V.G Naik. 2018. Association mapping of quantitative resistance to charcoal root rot in mulberry germplasm. PLoS ONE 13(7): e0200099. doi.org/10.1371/journal.pone.0200099.

Raghunath, M.K., K.N. Nataraja, J.S. Meghana, R.S. Sanjeevan, M.V. Rajan and S.M.H. Qadri. 2013. *In vitro* plant regeneration of *Morus indica* L.cv. V-1 using leaf explants. Am. J. Plant Sci. 4(10): 2001–2005.

Ravi, V., J.P. Khurana, A.K. Tyagi and P. Khurana. 2006. The chloroplast genome of mulberry: complete nucleotide sequence, gene organization and comparative analysis. Tree Genet. Genomes 3(1): 49–59.

Reynolds, M. and P. Langridge. 2016. Physiological breeding. Curr. Opin. Plant. Biol. 31: 162–171.

Rohela, G.K., P. Jogam, A.A. Shabnam, P. Shukla, S. Abbagani and M.K. Ghosh. 2018. *In vitro* regeneration and assessment of genetic fidelity of acclimated plantlets by using ISSR markers in PPR-1 (*Morus* sp.): An economically important plant. Sci. Hort. 241: 313–321.

Rukmangada, M.S., R. Sumathy, V. Sivaprasad and V.G. Naik. 2018a. Genome-wide identification and characterization of growth-regulating factors in mulberry (*Morus* spp.). Trees 32(6): 1695–1705.

Rukmangada, M.S., S. Ramasamy, V. Sivaprasad and G.N. Varkody. 2018b. Growth performance in contrasting sets of mulberry (*Morus* Spp.) genotypes explained by logistic and linear regression models using morphological and gas exchange parameters. Sci. Hort. 235: 53–61.

Saeed, B., M. Das, Q.M.R. Haq and P. Khurana. 2015. Overexpression of beta carotene hydroxylase-1 (bch1) in mulberry, *Morus indica* cv. K-2, confers tolerance against high-temperature and highirradiance stress induced damage. Plant Cell Tissue Organ Cult. 120(3): 1003–1015.

Saeed, B., V.K. Baranwal and P. Khurana. 2016. Comparative transcriptomics and comprehensive marker resource development in mulberry. BMC Genomics 17: 98. doi:10.1186/s12864-016-2417-8.

Sajeevan, R.S., K.N. Nataraja, K.S. Shivashankara, N. Pallavi, D.S. Gurumurthy and M.B. Shivanna. 2017. Expression of *Arabidopsis* SHN1 in Indian mulberry (*Morus indica* L.) increases leaf surface wax content and reduces post-harvest water loss. Front. Plant Sci. 8: 418. doi:10.3389/fpls.2017.00418.

Sarkar, T. 2014. Development of transgenic resistance to abiotic stress in groundnut using *AtDREB1A* gene through *Agrobacterium* mediated genetic transformation. Ph.D. Thesis, Saurashtra University, Rajkot, Gujarat, India.

Sarkar, T., T. Radhakrishnan, A. Kumar, G.P. Mishra and J.R. Dobaria. 2014. Heterologous expression of AtDREB1A gene in transgenic peanut conferred tolerance to drought and salinitystresses. PLoS One 9(12): e110507.doi.org/10.1371/journal.pone.0110507.

Sarkar, T., T. Radhakrishnan, A. Kumar, G.P. Mishra and J.R. Dobaria. 2016. Stress inducible expression of AtDREB1A transcription factor in transgenic peanut (*Arachis hypogaea* L.) crop conferred tolerance to soil-moisture deficit stress. Front. Plant Sci. 7: 935. doi.org/10.3389/fpls.2016.00935.

Sarkar, T., T. Mogili and V. Sivaprasad. 2017. Improvement of abiotic stress adaptive traits in mulberry (*Morus* spp.): An update on biotechnological interventions. 3 Biotech. 7: 214. oi.org/10.1007/s13205-017-0829-z.

Sarkar, T., T. Mogili, S. Gandhi Doss and V. Sivaprasad. 2018. Tissue culture in mulberry (*Morus* spp.) intending genetic improvement, micropropagation and secondary metabolite production: a review on current status and future prospects. pp. 467–487. *In:* Kumar, N. (eds.). Biotechnological Approaches for Medicinal and Aromatic Plants. Springer, Singapore.

Senapati, M.R., B. Behera and S.R. Mishra. 2013. Impact of climate change on Indian agriculture and its mitigating priorities. American J. Environ. Protection. 4109–111.

Shang, J., J. Liang, Z. Xiang and N. He. 2017. Anatomical and transcriptional dynamics of early floral development of mulberry (*Morus alba*). Tree genet. Genomes 13(2): 40.

Shavrukov, Y., M. Baho, S. Lopato and P. Langridge. 2016. The TaDREB3 transgene transferred by conventional crossings to different genetic backgrounds of bread wheat improves drought tolerance. Plant Biotechnol. J. 14: 313–322.

Susheelamma, B.N., J.S. Kumar, T. Mogili, K. Sengupta, M.N. Padma and N. Suryanarayana. 1992. Evaluation techniques for screening for drought resistance in mulberry. Sericologia 32: 609–614.

Thinnaluri, M. 2015. Discovery of QTLs for drought tolerance traits using a biparental mapping population of mulberry segregating for root and water use efficiency (WUE). Ph.D. Thesis. University of Agricultural Sciences, GKVK, Bengaluru, India.

Thorpe, T.A. 2007. History of plant tissue culture. Mol. Biotechnol. 37(2): 169–180.

Thumilan, B.M., N.M. Kadam, J. Biradar, H.R. Sowmya, A. Mahadeva, J.N. Madhura, U. Makarla, P. Khurana and S.M. Sreeman. 2013. Development and characterization of microsatellite markers for *Morus* spp. and assessment of their transferability to other closely related species. BMC Plant. Biol. 13: 194. doi:10.1186/1471-2229-13-194.

Thumilan, B.M., R.S. Sajeevan, J. Biradar, T. Madhuri, K.N. Nataraja and S.M. Sreeman. 2016. Development and characterization of genic SSR markers from Indian mulberry transcriptome and their transferability to related species of Moraceae. PLoS One 11(9): e0162909. doi:10.1371/journal.pone.0162909.

Tikader, A. and C.K. Kamble. 2008. Mulberry wild species in India and their use in crop improvement a review. Aust. J. Crop Sci. 2(2): 64–72.

Tikader, A. and K. Vijayan. 2017. Mulberry (*Morus* Spp.) Genetic diversity, conservation and management. pp. 95–127. *In:* Ahuja, M. and S. Jain (eds.). biodiversity And Conservation of Woody Plants. Sustainable Development and Biodiversity, vol 17. Springer, Cham.

Vasil, I.K. 2008. A history of plant biotechnology: From the cell theory of Schleiden and Schwann to biotech crops. Plant Cell Rep. 27(9): 1423.

Venkateswarlu, M., U.S. Raje, N.B. Surendra, H.E. Shashidhar, M. Maheswaran, T.M. Veeraiah and M.G. Sabitha. 2006. A first genetic linkage map of mulberry (*Morus* spp.) using RAPD, ISSR, and SSR markers and pseudotestcross mapping strategy. Tree Genet. 3: 15–24.

Vijayan, K., S.P. Chakraborti and B.N. Roy. 2000. Plant regeneration from leaf explants of mulberry: Influence of sugar, genotype and 6-benzyladenine. Indian J. Exp. Biol. 38(5): 504–508.

Vijayan, K., P.P. Srivastava and A.K. Awasthi. 2004. Analysis of phylogenetic relationship among five mulberry (*Morus*) species using molecular markers. Genome 47: 439–448.

Vijayan, K., S.G. Doss, S.P. Chakraborti and P.D. Ghosh. 2009. Breeding for salinity resistance in (*Morus* spp.). Euphytica. 169(3): 403–411.

Vijayan, K. 2010. The emerging role of genomic tools in mulberry (*Morus*) genetic improvement.Tree Genet. Genomes 6: 613–625.

Vijayan, K., A. Tikader and J.A.T. da Silva. 2011. Application of tissue culture techniques propagation and crop improvement in mulberry (*Morus* spp.). Tree Forestry Sci. Biotechnol. 5(1): 1–13.

Vijayan, K., P.J. Raju, A. Tikader and B. Saratchnadra. 2014. Biotechnology of mulberry (*Morus* L.)—a review. Emir. J. Food Agr. 26(6): 472–496.

Vijayan, K., G. Ravikumar and A. Tikader. 2018. Mulberry (*Morus* spp.). Breeding for higher fruit production. pp. 89–130. *In*: Al-Khayri, J., S. Jain and D. Johnson (eds.). Advances in Plant Breeding Strategies: Fruits. Springer, Cham.

Wang, D., L. Zhao, D. Wang, J. Liu, X. Yu, Y. Wei and Z. Ouyang. 2018. Transcriptome analysis and identification of key genes involved in 1-deoxynojirimycin biosynthesis of mulberry (*Morus alba* L.). PeerJ 6: e5443. doi.org/10.7717/peerj.5443.

Wang, H., W. Tong, L. Feng, Q. Jiao, L. Long, R. Fang and W. Zhao. 2014. *De novo* transcriptome analysis of mulberry (*Morus* L.) under drought stress using RNA-Seq technology. Russ. J. Bioorg. Chem. 40(4): 423–432.

Yile, P. and K. Oshigane. 1998. Chromosome number of wild species in Morus cathayana Hemsl and Morus wittiorum Handel-Mazett distribution in China. J. Seric. Sci. Jpn. 67: 151–153.

Zhao, Z.Z., M. Xuexia, Z. Yong, W. Sibao, H. Jianhua, X. Hui, P. Yile and H. Yongping. 2007. A comparison of genetic variation among wild and cultivated *Morus* species (Moraceae: *Morus*) as revealed ISSR and SSR markers. Biodivers. Conserv. 16: 275–290.

CHAPTER 14

Arbuscular Mycorrhizal Symbiosis Contribute Significant Benefits to Growth and Quality of Mulberry Plants

Songmei Shi,[1,#] *Miao Wen,*[1,#] *Xinshui Dong,*[1] *Lu Zhang,*[1] *Sharifullah Sharifi,*[1] *Xiao Xu*[2,*] and *Xinhua He*[1,3,*]

INTRODUCTION

Mulberry (*Morus alba* L.), a fast-growing multipurpose plant for the main purpose of rearing silkworm (*Bombyx mori* L.), has been widely planted in Asia, Europe, North and South America, and Africa because of its easy adaptability to tropical, subtropical, temperate and frigid habitats (Özgen et al. 2009). Globally a total of 30 mulberry species and 9 varieties are mainly grown in Japan, India and China (Lu et al. 2017). At present ~ 3,000 mulberry germplasm resources comprising 15 species and 4 varieties are preserved in China (Pan 2000), which is the birthplace of the world's sericulture and silk industry with a history of 5,500 years (Lu et al. 2017). Recently, products from mulberry trees are also widely developed as fruits, foods, drinks and even medicines due to an increasing exploration of its health benefits such as the risk reduction of atherosclerosis, coronary heart disease, high blood glucose and some cancers (Katsube et al. 2006, 2009, Harauma et al. 2007, Huang et al. 2013). Recent studies have showed that mulberry can be applied in ecological restoration (Huang et al. 2013, Xing et al. 2018) or as a potential bio-energy plant owing to its fast growth and adaptability under the global environment change scenarios (Sekhar et al. 2017). However, most of mulberry planting areas are facing with soil barrenness and severe biotic and abiotic stress problems, which restrict the plant growth and leaf yield, in addition to adversely affecting the cocoon quality and economic output.

Arbuscular mycorrhizal fungi (AMF), beneficial soil microorganisms, forming mycorrhizal symbioses with roots of > 80% of terrestrial plants, contribute to the nutritional requirements and development of the host plant (Smith and Read 2008). AM fungal hyphae penetrate into the root cortical cells to form complex dendritic structures, which facilitate the uptake of mineral nutrients such as phosphorus (P) and nitrogen (N) from the soil (Wipf et al. 2019). In return, AMF obtain carbon (C) from the plant in the form of sugars and fatty acids to sustain mycorrhizal colonization (Smith and Read 2008, Jiang et al. 2017). Furthermore, AM symbiosis plays a vital role in sustaining

[1] Centre of Excellence for Soil Biology, College of Resources and Environment, Southwest University, Chongqing 400715, China.

[2] Key Laboratory of Southwest China Wildlife Resources Conservation, China West Normal University, Nanchong 637002, China.

[3] School of Biological Sciences, University of Western Australia, Perth, WA 6009, Australia.

[#] These two authors contributed to this paper equally.

[*] Corresponding authors: xinhua.he@uwa.edu.au; xuxiao_cwnu@163.com

plant diversity, increasing plant productivity, and maintaining ecosystem processes by promoting plant fitness through a range of mechanisms (van der Heijden et al. 1998). These processes include enhanced water/nutrient uptake and photosynthesis performance, enhancing resistances to drought, salt, heavy metals, protecting the host from pathogens, and improving soil structure (Ouziad et al. 2005, Evelin et al. 2019). A number of studies have also documented that associations between agronomic plant species and AMF are able to increase the efficiency of fertilizer use and plant growth (Porras-Soriano et al. 2009, Baum et al. 2015, Conversa et al. 2015). However, there are seldom studies on AMF in both the bulk soil and rhizosphere of mulberry plants. This paper reviews on the progresses in the mycorrhizal associations with mulberry plants in order to promote future mulberry plantation and relevant silk industry.

Diversity of Arbuscular Mycorrhizal Fungi in the Rhizosphere of Mulberry

Probably the first reported study on AMF-stimulated mulberry growth was from Kim et al. (1984). After growing on a phosphate deficient soil for a period of 6-months, the shoot length, stem diameter, leaf yield and P_2O_5 (but not N), in *Glomus mosseae* (now *Funneliformis mosseae*)—inoculated mulberry seedlings were significantly greater than uninoculated counterparts. The occurrence of AMF in the *Morus alba* K2 variety roots were highly colonized by AMF in the field (Rajagopal et al. 1989). About 82% of the field mulberry saplings in Three Gorges Reservoir Region of southwest China were also colonized by fungi in pots. Studies showed that the AMF spores germinated at 4 days and appressoria were formed at 15 days after inoculation, and the root AM cololization rates grew up to 47% at 70 days, 51% at 90 days, respectively (Shu et al. 2011). According to their morphological characteristics, a total of 16 AMF morphotypes, including 10 *Glomus* species, 5 *Acaulospora* species, 1 *Gigarspora* species, were identified from the mulberry rhizosphere soil in the rock desertification area of the Chongqing Municipality, southwest China (Shi et al. 2013). Using the high throughput 454-sequencing technology, the AMF composition and diversity were determined in the nearby karst soils on mulberry roots in Guizhou, southwest China by Xing et al. (2018), who identified a total of 8 genera including *Acaulospora*, *Ambispora*, *Archaeospora*, *Claroideoglomus*, *Diversisporav*, *Glomus* (dominant genus), *Paraglomus* and *Redeckera*. These results thus showed that mulberry is a highly AMF dependent plant species.

Studies also showed the selective affinity of AM fungi in mulberry. Through a two-year period of field experiment in a rainfed lateritic soil, the branch number, leaf area and moisture, leaf P uptake and yield of 2-year-old mulberry saplings were consistently greater under *Glomus fasciculatum* inoculation plus 10 kg phosphate ha⁻¹ yr⁻¹ fertilizer than *G. etunicatum* or *F. mosseae* inoculation plus either 10 or 25 kg phosphate ha⁻¹ year⁻¹ fertilizer (Setua et al. 1999). In contrast, greater improvement effects on the growth of 1-year-old mulberry samplings in southern Xinjiang ranked as: *F. mosseae* BGC XJ08A > *G. versiforme* BGC XJ08F > *F. mosseae* BGC XJ07A (Lu et al. 2014). Furthermore, the effects of the single or dual inoculation with *F. mosseae* and/or *G. intraradices* on the 90-day-old pot grown seedlings, were greater for their root colonization under *G. intraradices* than under *F. mosseae*, but the growth of seedlings was similar between the single and dual AMF inoculation (Lu et al. 2015). In addition, greater biomass production, root activity, chlorophyll, N and P content were observed in AMF seedlings than in non-AMF-inoculated counterparts. Meanwhile, positive responses noted of mulberry to AMF species ranked as *Funneliformis mosseae* > *Acaulospora scrobiculata* > *Rhizophagus intraradices* for plant's physiological and growth characteristics, whilst as *F. mosseae* > *A. scrobiculata* > *R. intraradices* for leaf quality (Shi et al. 2016). Among six tested AMF species *Acaulospora scrobiculata*, *G. aggregatum*, *G. fasciculatum*, *G. microcarpum*, *F. mosseae*, *Gigaspora margarita*, and *G. microcarpum*, followed by *G. fasciculatum* and *G. aggregatum*, seemed consistently more efficient in enhancing the uptake of P, potassium (K) and zinc (Zn), while the N uptake was enhanced by *G. aggregatum* (Subramanian et al. 2010). As a result, such differences of AMF functioning might be related to AMF species and abiotic growth conditions.

AMF Improves the Growth and Nutrition uptake of Mulberry

Compared to the 100% of chemical P fertilizer, no negative effects on the growth of 2-year-old field mulberry saplings in the Kashmir valley of India were observed when 50 or 75% P fertilizer and *F. mosseae* or *G. fasciculatum* inoculation was combined (Kour et al. 2009). Both the growth (leaf numbers) and fresh biomass production of 10-year-old mulberry plants were similar between the 100% P and the *G. fasciculatum* +50% P fertilizer (Mamatha et al. 2002). The responses of 4-year-old mulberry saplings to AMF inoculation with *F. mosseae* were evaluated in combination with or without phosphate fertilizer (0, 30, 60 and 120 kg phosphate ha^{-1} yr^{-1}) (Katiyar et al. 1995). Not only the plant growth, leaf chlorophyll and yield, but also the silkworm growth (moulting test), were similar between AMF+30 kg phosphate ha^{-1} yr^{-1} and non-AMF+120 kg phosphate ha^{-1} yr^{-1} fertilizer, which indicated a 75% of the total P contribution by *F. mosseae* or *G. fasciculatum*. The improved nutritional status in mycorrhizal associated plants contributes to an enhanced plant biomass production (Treseder 2013).

A mixed inoculum of diverse beneficial micro-organisms with AMF had been applied for promoting the growth of both mulberry and silkworm. For instance, the co-inoculation of two AMF species (*G. fasciculatum* and *F. mosseae*) with other three beneficial microorganisms (N_2-fixing *Azotobacter* sp., phosphate solubilizing bacterium *Bacillus megaterium* and fungus *Aspergillus awamori*) did increase the uptake of N, P and K in 1-year-old mulberry in India (Baqual et al. 2005, Baqual and Das 2006). Meanwhile, the dual inoculation of *F. mosseae* and bacterial biofertilizer (*Azotobacter chrococcum*) saved 50% of chemical fertilizer utilization with an improved mulberry leaf quality (also cocoon) under semi-arid fields in India (Ram Rao et al. 2007). The combined application of biofertilizers (AMF, bacterial biofertilizer, phosphate solubilizing bacterium and N_2-fixing bacteria) could be a promising management strategy to sustain both the production and quality of mulberry leaves as well as cocoons.

AMF Significantly Promotes the Photosynthetic Capacity of Mulberry

AM symbiosis improves plant P and N acquisition because the extraradical mycelium grows beyond the nutrient depletion zone of the root system in exchange of about 20% of photosynthetic carbohydrates from the host plant to maintain the AMF growth (Bago et al. 2000, Parniske 2008). The photosynthates derived from plant photosynthesis for the performance of AMF symbiosis is often referred to as the "cost", and the nutrients obtained by plants through such symbiosis is often referred to as the "benefit" (Koide and Elliot 1989). Numerous studies have found a leaf chlorophyll increase in AMF-inoculated mulberry (Baqual et al. 2005, Baqual and Das 2006, Shi et al. 2013, 2016, Chen et al. 2014, Lu et al. 2015). In general, leaf chlorophyll is closely related to photosynthetic ability in plants. Since greater chlorophyll content is in the AMF-associated plants, it seems that AMF plants maintain better photosynthetic characteristics than non-AMF counterparts, thereby enhancing photosynthetic production for sustaining the mycorrhizal performance and thus the plant growth (Smith and Read 2008). This phenomenon was also observed in mulberry plants. Colonization with AMF (*Acaulospora scrobiculata, F. mosseae*, and *Rhizophagus intraradices*) significantly enhanced total chlorophyll content by 16%–25%, and leaf net photosynthetic rate by 52–99% (Shi et al. 2016). Enhancement in photosynthesis by AMF was attributed (Mahmood et al. 2011) to the increase of nutrition uptake especially P and magnesium (Mg), and the latter is an important component of chlorophyll molecule. P is used for energy supply (ATP and NADPH) in photosynthesis, regeneration of the CO_2 acceptor ribulose biphosphate, and regulating the ratio of starch:sucrose biosynthesis (Rychter and Rao 2005). In addition, the enhancement in photosynthesis is also attributed to the increasing total leaf area by AM symbiosis (Shi et al. 2016).

AMF Significantly Increases the Yield and Quality of Mulberry

In the mulberry industry, the focus has mainly been on leaf yield and quality, which directly affect the growth and development of silkworm and production of cocoon. Numerous studies have displayed that both the leaf yield and quality of mulberry were increased by AMF inoculation under diverse conditions. The yield (leaf area, number and biomass) and leaf quality, including leaf moisture, total N, all essential amino acids, soluble protein, sugar, and fatty acid consistently increased under 6-month-old mycorrhizal inoculated mulberry seedlings (Shi et al. 2016). A qualitative comparison of protein in leaf extracts from non-mycorrhizal and AMF inoculated mulberry cultivar MR2 plants showed that several additional polypeptides appeared in mycorrhizal extracts (Subramanian et al. 2010). For instance, the polypeptide 19 kDa was expressed in *F. mosseae, G. fasciculatum* and *Gigaspora margarita* inoculated leaves, while 39.0 kDa was expressed in *G. aggregatum, F. mosseae* and *G. fasciculatum* inoculated mulberry leaves. Mulberry plants inoculated with *G. aggregatum* and *G. fasciculatum* showed an increase in sugars, total proteins, amino acids and phenols. But the leaf lipid contents were greater in plants inoculated with *G. fasciculatum* than with *G. microcarpum* (Subramanian et al. 2010). The dual inoculation of AMF and bacterial biofertilizers also did increase the leaf moisture retention capacity, protein and amino acid content, sugar and starch content, components N as well as nitrate reductase activity, and leaf yield under semi-arid fields in India (Ram Rao et al. 2007).

AMF Improves the Stress Tolerance Capacity of Mulberry

Mycorrhizae can increase the effective absorptive surface area of the mycorrhizal colonized plant. In nutrient-poor or moisture-deficient soils, nutrients being taken up by the extramatrical hyphae can lead to an improved plant growth and amelioration to abiotic environmental stresses.

Drought Resistance

Drought limits plant growth and productivity. AMF are known to confer in mulberry host plant a greater drought tolerance (Ye et al. 2012, Huang et al. 2013, Tang et al. 2013, Sekhar et al. 2017). In a pot experiment the effects of *Gigaspora rosea* and 0, 10%, 15%, 20%, 25% PEG6000 on drought were examined with the 5-month-old mulberry seedlings (Ye et al. 2012). Results showed that mycorrhizal mulberry could endure water stress under 20% PEG6000 through extraradical hyphae, which could absorb water from contiguous soils, and then deliver it to the cortex cells by intraradical mycelia (Ye et al. 2012). AMF-inoculated mulberry plants had significantly high chlorophyll content and net photosynthetic rate, stomatal conductance and transpiration rates, greater root vigor, higher enzyme activity of catalase (CAT), peroxidase (POD) and superoxide dismutase (SOD) as well as proline and sugar contents than non-AMF-inoculated plants (Tang et al. 2013). After two years of transplantation in a karst rocky desertification area in Chongqing, AMF *Gigaspora rosea* inoculated mulberry plants had 1.23 fold higher root numbers and 2.6 fold higher root dry weights. Plant height, stem diameter, leaf number and the 5th leaf area were also significantly increased by 43%, 72%, 110% and 117%, respectively, after inoculation with *Gigaspora rosea* (Chen et al. 2014). The contents of chlorophyll and osmosis substances (proline, soluble sugar and soluble protein), and the leaf antioxidant enzyme activity of CAT, POD and SOD were also significantly higher in AMF *Gigaspora rosea* inoculated mulberry than in non-AMF seedlings (Cao et al. 2017). The potential mechanisms of drought tolerance in mycorrhizal mulberry plants could be as follows: (1) an improved root nutrient and water uptake through an enhanced root vigor and an expanding root absorption area with intensive hyphal mycelia; (2) an enhanced vegetative growth by higher chlorophyll content and photosynthesis rate, water use efficiency; and (3) an increased osmotic adjustment and antioxidant enzyme activity.

Salinity Resistance

Salinity is one of the most serious factors limiting plant performance (Ashraf and Foolad 2007). Mulberry grows well in slightly acidic soils and is hard to grow in salt-affected soils. The slow growth of mulberry leaves with poor quantities in salt-affected areas is of a serious concern to sericulture. Various studies have shown that AMF could mitigate such deleterious salinity effects on a number of plant species including *Catharanthus roseus, Elaeagnus angustifolia, Trigonella foenum-graecum* and *Solanum lycopersicum* (Evelin and Kapoor 2014, Frosi et al. 2017, Khalloufi et al. 2017, Garg and Bharti 2018). However, there is limited knowledge of AMF status in the rhizosphere of mulberry plants under salt stress. The growth parameters (stem height, girth, node numbers, average weight) and AMF root colonization and spores of mulberry (*Morus alba* var. sujanpuri) showed a consistent decrease when watered with 0.05% NaCl (Kashyap et al. 2004). However, the inoculation of a mixture of AMF species (*F. mosseae, Glomus microcarpum, G. macrocarpum, G. fasciculatum, Gigaspora margarita* and *G. heterogama*) significantly enhanced the survival percentage of saplings from 25 to 45% under salt stress and to 50% under the dual inoculation of AMF and *Azotobacter* (Kashyap et al. 2004). The AM-inoculated mulberry plants thus showed remarkable biochemical changes under saline conditions, but the exact mechanisms involved in such changes leading to tolerance are still not fully understood. However, the role of AMF (*Glomus* and *Gigaspora* species) colonized root extracts was found equivalent to that of the indole butyric acid (1.0 mg IBA l^{-1}) in promoting rhizogenesis, root growth and proliferation of *M. alba* with addition of 0.1 to 0.4% (w/v) NaCl (Sharma et al. 2005).

Toxic Metals Resistance

The products made from mulberry leaves and fruits have acquired ever-increasing attention in the food, drink and pharma industry. Nevertheless, there are potential food security concerns due to toxic heavy metal accumulation in mulberry products. For instance, the average mineral concentrations (mg kg^{-1} DW fruit) in fully-ripened fruits of four mulberry species (*Morus alba*, *M. nigra*, *M. macroura*, and *M. laevigata*) were as Ca ($3,680 \pm 540$), Cu (15 ± 1), Fe (792 ± 261), K ($2,774 \pm 230$), Mg ($1,730 \pm 521$), Mn (26 ± 8), Na ($2,776 \pm 230$), P ($2,673 \pm 203$) and Zn (817 ± 266), respectively (Mahmood et al. 2011). Thus the *Morus* fruits could be particularly explored as a rich source of Fe and Zn, which generally are in a short supply in human diets. However, such fruits also had a concentration of up to 423 ± 220 mg Al kg^{-1} DW fruit. The heavy metal (Cd, Co, Cr, Cu, Mn, Ni, Pb, and Zn) concentrations both in soil and mulberry plants (leaf and fruit) grown at 20, 100, 400 and 1,000 m roadside distances in the Upper Coruh Valley of Turkey analysed had highest all metal values at the 20 m distance but for Pb it was high at 100 m distance and for Ni at all above-mentioned distances (Pehluvan et al. 2012). Such high metal concentrations in the mulberry fruits are greater than their limited standards as worked out by the Turkish food security. These results showed a strong relationship of the heavy metal concentration between the bulk soil and mulberry fruits (Pehluvan et al. 2012). Negative associations were also observed between plant growth, leaf nutritional quality, and leaf Cd concentration (Prince et al. 2002). The nutritional qualities of leaf total protein, carbohydrate, and chlorophyll too decreased, but the leaf free amino acid, total N, Cd concentration (from the external addition of 5 to 20 μg Cd/g) increased (Prince et al. 2002).

Studies have shown that AM fungi can play an important role in improving plant health even in metal contaminated soils. For instance, inoculation of *Gigaspora rosea* could significantly improve the biomass production and accumulations of mineral nutrients (N, K, Ca, Mg) in root, stem and leaf of 60-day-old mulberry seedlings under 0, 5, 20 and 40 mg Cd/kg (Wang et al. 2017). The Cd in the AM mycelia and plant parts (leaf, stem and root) showed an increased trend with an increase in the tested Cd concentrations. The portion of Cd accumulation ranked as root (92.9%) > stem (4.4%) > leaf (2.7%) in the 60-day-old mycorrhizal mulberry seedlings. However, Cd accumulation with the increase of external Cd addition though significantly increased in the underground roots, but decreased in the aboveground tissues. In addition, the Cd concentration was negatively correlated with the Mg concentration in the mycelia (Wang et al. 2017).

Disease Resistance

Mulberry is a perennial plant that is widely exposed to soil-borne diseases. The diseases generally cause 5 to 10% of leaf biomass loss by defoliation and an additionally economic loss of 20 to 25% by deterioration in leaf quality (Philip et al. 1994, Beevi and Qadri 2010). Hence soil disease is one of the main direct limiting factor for mulberry plantation and silk industry. AMF is particularly promising in the pathogen biocontrol by altering the morphology, nutritional, disease defense status, and rhizosphere antagonistic microflora composition, hence enhancing the plant growth and development (Whipps 2004). However, at present limited information is available on how AMF could contribute to plant disease resistances.

Research Prospects

Accumulated evidence has established about the multiple benefits of AMF symbiosis to the leaf growth and quality of mulberry plants. Here we present the following aspects for future promising mycorrhizal studies in enhancing mulberry plantations.

Mycorrhizal Fungal Diversity and Community Composition in Mulberry Plantations

Mulberry has been domesticated over thousands of years ago. It has adapted to climatic conditions of a wide area of tropical, subtropical and different temperature zones in Asia, Europe, North and South America, and Africa. Mulberry plants present high mycorrhizal dependency and AMF communities are considered as a vital indicator of soil quality in the mulberry plantations. In general a higher AMF richness occurs in the rainy season between July and October (Wang et al. 2015). It is unknown if the composition and diversity of the indigenous AMFs vary with the plantation age of mulberry trees, either seasonally or spatiotemporally in mulberry plantations. Although Xing et al. (2018) and Shi et al. (2013) have investigated the AMF composition and diversity of mulberry in the karst area of Guizhou and Chongqing in southwest China, mycorrhizal colonization and species richness vary with habitats. A total of 15 species and 4 varieties with about 3,000 mulberry germplasms have been reserved in China. No studies have been addressed whether different mulberry plants could anchor distinctive AMF communities and display preferences for some AMF species, and whether a specific AMF species could contribute a unique function to mulberry plantation. Therefore, it is imperative to screen the dominant AMF species in mulberry plantations for their practical field inoculation in order to improve mulberry leaf growth and fruit quality.

The Molecular Mechanisms of Mycorrhizal Mulberry to Stress Resistances

AMF acts as biofertilizer and bioprotector in order to confer better mulberry plant growth, higher nutrient uptake, greater tolerance to abiotic and biotic stresses including drought, salt, toxic metal and disease. An enhanced stress tolerance of mulberry by mycorrhization reflects complex biological processes that involve improvement of water and nutrient uptake through extraradical hyphal exploitation, ameliorate root vigor and intensive systems, enhance photosynthetic activity, have greater osmotic adjustments (through proline, soluble sugar and soluble protein accumulations), and scavenging reactive oxygen using antioxidant enzymes (POD, SOD and CAT), etc. However, the physiological and molecular mechanisms of these processes are not understood. Stable isotope techniques could be applied to study nutrient and water translocations mediated by mycorrhizal hyphae or networks. Molecular approaches and genetic manipulations may also provide new insights into the alleviating role of mycorrhizal symbiosis under abiotic and/or biotic stresses. Furthermore, the transcriptome and proteomic analyses could identify the responsible genes that contribute benefits to tolerate abiotic and biotic stresses for mycorrhizal mulberry plants. The target genes can be also engineered and transferred into plants through transgenic manipulations to tolerate harsh environments for new generations of mulberry plants.

Acknowledgements

Researches in the authors' laboratory are jointly supported by projects from the Science and Technology Department of Sichuan Province (2018JZ0027), National Youth Natural Science Foundation (4111800096), National Key Research and Development Program (2016YFD0200104), and the 100 Talents program of the Chongqing Municipality (2015-2), China. The authors have no conflicts of interest to declare.

References

Ashraf, M. and M.R. Foolad. 2007. Roles of glycine betaine and proline in improving plant abiotic stress resistance. Environ. Exp. Bot. 59: 206–216.

Bago, B., P.E. Pfeffer and Y. Shacharhill. 2000. Carbon metabolism and transport in arbuscular mycorrhizas. Plant Physiol. 124: 949–958.

Baqual, M.F., P.K. Das and R.S. Katiyar. 2005. Effect of arbuscular mycorrhizal fungi and other microbial inoculants on chlorophyll content of mulberry (*Morus* spp.). Mycorrhiza News 17: 12–14.

Baqual, M.F. and P.K. Das. 2006. Influence of biofertilizers on macronutrient uptake by the mulberry plant and its impact on silkworm bioassay. Caspian J. Environ. Sci. 4: 98–109.

Baum, C., W. El-Tohamy and N. Gruda. 2015. Increasing the productivity and product quality of vegetable crops using arbuscular mycorrhizal fungi: A review. Sci. Hortic. 187: 131–141.

Beevi, N.D. and S.M.H. Qadri. 2010. Biological control of mulberry root rot disease (*Fusarium* spp.) with antagonistic microorganisms. J. Biopest. 3: 90–92.

Cao, M., M. Yu, M. Lu, Y.Y. Jiang, X.Z. Huang and X.H. Yang. 2017. Ecological adaptability of mulberry saplings inoculated with *Gigaspora rosea* under stress conditions in rocky desertification areas and riparian zones. Science of Sericulture 43(2): 189–195 (in Chinese with English abstract).

Chen, K., S.M. Shi, X.H. Yang and X.Z. Huang. 2014. Contribution of arbuscular mycorrhizal inoculation to the growth and photosynthesis of mulberry in karst rocky desertification area. Appl. Mech. Mater 488: 769–773.

Conversa, G., A. Bonasia, C. Lazzizera and A. Elia. 2015. Influence of biochar, mycorrhizal inoculation, and fertilizer rate on growth and flowering of Pelargonium (*Pelargonium zonale* L.) plants. Front Plant Sci. 6: 429.

Evelin, H. and R. Kapoor. 2014. Arbuscular mycorrhizal symbiosis modulates antioxidant response in salt-stressed *Trigonella foenum-graecum* plants. Mycorrhiza 24: 197–208.

Evelin, H., T.S. Devi, S. Gupta and R. Kapoor. 2019. Mitigation of salinity stress in plants by arbuscular mycorrhizal symbiosis: Current understanding and new challenges. Front Plant Sci. 10: 470.

Frosi, G., V.A. Barros, M.T. Oliveira, M. Santos, D.G. Ramos, L.C. Maia and M.G. Santos. 2017. Arbuscular mycorrhizal fungi and foliar phosphorus inorganic supply alleviate salt stress effects in physiological attributes, but only arbuscular mycorrhizal fungi increase biomass in woody species of a semiarid environment. Tree Physiol. 38: 25–36.

Garg, N. and A. Bharti. 2018. Salicylic acid improves arbuscular mycorrhizal symbiosis, and chickpea growth and yield by modulating carbohydrate metabolism under salt stress. Mycorrhiza 28: 727–746.

Harauma, A., T. Murayama, K. Ikeyama, H. Sano, H. Arai, R. Takano, T. Kita, S. Hara, K. Kamei and M. Yokode. 2007. Mulberry leaf powder prevents atherosclerosis in apolipoprotein E-deficient mice. Biochem. Biophys. Res. Commun. 358: 751–756.

Huang, X.H., Y. Liu, J.X. Li, X.Z. Xiong, Y. Chen, X.H. Yin and D.L. Feng. 2013. The response of mulberry trees after seedling hardening to summer drought in the hydrofluctuation belt of the Three Gorges Reservoir Area. Environ. Sci. Pollut. Res. 20: 7103–7111.

Jiang, Y., W. Wang, Q. Xie, N. Liu, L. Liu, D. Wang, X. Zhang, C. Yang, X. Chen and D. Tang. 2017. Plants transfer lipids to sustain colonization by mutualistic mycorrhizal and parasitic fungi. Science 356: 1172–1175.

Kashyap, S., S. Sharma and P. Vasudevan. 2004. Role of bioinoculants in development of salt-resistant saplings of *Morus alba* (var. sujanpuri) *in vivo*. Sci. Hortic. 100: 291–307.

Katiyar, R.S., P.K. Das, P.C. Chowdhury, A. Ghosh, G.B. Singh and R.K. Datta. 1995. Response of irrigated mulberry (*Morus alba* L.) to VA mycorrhizal inoculation under graded doses of phosphorus. Plant Soil 170: 331–337.

Katsube, T., N. Imawaka, Y. Kawano, Y. Yamazaki, K. Shiwaku and Y. Yamane. 2006. Antioxidant flavonol glycosides in mulberry (*Morus alba* L.) leaves isolated based on LDL antioxidant activity. Food Chem. 97: 25–31.

Katsube, T., Y. Tsurunaga, M. Sugiyama, T. Furuno and Y. Yamasaki. 2009. Effect of air-drying temperature on antioxidant capacity and stability of polyphenolic compounds in mulberry (*Morus alba* L.) leaves. Food Chem. 113: 964–969.

Khalloufi, M., C. Martínez-Andújar, M. Lachaâl, N. Karray-Bouraoui, F. Pérez-Alfocea and A. Albacete. 2017. The interaction between foliar GA3 application and arbuscular mycorrhizal fungi inoculation improves growth in salinized tomato (*Solanum lycopersicum* L.) plants by modifying the hormonal balance. J. Plant Physiol. 214: 134–144.

Kim, J.C., Y.H. Choi, J.Y. Moon and J.U. Kim. 1984. Growth stimulation of mulberry trees in unsterilized soil under field conditions with VA mycorrhizal inoculation. Korean J. Sericultural. Sci. 26: 7–10.

Koide, R. and G. Elliot. 1989. Cost, benefit and efficiency of the vesicular-arbuscular mycorrhizal symbiois. Funct. Ecol. 3: 252–255.

Kour, R., M.R. Mir, N.A. Mir, M. Khan, G. Darzi and M. Farooq. 2009. Impact of arbuscular mycorrhiza fungal inoculation on growth and development of Mulberry (*Morus* spp.) saplings under Kashmir conditions. Appl. Biol. Res. 11: 49–52.

Lu, C., F.D. Ji, F.R. Zhu, A.C. Zhao, G.Q. Luo and C. Su. 2017. Mulberry Cultivation Varieties in China. Chongqing: Southwest Normal University 3–11 (in Chinese with English abstract).

Lu, H., R.X. Ye, J. Qin, S.C. Zuo and L.L. Wu. 2014. Improvement effects of three arbuscular mycorrhizal fungal strains on growth of mulberry trees planted in soil of southern Xinjiang. Science of Sericulture 40: 804–810 (in Chinese with English abstract).

Lu, N., X. Zhou, M. Cui, M. Yu, J. Zhou, Y. Qin and Y. Li. 2015. Colonization with arbuscular mycorrhizal fungi promotes the growth of *Morus alba* L. seedlings under greenhouse conditions. Forests 6: 734–747.

Mamatha, G., D. Bagyaraj and S. Jaganath. 2002. Inoculation of field-established mulberry and papaya with arbuscular mycorrhizal fungi and a mycorrhiza helper bacterium. Mycorrhiza 12: 313–316.

Ouziad, F., U. Hildebrandt, E. Schmelzer and H. Bothe. 2005. Differential gene expressions in arbuscular mycorrhizal-colonized tomato grown under heavy metal stress. J. Plant Physiol. 162: 634–649.

Özgen, M., S. Serçe and C. Kaya. 2009. Phytochemical and antioxidant properties of anthocyanin-rich *Morus nigra* and *Morus rubra* fruits. Sci. Hortic. 119: 275–279.

Pan, Y. 2000. Progress and prospect of germplasm resources and breeding of mulberry. Acta Sericologic Sinica 26: 1–8 (in Chinese with English abstract).

Parniske, M. 2008. Arbuscular mycorrhiza: the mother of plant root endosymbioses. Nat. Rev. Microbiol. 6: 763.

Pehluvan, M., H. Karlidag and M. Turan. 2012. Heavy metal levels of mulberry (*Morus alba* L.) grown at different distances from the roadsides. J. Anim. Plant Sci. 22: 665–670.

Philip, T., V.P. Gupta, Govindaiah, A.K. Bajpai, R.K. Datta and T. Philip. 1994. Diseases of mulberry in India—Research priorities and management strategies. Int. J. Trop. Plant Dis. 12: 1–21.

Prince, W.S., P.S. Kumar, K. Doberschutz and V. Subburam. 2002. Cadmium toxicity in mulberry plants with special reference to the nutritional quality of leaves. J. Plant Nutr. 25: 689–700.

Rajagopal, D., S. Madhavendra and K. Jamil. 1989. Occurrence of vesicular-arbuscular mycorrhizal fungi in roots of *Morus alba* L. Curr. Sci. 58: 687–689.

Ram Rao, D.M., J. Kodandaramaiah, M.P. Reddy, R.S. Katiyar and V.K. Rahmathulla. 2007. Effect of VAM fungi and bacterial biofertilizers on mulberry leaf quality and silkworm cocoon characters under semiarid conditions. Caspian J. Environ. Sci. 5: 111–117.

Rychter, A.M. and I. Rao. 2005. Role of phosphorus in photosynthetic carbon metabolism. Handbook of Photosynthesis 2: 123–148.

Sekhar, K.M., K.S. Reddy and A.R. Reddy. 2017. Amelioration of drought-induced negative responses by elevated CO_2 in field grown short rotation coppice mulberry (*Morus* spp.), a potential bio-energy tree crop. Photosynth. Res. 132: 151–164.

Setua, G., R. Kar, J. Ghosh, K. Das and S. Sen. 1999. Influence of arbuscular mycorrhizae on growth, leaf yield and phosphorus uptake in mulberry (*Morus alba* L.) under rainfed, lateritic soil conditions. Biol. Fertil Soils 29: 98–103.

Sharma, S., S. Kashyap and R. Vasudevan. 2005. *In vitro* rhizogenesis of *Morus alba* by mycorrhizal extracts under saline stress. Eur. J. Hort. Sci. 70: 79–84.

Shi, S.M., K. Chen, B. Tu, X.H. Yang and X.Z. Huang. 2013. Diversity of AMF in mulberry rhizosphere in a rock desertification area and vigorous mulberry seedling culture. Journal of Southwest University 35(10): 1–8 (in Chinese with English abstract).

Shi, S.M., K. Chen, Y. Gao, B. Liu, X.H. Yang, X.Z. Huang, G.X. Liu, L.Q. Zhu and X.H. He. 2016. Arbuscular mycorrhizal fungus species dependency governs better plant physiological characteristics and leaf quality of mulberry (*Morus alba* L.) seedlings. Front. Microbiol. 7: 1030.

Shu,Y.F., J. Ye, C.Y. Pan, X.H. Yang, X.Z. Huang and J. Qin. 2011. Developmental features of mycorrhiza and its promotion effect on growth of mulberry saplings in Three Gorges Reservoir Region. Science of Sericulture 37(6): 0978–0984 (in Chinese with English abstract).

Smith, S.E. and D.J. Read. 2008. Mycorrhizal symbiosis (Third edition). Academic Press.

Subramanian, K., E. Sanniyasi and P. Mylsamy. 2010. Effect of VAM fungi on growth and physiological parameters of mulberry (*Morus alba* L.) cultivars in South India. Biosci. Biotechnol. Res. Asia 7: 793–806.

Sukumar, J., S. Padma, Tomy Philip, V.P. Gupta, Govindaiah, A.K. Bajpai, R.k. Datta and T. Philip. 1994. Diseases of mulberry in INdia—Research priorities and management strategies. International Journal of Tropical Plant Diseases 12: 1–21.

Tang, X., D.J. Liu, B. Tu, X.H. Yang and X.Z. Huang. 2013. Promotion effect on growth of mycorrhiza-inoculated mulberry saplings and physiological and biochemical mechanism to drought tolerance. Journal of Southwest University 35(8): 19–26 (in Chinese with English abstract).

Treseder, K.K. 2013. The extent of mycorrhizal colonization of roots and its influence on plant growth and phosphorus content. Plant Soil 371: 1–13.

van der Heijden, M.G.A., J.N. Klironomos, M. Ursic, P. Moutoglis, R. Streitwolf-Engel, T. Boller, A. Wiemken and I.R. Sanders. 1998. Mycorrhizal fungal diversity determines plant biodiversity, ecosystem variability and productivity. Nature 396: 69–72.

Wang, C., Z. Gu, H. Cui, H. Zhu, S. Fu and Q. Yao. 2015. Differences in arbuscular mycorrhizal fungal community composition in soils of three land use types in subtropical hilly area of Southern China. PloS one 10: e0130983.

Wang, K.Y., Y.Y. Jiang, W.J. Song, J.Y. Liu, X.Z. Huang and X.H. Yang. 2017. Effects of interaction between AMF and Cd on mulberry growth and absorption and migration of inorganic elements. Mycosystema 36(7): 996–1009 (in Chinese with English abstract).

Whipps, J.M. 2004. Prospects and limitations for mycorrhizas in biocontrol of root pathogens. Can. J. Bot. 82(8): 1198–1227.

Wipf, D., F. Krajinski and P.E. Courty. 2019. Trading on the arbuscular mycorrhiza market: from arbuscules to common mycorrhizal networks. New Phytol. 223: 1127–1142.

Xing, D., Z.H. Wang, J.J. Xiao, S.Y. Han, C.B. Luo, A.M. Zhang, L.L. Song and X.B. Gao. 2018. The composition and diversity of arbuscular mycorrhizal fungi in karst soils and roots collected from mulberry of different ages. Ciencia Rural 48: 1–14.

Ye, J., B. Tu, S.M. Shi, X.H. Yang, X.Z. Huang and J. Qin. 2012. Ecological response at the cytological level of the root primary structure in mycorrhiza-inoculated mulberry saplings to gradient water stress. Journal of Southwest University 34: 67–72 (in Chinese with English abstract).

CHAPTER 15

Mulberry (*Morus* sps.) Cultivation for Sustainable Sericulture

Dandin SB[1],* and *Vijaya Kumari N[2]*

INTRODUCTION

Mulberry, the only food source to the Mori Silkworm belongs to the genus *Morus* which was established by *Linnaeus* (1773) and reported to include about 70 species, distributed in both new and old world. The genus has a multi-centric origin covering south east and central Asia, Caucasious region in the old world, central and south America, and Europe in the new world. The species of the genus *M. laevaegata* and *M. serrata* are occurring sea level to about 1700 mtr above Mean Sea Level, respectively, and found in all types of forests, i.e., from dry deciduous to tropical rain forests. Some of the species like *M. indica* occur naturally as small trees and *M. laevaegata* is a gigantic lofty tree reaching a height about 50 mts. The shape of the trees vary from a woody climber (*M. sparciflorus*) to a huge tree with a girth of about 18 mtr (*M. serrata*). Other unique features associated with genus *Morus* are firstly, *M. serrata* is the oldest angiosperm species of more than 4000 year old and found in Joshi mutt of Chamoli district in Uttarakhand, India. Secondly, chromosome number ranges from diploid (2n = 2x = 28) in *M. indica* to decosoploid level of 2n = 22x = 308 chromosomes in (*M. nigra*). Thirdly, nature of sex in plants of the genus was demonstrated to the world by Camerarius (1694) in his classical work De sexu plantarum epistle. Fourthly, genus shows highest variability because of its wide spread occurrence right from temperate, subtropical and tropical regions of the world (Figures 1–3).

Genus *Morus* is of high economic importance because of its multipurpose use as a source of four "F" s, i.e., Feed, Fodder, Fuel and Fiber and meeting the diversified human needs. *M. alba (M. indica), M multicaulis (M. latifolia), M. bombysis* and *M. luhu* are cultivated as high bush to small tree in large areas of China, India, Vietnam, Thailand, USSR, Brazil, etc., for the production of foliage for rearing silkworms (*Bomboyx mori* L.) which feeds only on the leaves of mulberry and produces the most valued silk of commerce. The foliage of few mulberry species is also used to feed live stock such as sheep, goat, cows, etc. Four species of the genus *Morus* namely *M. indica, M. laevigata, M. serrate* and *M. nigra* are the major source of valuable timber. Mulberry wood is valued because of its fine grains, flexibility and taking fine polish, etc. Because of these characters, the wood is used in manufacture of cricket bats, tennis rackets, tool handles and in other turnery works. The wood of *M. laevigata* is also used in building houses, boats and other industrial purposes. Fruits of *M. nigra, M. alba, M. multicaulis* and *M. leavigata* are delicious and used both for table dessert and in preparation of jelly, jam and juice. Of late, the tea prepared out of mulberry leaves is becoming very popular as an anti-diabetic health drink (Mala and Sharma

[1] Consultant, Biodiversity International, Rome; Former Vice-Chancellor, UHS Bagalkot, Former Director CSB, Bangalore.
[2] Dept. of Biosciences and Sericulture, Sri Padmavati MahilaVisvavidyalayam, Women's University, Tirupati.
* Corresponding author: dandinbnm@gmail.com

Figure 1. Mulberry varieties for Southern Zone (Source: Hand Book of Sericulture Technologies 2010).

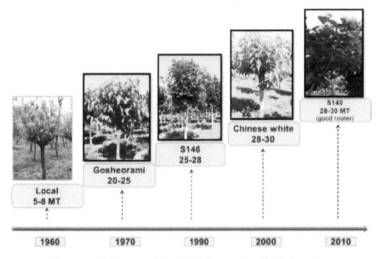

Figure 2. Mulberry varieties for Eastern and North-Eastern Zone.

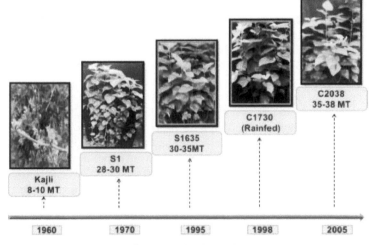

Figure 3. Mulberry varieties for North Western Zone.

Table 1(a). Mulberry varieties developed for Southern States of India.

Sl. No	Mulberry Variety	Potential Leaf yield (Mt./ha/yr.)	Recommended for
1.	K2	32	Irrigated conditions
2.	S-54 (1984)	35	Irrigated conditions
3.	S36 (1986)	45	Young age rearing
4.	S13 (1990)	14	Rainfed condition
5.	S34 (1990)	14	Rainfed condition
6.	RFS 135 (1990)	35	Rainfed condition
7.	V1 (1997)	60	Young and late age silkworm rearing
8.	RFS 175	45	Rainfed condition
9.	Sahana	25	under shade in a coconut garden
10.	AR 11 (1999)	10	Rainfed and semi-arid condition
11.	AR 12 (2000)	25	Alkaline soils
12.	G4 (2003)	60	Late age silkworm rearing
13.	G2 (2003)	40	Young age silkworm rearing
14.	RC 1 (2002)	26	Sub optimal conditions
15	RC2 (2002)	26	Sub optimal conditions

(Source: Dandin and Giridhar (2010), Hand Book of Sericulture Technologies Central Silk Board, Bangalore 2010)

2010). The roots and bark of the mulberry are used in curing some of the health ailments. Hence, mulberry serves as a source of many beneficial products in the service of mankind and being worshiped in some parts of the world as tree—"Kalpavruksha".

Development of New Mulberry Varieties

Realising the importance of quality mulberry leaf for bivoltine silkworms, research works have been initiated at CSRTI, Pampore, Berhampore and Mysore. The idea of popularising temperate mulberry varieties namely; Kosen, KNG Ichinose, etc., have not yielded any encouraging results due to poor rooting and response to repeated pruning. The only noteworthy success was development of BC-259 at CSRTI, Berhampore through back crossing of Kosen into local variety. Later three varieties namely; Mandalaya (Ber S1) at CSRTI Berhampore, Kanva-2 (M5) at CSRTI Mysore and Check Majra, (136) at RSRTI Jammu were developed with higher yield of 20 Mt, 25 Mt and 15 Mt/ha/yr respectively. These varieties were only the natural selections out of open pollinated hybrids.

Later the efforts made by CSRTI Mysore between 1970–80 through mutation breeding, four selections namely; S-54, S-41, S-36 and S-30 were isolated with a good quality leaf and higher yield. S-54 was high yielding with spreading type of branches whereas, S 41 was fast growing type with very high rooting ability. However, the leaf moisture and protein content were low. Among S 30 and S 36, variety S 36 is found superior in leaf quality especially for rearing young age silkworms and even till today the variety is preferred over other varieties for rearing Chawki worms, especially bivoltine. Efforts made simultaneously for developing rain fed varieties have resulted in developing varieties RFS 135 and RFS 175 which proved better over other varieties. During the same period, CSRTI Berhampore has developed verities BER-776, BER-769, etc., and conducted large scale field trials.

New mulberry breeding programmes initiated during 1980's and 90's have yielded excellent results and variety V-1 was developed by cross intraspecific hybridization which was found superior over all the other existing mulberry varieties. This has laid real foundation for the success of bivoltine in the country in general and in Southern Indian states in particular. G 2 and G 4 were also isolated during the same period and found equally

Table 1(b). Mulberry varieties for temperate conditions.

Mulberry Variety	Leaf Yield
Tr-10, S146, S1635 and Vishala	14 to 19 MT/ha/yr (Under rainfed conditions) Recommended for Subtropical region
Goshoerami, Ichinose, KNG and Tr-10	22 to 26 MT/ha/yr (Under rainfed conditions). Recommended for temperate region

(Source: Dandin and Giridhar (2010), Hand Book of Sericulture Technologies Central Silk Board, Bangalore 2010)

Table 2. Morphophysiological characteristics of Mulberry varieties cultivated under irrigated conditions.

Varieties	Kanva-2	S-36	S-135	S-175	V-1	G-4
Parentage	OPH from local	Mutant of Ber. local	OPH from Kanvaaa-2	OPH from Kanvaaa-2	S-30 x Ber. C 776	M. multicaulis x S-13
Year of Release	1966	1986			1977	2014
Plant type	Erect	Semi-erect	Erect	Erect	Erect	Erect
Inflorescence	Unisexual	Unisexual	Unisexual	Unisexual	Unisexual	Unisexual
Flower/Sex	Female	Female	Male	Male	Male	Female
Ploidy	Diploid	Diploid	Diploid	Diploid	Diploid	Diploid
Rooting ability	> 80	> 48	> 80	> 85	> 90	> 90
Inter nodal distance (cm)	4.5	3.53	4.4	4.6	5.2	3.9
Leaf nature	Unlobed Cordate Medium Size	Unlobed Cordate Large Size	Unlobed Cordate Large Size	Unlobed	Unlobed Ovulate, Large, Dark Green	Entire, cordate wavy, dark green
Fresh weight of 100 leaves (g)	370–390	395–425	490–520	525–560	530–560	416–425
Leaf Moisture content (%)	65–70	70–74	73–74	72–75	75–78	75–77
Leaf MRC (%)	61–63	75–78	74–78	72–76	78–82	74–76
Proteins %	18–21	23–25	25–27	20–25	26–28	24–26
Sugars %	9–12	10–13	11–14	12–15	16–18	16–19
Carbohydrates %	20–24	23–25	24–28	24–28	28–30	27–30
Leaf yield (MT/ha/year)	30–35	40–45	40–42	50–55	60–65	61–65
Recommended for	Irrigated condition	Irrigated condition	Irrigated condition	Irrigated condition	Irrigated condition	Irrigated condition
Specially recommended for	Water scarcity areas also	Chawki rearing	Water scarcity areas	Water scarcity areas	Chawki and late age rearing	Late age rearing

(Source: Dandin and Giridhar (2010), Hand Book of Sericulture Technologies Central Silk Board, Bangalore 2010)

good, but they could not be popularized because of the fast spread and popularity of V-1 variety (Sivaprasad 2017). Comparative yield performance of all newly developed varieties is shown in Tables 1 to 4 and Figures 1–3.

Package of Practices for Mulberry Cultivation

Among all the factors contributing to success of cocoon crops, share of leaf improvement and quality (Figures 4, 5) is about 38.4% followed by environment factors which is 37.0%. Mulberry cultivation is the first and foremost

Table 3. Morpho-physiological characteristics of Mulberry varieties cultivated under rainfed soil moisture stress conditions.

Varieties	Mysore local	Kanva-2	S-13	S-34	MSG-2
Parentage	Locally adapted	OPH from local	OPH from Kanva-2	S-30 x Ber. C 776	BR-4 x S13
Year of Release	----	1966	1990	1990	2015
Plant type	Erect	Erect	Erect	Erect	Erect
Inflorescence	Unisexual	Unisexual	Unisexual	Unisexual	Unisexual
Flower/Sex	Monoecious-Female and male on separate plants	Female	Male	Male	Male
Ploidy	Diploid	Diploid	Diploid	Diploid	Diploid
Rooting ability %	> 90	80	> 80	> 75	> 80
Inter nodal distance (cm)	3.8	4.5	3.7	3.8	3.9
Leaf nature	Heterophyllous-Lobed (1–5 lobes)	Unlobed Cordate Medium Size	Unlobed, Ovate, Medium Size	Unlobed, Cordate Medium Size	Unlobed, Cordate Medium Size
Fresh weight of 100 leaves (g)	245	290	365	642	420
Moisture content %	60–65	68–72	73–75	73–76	75–77
Leaf MRC %	61–62	61–63	75–78	74–77	73–75
Proteins %	14–16	16–20	21–23	20–24	22–25
Sugars %	8–10	13–14	13–17	12–16	15–18
Leaf yield (MT/ha/year)	4–6	7–9	13–16	13–16	22–23
Recommended for	Rainfed areas	Rainfed conditions	Rainfed red soil	Black soil	Rainfed Red soil
Specially recommended for	Water scarcity areas also	Water scarcity areas also	Chawki and late age rearing under rainfed conditions	Late age rearing	Chawki and late age rearing under rainfed conditions

(Source: Dandin and Giridhar (2010), Hand Book of Sericulture Technologies Central Silk Board, Bangalore 2010)

activity to be taken up for starting of sericulture enterprise and about 60% of the initial cost towards capital expenditure is spent on establishment of healthy mulberry garden (Figure 6). The third important aspect to be considered is that, the establishment of good mulberry garden sustains the cocoon production at least upto 20 years. Hence, importance of improved mulberry along with all the good agriculture practices needs no emphasis. Mulberry cultivation involves four to five important aspects namely; selection of the land, choice of mulberry variety, spacing and planting system followed by regular cultivation practices such as, fertilization, irrigation, crop protection and lastly pruning and leaf harvest. All the above aspects need to be attended carefully for production of healthy leaf and in turn cocoons with sustainable income over the years for sericulture in general and bivoltine cocoon production in particular. Bivoltine silkworm rearing is basically practiced in temperate countries namely Japan, China, South Korea and few Russian countries and was practiced in Jammu and Kashmir valley since centuries due to prevailing temperate climatic conditions. Since the native varieties/species of mulberry were not suitable for bivoltine silkworm rearing, varieties like *Kosen, Ghoshoerami, Tserkasaguwa, Ensatakasuke, Ichii, Ichinose*, Kasyonizumigeshi, etc., were introduced from Japan to the valley as early as 1900 or even early. On the contrary, the Southern and Eastern Indian states were practising only multivoltine based sericulture using Nistari and Pure Mysore races with local mulberry varieties namely; Khajali in West Bengal, and 'Mysore local' in the then Mysore provinces and adjoining areas of Tamil Nadu and Andhra Pradesh (Dandin et al. 2000).

Table 4. Morphophysiological characteristics of Mulberry varieties cultivated under Specific climatic conditions.

Recommended For	Alkaline soils	Intercrop under coconut shade	Semi-arid areas	For hilly ar Tamilnadu	Resource constraints (50% irrigation and 50% fertilizers recommended for optimal conditions)		Sub-optimal irrigated condition (60% irrigation recommended)
Varieties	AR-12	Sahana	AR-11	MR-2	RC-1	RC-2	AGB-8
Parentage	S-41 (4x) x Ber. 776 (2x)	Kanva-2 X Kosen	OPH from Kanva-2	OPH from exotic variety	Punjab local x Kosen	Punjab local x Kosen	(Sujanpur-5 x Phillipines) × (K-2 x Black cherry)
Plant type	Semi-erect	Erect	Erect	Erect	Semi-erect	Erect	Erect
Inflorescence	Unisexual	Unisexual	Unisexual/ Monoecious	Unisexual	Unisexual	Unisexual	Unisexual
Flower/Sex	Female	Female	Predominantly male	Male	Female	Female	Male
Ploidy	Triploid	Diploid	Diploid	Diploid	Diploid	Diploid	Diploid
Rooting ability (%)	> 90	> 65	> 80	> 90	> 90	94	> 90
Inter nodal distance	3.75	4.8	3.8	4.0	4.6	4.4	5.2
Leaf nature	Entire, Cordate, Dark Green	Unlobed, Cordate, Glossy	Unlobed, small, thick semi-glossy, Cordate	Hetero-phyllous	Two to three lobed, cordate, glossy	Unlobed, Medium	Unlobed, large Cordate, smooth and glossy
Fresh weight of 100 leaves (g)	425	379	290	374	410	405	605
Moisture content %	71–74	71–73	67–70	73–76	72–73	72–73	70–73
Leaf MRC %	64–66	73–75	56–58	73–76	74–77	70–73	71–74
Proteins %	19–21	20–22	15–17	24–28	21–23	20–23	21–23
Sugars %	13–14	11–13	11–12	14–17	14–16	12–16	14–17
Leaf yield (MT/ ha/year)	20–24	28–30	9–10	35–40	25–26	40–45	45–47
Recommended for	Irrigated	Irrigated	Rainfed	Hilly areas	Sub-optimal conditions of 50% irrigation and fertilizers		Sub-optimal conditions of 50% irrigation
Specially recommended for	Alkaline soil up to pH 9.4 with SAR 30 %	Intercropping with coconut plantation	Semi arid areas with pH 9.0	Mildew infested areas	Suitable for poor soil and poor input application condition.		Suitable for soil moisture stress prone areas semi-arid conditions

(Source: Dandin and Giridhar (2010), Hand Book of Sericulture Technologies Central Silk Board, Bangalore 2010)

Cultivation Practices for Southern India

Supplementing the improved mulberry varieties, with efficient package of practices for mulberry cultivation were also started simultaneously. Wider spacing (improved productivity), tree cultivation (sustainable leaf production) drip and protective irrigation (water conservation), nutrient supplements (Poshan), organic leaf production systems, integrated nutrient management (bio-fertilizers, vermicomposting, green manuring, trenching & mulching) have also been developed. These resulted in minimizing the application of chemical fertilizers and improved soil health and fertility. Eco- and user-friendly pest and disease management practices in mulberry such as Navinya (root rot), Nemahari (root knot) and Nursery guard (nursery diseases) are the other notable achievements, which played a vital role in increasing the mulberry productivity. The utilization of biocontrol agents: predatory beetles (*Cryptolaemus montrouzi*eri or *Scymnus coccivora*) for tukra, egg parasitoid (*Trichogramma chilonis*) for controlling leaf roller and Bihar hairy caterpillar, exotic parasitoid (*Acerophagus papaya*) for the management of papaya mealy bug

Table 5. Morphophysiological characteristics of Mulberry varieties cultivated under Specific Conditions for Chawki rearing at CRCs.

Varieties	S-36	V-1	G-2
Parentage	Mutant of Ber. local	S-30 x Ber. C 776	M. multicaulis x S-34
Year of Release	1986	1977	
Plant type	Semi-erect	Erect	Semi-erect
Inflorescence	Unisexual	Unisexual	Unisexual
Flower/Sex	Female	Male	Female
Ploidy	Diploid	Diploid	Diploid
Rooting ability	48	> 90	> 90
Inter nodal distance (cm)	3.6	5.2	4.3
Leaf nature	Unlobed Cordate Large Size	Unlobed Ovulate, Large, Dark Green	Entire, cordate, wavy, dark green
Fresh weight of 100 leaves	90–95	67–70	125–132
Moisture content %	78–79	79–81	79–81
Leaf MRC %	80–83	82–84	81–83
Proteins %	23–25	25–26	25–26
Sugars %	10–12	12–13	12–13
Carbohydrates %	23–28	24–26	23–25
Leaf yield (MT/ha/year)	27–28	30–32	36–39
Recommended for	Irrigated condition	Irrigated condition	Chawki rearing
Specially recommended for	Chawki rearing	Chawki and late age rearing	Exclusively chawki gardens of CRCs

menace, and *Chrysoperla sellomi* for the management of thrips, resulted in preventing huge losses due to mulberry pests. Important technologies which have helped in getting the sustainable leaf yield and better quality as well as environmental safety are summarised in Table 6.

Cultural Practices for Eastern and North Eastern India

Similarly, for Eastern and North Eastern states, CSRTI, Berhampore, India has developed suitable cultivation practices to get good leaf yield of better quality. The recommended practices are as follows:

- Under irrigated condition: Mulberry leaf yield increased from 10–12 Mt/ha/year (Khajali/Bombai local) to 28–29 Mt/ha/yr (S1) and to 40–45 Mt/ha/yr (S1635). At present, the improved variety C2038 (54–55 Mt/ha/yr) is under All India Coordinated Experiment on Mulberry (AICEM) (Phase–III).

- Sub-tropical hills of Darjeeling and Sikkim: $BC_2 59$ (16–17 Mt/ha/yr) out yielded Kosen (7–8 Mt/ha/yr) with productivity increase of 9 Mt/ha/yr.

- Flood prone/water logged areas in Murshidabad, Nadia and Birbhum districts of West Bengal, Gumla district of Jharkhand; also Palakkad district of Kerala: C-2028 (35–36 Mt/ha/yr).

- Out of 164 germplasm Bank mulberry accessions: C-1540 and C-1726 have been shortlisted as cold tolerant variety as early sprouters (17–27 days), late senescence (76–85 days) with maximum leaf yield (4507 kg/ha, 4215 kg/ha/crop) for breeding program.

- Mulberry genotypes for low input soils: On farm testing at Regional Sericultural Research Station, Koraput, Ranchi and Jorhat. Leaf yield gain is around 18–50% over S1635.

- Soil test based NPKS fertilizers: Its application reduced economic burden to the farmers and maintained soil health.

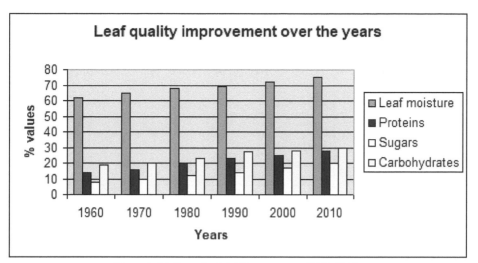

Figure 4. Mulberry leaf yield improvement (1960–2010).

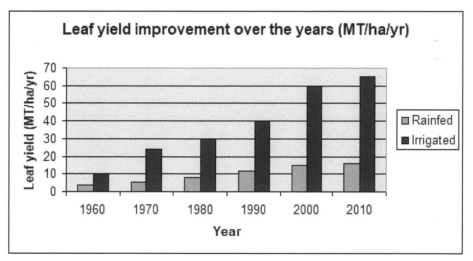

Figure 5. Leaf quality improvement (1960–2010) (Source: Dandin and Giridhar (2010), Hand Book of Sericulture Technologies Central Silk Board, Bangalore 2010).

- Biofertilizer: NITROFERT @ 20 kg/ha/yr for irrigated and 10 kg/ ha/yr for rainfed garden and PHOSPHOFERT 75 kg/ha once in 4 yr in irrigated and 40 kg/ha once in 4 yr in rainfed condition curtailed 50% chemical nitrogenous fertilizer and 70–80% phosphatic fertilizer use.
- Foliar Spray of Morizyme-B: 0.1% after 15–20 days of pruning during winter season increases mulberry leaf yield, leaf protein and sugar by 25–30%, 30% and 31%. respectively.

Cultivation Practices for Chawki Rearing

Need for exclusive Chawki garden

Chawki rearing is a vital aspect of sericulture industry for development of healthy worms and ultimately the enhanced cocoon crop. Of the different factors responsible for healthy Chawki rearing, leaf quality plays an

Table 6. Technology and its advantages in mulberry production and protection.

Sl. No.	Technology	Advantages
1.	Paired row system of plantation	It facilitates movement of power tillers and tractors and curtails expenditure on intercultural operations by at least by half. The mechanized cultivation improves the quality of leaves and curtails expenditure on intercultural operations by at least by half
2.	Drip irrigation	Up to 40% Water saving in mulberry without affecting the leaf yield. (Furrow irrigation-87692 gallons/week and drip irrigation 51281 gallons/week)
3.	Trenching and mulching	Fertility building by subsoil bio massing/manuring and soil moisture conservation by organic mulching
4.	Bio-fertilisers	inoculation of *Azotobacter* and VA-mycorrhiza helps in reduction of N and P fertilizer application by 50% in mulberry
5.	Poshan:	It is a multi-nutrient formulation, corrects the physiological disorders caused due to deficiencies of both macro and micro nutrients
6.	Morizyme-	It is a plant Growth Regulator. The Foliar Spray of this increases mulberry leaf yield by 25–30%, leaf protein by 30% and sugar by 31% during winter season. Commercialized through NRDC, and Licensed to 4 entrepreneurs
7.	Navinya	Controls root rot disease effectively. It rejuvenates the affected mulberry bushes within 30 days after treatment. It is eco-friendly, no residue, non-toxic, and water soluble.
8.	Nemahari	Plant-based formulation for management of root knot disease of mulberry. It controls 82% disease and reduces 83% nematode population in the soil. Target specific, inhibits nematode and its multiplication in soil and mulberry roots.
9.	Chetak	A herbal based formulation for control of all major foliar and root diseases.

important role. Compared to the quality of mulberry required for late age worms viz., Moisture-70%; Protein-21%; Sugar-11%, requirement for Young age silkworms is different. Young worms require better quality leaf with leaf moisture—> 80%; Proteins—25%; Sugars—13% and Succulent—soft leaves. To meet this specific leaf quality requirement and economic leaf yield suitable only for Chawki worms large scale experiments were conducted in exclusive garden established for this purpose (Figure 6) and based on the results three varieties namely S.36, V-I and G-2 were finally recommended. Comparative account of these three varieties is given in Table 7a,b.

- Select healthy saplings of 3–4 months old plant in erect position, cut all the saplings at 10–15 cm height from ground level using secateurs.
- Allow all the branches to grow
- Apply basal dose of 20:20:20 kg NPK/acre/year, after 2 months—120 kg 17:17:17 complex/acre.
- During the first 6 months do not harvest or cut the branches.

  After 6 months, select 3 strong & healthy branches for each plant and cut them at 25 cms height from ground level.

  Remove all the other weak and minor branches.

- After 25 days of pruning, remove all the weak branches at the bottom. Rearing can be taken 40–45 days after pruning
- After every rearing, cut all the branches at a height of 30 cm, and remove the weak and dry branches. Keep only 10–15 healthy branches to grow.

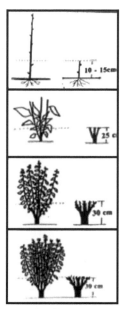

Figure 6. Establishment and Training of the New Mulberry Garden.

Cultivation Requirements for Chawki Garden

Land and soil	Soil pH-6.5–7.5 Organic Carbon—> 0.65% Flat land, Porous, Good drainage, Sandy loam Fertile soil
Plantation area	2.5–3.2 acres—Four blocks (0.625 to 0.85 acres each)
Mulberry variety	S-36/V-1/G-2
Plant spacing	90 x 90 cm or (150 + 90 cm) x 60 cm Mechanization facility
Manure (FYM)	50 MT/ha/year (Two split doses (after 1st and 5th crops)
Chemical fertilizers (N:P:K kg/ha/year)	260:140:100 kg in 8 equal split doses
Straight fertilizers	
Ammonium sulphate	66 kg crop/acre/year
Single super phosphate	44 kg crop/acre /year
Muriate of Potash	11.5 kg crop/acre/year

Table 7a. Agronomical descriptions of varieties suitable for chawki rearing.

	G 2	S 36	V 1
Chawki leaf yield (mt/ha/yr.)	38.00	27.94	31.86
No. of shoots/plant	10–12	8–10	10.12
Total shoot length (cm)	1170	924	1280
Inter-nodal distance (cm)	4.50	3.53	5.20
Leaf size (cm)	L-21-24 B-16-19	L- 20-23 B-16-19	L- 20-23 B-13-15
Leaf shape	Entire, Cordate	Entire, Cordate	Entire, Ovate
Leaf surface	Smooth, glossy, wavy	Smooth glossy, pale green	Smooth glossy, dark green
100 leaf weight (g)	460	465	560
Sprouting (30th day)	96%	70%	90%
Rooting (90th day)	94%	48%	94%

Table 7b. Leaf quality parameters.

	G 2	S 36	V 1
Nitrogen (%)	4.21	4.05	4.19
Phosphorus (%)	0.32	0.32	0.33
Potassium (%)	2.20	2.21	2.14
Protein (%)	26.31	25.31	26.19
Total Sugars (%)	12.85	11.89	12.13
Reducing sugars (%)	1.46	1.45	1.41
Starch (%)	11.06	11.96	11.90
Total carbohydrates (%)	23.91	23.85	24.03
Fibre content (%)	10.74	10.73	10.73
Total chlorophyll content [mg/g FW]	3.654	3.252	3.726
Moisture content (%)	80.30	79.90	80.10
Moisture retention after 6 hours of harvest (%)	83.40	82.90	83.20

(Source: Dandin and Giridhar (2010), Hand Book of Sericulture Technologies Central Silk Board, Bangalore 2010 and Sivaprasad et al. 2016)

Complex fertilizers	
1) 19:19:19	37 kg + 30 kg AS
2) 17:17:17	40 kg + 30 kg AS
3) 15:15:15	45 kg + 30 kg AS
Foliar supplement	
Poshan	7 ml per liter
Seri boost	2.5 ml per liter
Irrigation frequency	3.75-hectare cm
Harvesting – Leaf and: shoot-lets	Leaf harvest after bottom pruning and top clipping and shoot let harvest
No. of crops/year	32 crops-8 crops per plot x 4 plots

Schedule of Leaf Harvest

Schedule	Block-1	Block-2	Block-3	Block-4
Pruning 1st Harvest:	10.6.16 21.7.16	20.6.16 31.7.16	30.6.16 11.8.16	11.7.16 21.8.16
Topping 2nd Harvest	26.7.16 1.9.16	5.8.16 11.9.16	16.8.16 22.9.16	26.8.16 1.10.16
Pruning 3rd Harvest:	7.9.16 16.10.16	17.9.16 27.10.16	28.9.16 7.11.16	7.10.16 17.11.16
Topping 4th Harvest	22.10.16 1.12.16	1.11.16 11.12.16	13.11.16 22.12.16	22.11.16 1.1.17
Pruning 5th Harvest	7.12.16 27.1.17	17.12.16 6.2.17	27.12.16 16.2.17	6.1.17 26.2.17
Topping 6th Harvest	1.2.17 11.3.17	12.2.17 22.3.17	22.2.17 1.4.17	2.3.17 12.4.17
Pruning 7th Harvest	16.3.17 26.4.17***	27.3.17 7.5.17	6.4.17 16.5.17	17.4.17 27.5.17
Topping 8th Harvest	1.5.17 6.6.17	12.5.17 17.6.17	21.5.17 26.6.17	1.6.17 6.7.17

(Source: Commercial Chawki rearing, CSRTI, Mysuru (2016))

Mulberry Crop Protection

Diseases

Diseases are one of the limiting factors for successful mulberry cultivation. Many diseases caused by fungi, bacteria, viruses and nematode affect mulberry. These diseases are either air borne or soilborne in nature, which reduce about 10–20% leaf yield besides deteriorating the leaf quality (Sengupta and Govindaiah 1999, Shekar and Sathyaprasad 2010, Philip et al. 1997).

Foliar Diseases

1. **Leaf spots:** More than 25 pathogens are reported to cause leaf spot disease in mulberry. Among them, brown leaf spot is the most devastating one throughout the country. The disease is caused by a fungus, *Cercospora moricola* Cooke. belongs to the family *Moniliacea*e; order *Moniliales* of class *Deuteromycetes.*

Occurrence: More prevalent during rainy and winter seasons.

Crop loss: 10–12%.

Symptoms: Brownish necrotic, irregular spots appear on the leaf surface. Spots enlarge, coalesce and leave characteristic `shot hole'. Leaves become yellow & wither off during severe stage.

Control: Adoption of wider spacing. Foliar spray of 0.2% Bavistin (Carbendazim 50% WP) solution (dissolve 2 g Bavistin in 1 litre water). 150–180 liters of fungicide solution is required for one-acre garden. Fungicidal spray should be taken up immediately after the initial appearance of the disease (30–35 days after pruning/leaf harvesting). Fungicide treated leaves can be fed to silkworm safely after 2–3 days of spraying.

2. **Powdery mildew:** Powdery mildew is caused by a fungus *Phyllactinia corylea (Pers.) Karst.* belonging to the family *Erysiphaceae, order Erysiphales of class Ascomycetes. It is also known as Phyllactinia moricola (P. Henn.) Homma.* This disease occurs during winter and rainy seasons.

Crop loss: 5–10% and diseased leaves have less moisture, protein contents and higher starch as well as cellulose as compared to healthy leaves.

Symptoms: White powdery patches appear on the ventral surface of leaves and turn blackish brown in later stages.

Control: Foliar spray of 0.2% Karathane (Dinocap)/Bavistin. The treated leaves can be fed to silkworm safely after 2–3 days of spraying.

3. **Leaf rusts:** Mulberry is attacked by several rust pathogens. Among them, black rust and red rust are more prevalent in India and belong to the family *Uredinaceae;* order *Uredinales; class Basidiomycetes.*

(i) Black rust: The pathogen of black rust is *Cerotelium fici* Cast. (Arth.) It is also known as *Peridiospora mori* Barclay. The disease was first reported from Maharashtra in 1907 and later from other parts of India. The disease occurs both in post rainy and winter seasons and is more serious in South India and West Bengal. Other than quantitative loss, the moisture, crude protein and sugar contents are also reduced. It causes 10–15% crop loss

Symptoms: Circular pinhead size brown eruptive lesions appear on the leaves. Later, leaves become yellow and wither off.

(ii) Red rust: Red rust disease caused by a Basidiomyceteous fungus, *Aecidium mori* (Sydow & Butler) Barclay, is observed in northern parts of the country, like Himachal Pradesh, Uttar Pradesh, West Bengal, North-Eastern region and also in Karnataka. It appears during rainy and winter seasons and reduces about 8–10% leaf yield. The disease affects young buds, leaves, petioles and shoots. The affected buds become swollen and curl up in abnormal shapes with many protruded golden yellow spots. On both the surfaces of the affected leaves, numerous small, round, shiny-yellow coloured protruded spots appear in the form of powdery substance. If the shoot and petiole are affected, the fungus spreads through vascular bundles. It spreads through aeciospores that are dispersed by wind and water current. Aeciospores produce germ tubes, which enter the leaf tissues either directly through leaf epidermis or stomata.

Control: Adoption of wider spacing. Avoiding the delay in harvesting of leaves; Foliar spray of 0.2% Kavach (Chlorothalonil 75% WP). The treated leaves can be fed to silkworm safely after 2–3 days of spraying.

4. **Leaf blights:** Both fungi as well as bacteria cause leaf blight in mulberry. The fungal leaf bight is prevalent mainly in tropical regions whereas bacterial blight is worldwide in occurrence.

   (a) Fungal leaf blight: Fungal leaf blight is caused by the various species of *Alternaria, Fusarium, Helminthosporium* and Colletotrichum. Among them, Alternaria and Fusarium species are most commonly observed to cause leaf blight in mulberry. Fungal leaf blight causes leaf yield loss up to 10% in addition to alterations of leaf quality.

   (b) Bacterial leaf blight: The disease is caused by a bacterium, *Pseudomonas syringae* pv. *mori* (Boyer and Lambert) belonging to the order *Pseudomonadales* of class *Schizomycetes.* There are four types of bacterial diseases reported in mulberry *viz.,* leaf blight (*P. syringae* pv. *mori*), leaf spot (*X. campestris* pv. *mori*), shoot soft rot (*E. carotovora.* var. *carotovora*) and leaf scorch (Fastidious Xylem inhabiting bacteria). Among them, in India, bacterial blight caused by *P. syringae* pv. *mori* is most serious. It was first observed in hilly region of Uttar Pradesh in 1966. Since then, the disease is reported from other states like Karnataka, Tamil Nadu, West Bengal, Andhra Pradesh and Kerala.

Occurrence: Rainy and winter seasons. This disease causes 5–10% crop loss: Symptoms: Numerous blackish brown irregular water-soaked patches appear on the leaves resulting in curling & rotting of leaves.

Control: Spraying of Streptomycycin (0.2%) or Dithane M-45 (0.2%).

Soil Borne Diseases

Soilborne diseases pose a serious problem for mulberry cultivation during nursery plantation and established gardens, which cause severe loss in revenue generation of mulberry growers as compared to foliar diseases. Various soilborne diseases affect mulberry. Among them, root knot and root rot affect the established plantation resulting in severe loss in leaf yield apart from deterioration in leaf quality, which is a prerequisite in successful sericulture to get the good quality of cocoons. Besides, stem-canker, cutting rot, collar rot and die-back, affect the initial establishment and survivability of mulberry plantation in nursery.

Nematode/Root knot disease

More than 42 species of nematodes belonging to 24 genera are reported to cause the different diseases in mulberry all over the world. *Meloidogyne incognita* (Kofoid & White) Chitwood, a non-segmented worm belonging to the Family Heteroderoidea, Order Tylenchida of Class Secerentia under Phylum Nematoda causes root knot disease which is very serious and chronic to the mulberry. The disease is widespread and more prevalent in sandy soils under irrigated farming system. *M. incognita* has a wide range of host plants, which infects more than 2000 species of plants including almost all agricultural, horticultural, oilseeds, ornamentals, plantation crops, etc. It is more dangerous to mulberry not only because of the direct damage to the crop but also it predisposes the plant to various soilborne plant pathogens (Philip et al. 1995). The disease spreads primarily through contaminated soil, farm implements and run-off irrigation water from infected plot to the other plot. Plantation of infected saplings, cultivation of other susceptible crops along with mulberry, and growth of some susceptible weeds in and around the mulberry gardens acts as the secondary sources of infection

Crop loss: More than 12%.

Symptoms: Yellowing and necrosis of leaves, stunted growth of plant and presence of galls on root system.

Control: The nematode population in the infected gardens can be reduced by deep digging/ploughing to a depth of 30–40 cm during summer. The higher temperature and lower humidity developed by this practice kill the nematode eggs and larvae present in the soil because they are sensitive to heat. Other control measure include:

(a) Intercropping of plants like marigold (*Tagetus patula*), sesame (*Sesamum indicum*) or sun hemp (*Crotalaria spectabilis*) at 30 cm distance in between mulberry rows reduces 60–65% root galls and egg masses of nematode thus resulting in marginal increase of 8–10% in leaf yield. Mulching of these plants after full growth enriches the soil fertility, in addition to controlling the disease.

(b) Application of Furadan at the rate 16 kg/acre yr (0.5 g/plant/dose) in 4 splits around the basin of the plant followed by regular irrigation. Safe period of 45 days is essential.

(c) Soil application of Neem oil cake at the rate 800 kg (20 g/plant/dose).

(d) Biological method: Soil application of Bionema (*Verticillium chlamydosporium*) along with Neem oil cake and FYM (1:24:200) around the root-zone at the rate 200 g/ plant in 3 doses/yr followed by regular irrigation. *V. chlamydosporium* does not show any residual toxicity on mulberry and silkworms.

Root rot diseases

Root rot is the most dangerous disease due to its epidemic nature and potentiality to kill the plants completely and poses a serious problem during mulberry cultivation in almost all the sericultural countries. Various types of the mulberry root rot diseases have been reported from all over the world. These are dry root rot, black root rot, charcoal root rot, violet root rot, white root rot, *Armilaria* root rot and bacterial root rot. Among them, the dry, black and charcoal root rots (caused by *Fusarium*) are reported in India. For example, in India, earlier it has been reported that violet and white root rots damage the mulberry (Rangaswami et al. 1976), but various surveys indicate that the disease is dry root rot caused by *Fusarium solani* (Mart.) Sacc. and *F. oxysporum* Schlecht. (Philip et al. 1995, 1997). Both these pathogens belong to the Family Tuberculariaceae, Order Moniliales of Class Deuteromycetes under Division Mycota.

Occurrence: Throughout the year in all types of soils.

Crop loss: 12–14%

Symptoms: Sudden withering and defoliation of leaves and rotting of root followed by death of plants in isolated patches.

Control: Root dipping of saplings in 0.1% Bavistin solution (1 g dissolve in 1 litre water) and planting in pits dusted with 10 g of Dithane M-45 followed by irrigation. Another approach is by integrated method.

Integrated method: The diseased plants are to be uprooted and burnt. About 10 g of Dithane M-45 per pit should be applied and plant the new saplings after soaking in Dithane M-45 (0.1%) solution for 30 minutes. After 15–20 days, the Raksha mixture at the rate of 500 g/plant in the root zone has to be applied and irrigated. The Raksha mixture can be prepared by mixing of 1 kg Raksha with 50 kg farmyard manure (sufficient for 100 plants) and to this added 8-10 liters of water to maintain the 15–20 % moisture. The mixture should be stored under shade for about one week by covering with gunny cloth for enhancing the multiplication of *Trichoderma* colonies. Application of Raksha should be continued for one year at an interval of 3 months. *T. harzianum* does not show any residual toxicity on mulberry and silkworms.

Diseases of Nursery

In India, mulberry is propagated through stem cuttings either to raise potted saplings or direct plantation of cuttings in the main field, whereas in other countries the crop is propagated through seeds. Further, the breeders use the seedlings as stock for grafting in all sericultural countries including India for the conventional breeding programme. Both of the propagated materials are attacked by several soilborne pathogens, which reduce the survival percentage of saplings/seedlings considerably in addition to spreading the diseases in the disease free areas. Some of the common nursery diseases causing agents and their management are given below:

Fusarium solani (rotting of whole cuttings and decaying of bark), *Botryodiplodia theobromae* (presence of greenish-black eruptions on cuttings), *Phoma sorghina/P. morourm* (rotting of cuttings near soil line), *Botryodiplodia theobromae* (sudden withering and death of spalings starting from apex to down wards).

Crop loss: 30–35 % mortality of cuttings and saplings during initial establishment.

Management: Application of biofungicide Nursery-guard (*Trichoderma pseudokoningii*) after mixing with FYM (1:60) at the rate 2 kg/m^2 before plantation later plant the cuttings after soaking in Dithane M-45 (0.1%) solution for 30 minutes.

Bio-control measures are becoming more popular in the recent years because of low or no residual toxicity. Important bio-control agents used for the control of different mulberry diseases is given in Table 8.

Pests

Like most of the other economic plantations and field crops, mulberry is prone to the attack of a varied pest complex (more than 300 insect and non-insect pests) belonging to a large number of insect orders. Information on different pests of mulberry are given below:

Mealy bug causing Tukra

It is a major pest popularly known as 'hard to kill' pest. Commonly also called as 'Pink mealy bug', the scientific name is *Maconellicoccus hirsutus,* belongs to the order *Homoptera* and family *Pseudococcidae.*

Period of occurrence: The pest has been reported to occur throughout the year but the incidence is high in summer months.

Type of damage and symptoms: The leaf yield is tremendously reduced and low in nutritive value.

The leaves are wrinkled, thickened, become dark green in colour and then turn into yellowish. Heavily infested plants have shortened internodes resulting into rosette or a 'bunchy top' appearance. The symptom is generally called as 'tukra'. A heavy, black sooty mold may develop on an infested plant leaves and stem with heavy honey dew secretions by this mealy bug. Leaf yield loss upto 4500 kg/ha/yr has been reported.

Table 8: Bio-control agents for mulberry disease management.

Sl. No.	Mulberry disease	Casual organism	Bio-control agent
1.	Leaf spot disease	*Cercospora moricola*	*Trichoderma harzianum* (Th-1) and *Trichoderma pseudokoningii*
2.	Powdery mildew	*Phyllactinia corylea*	Lady bird beetles (*Illeis cincta* Fab and *I. indica*), (Fungal hyperparasite *Cladosporium* spp.)
3.	Leaf Rust	*Aecidium mori*	*Trichoderma harzhianum* and *T. pseudokoningii*
4.	Bacterial blight	*Pseudomonas syringae* pv. *mori*	*Pseudomonas fluorescences* and *Trichoderma harizhianum*
5.	Stem canker	*Botryodiplodia theobromae*	*Trichoderma pseudokoningii*
6.	Root rot	*Macrophomina phaseolina*	Raksha containing *T. harizhianum* integrated with another bioformulation of *Pseudomonas flourescens*
7.	Root knot	*Meloidogyne incognita*	*Verticillium chlamidosporium, Pacilomyiclila cinus, Arthrobotry soligospora, Purpureocillium lilacinus, Bacillus firmus, Pasteuria penetrans* and *Pseudomonas flourescens*

(Source: Hand book of Sericulture Technologies 2010)

Management

1. Mechanical

(a) Removal of infested portion by top clipping and burning the clip as and when infestation noticed.

(b) Prevent further mealy bug infestation to better maturity in leaf suitable for late fifth instar silkworm larva.

(c) After every leaf harvest remove the left-over leaf and apical buds, This will prevent future infestation, activation of axillary buds and better development of secondary shoots.

2. Chemical: Spray of 0.2% DDVP (76% EC) (2.63ml/l) 2 to 3 times with 10 days interval. Safe period: 15 days.

3. Biological: Release of predators either *Cryptolaemus montrouzieri*—250 adults/acre in two split dosages: Oct–Nov. and Jan–Feb. or *Scymnus coccivora*—500 adults/acre. The biocontrol agents should be released in late evening hours.

Mulberry Leaf Roller

The leaf roller, *D. pulverulentalis* (Hampson) (Lepidoptera: Pyralidae) is reported to infest mulberry plantation in Karnataka, Andhra Pradesh and Tamil Nadu. The infestation causes considerable reduction in leaf yield resulting in economic loss to sericulturists. The infestation is observed mainly in the chawki garden (15 days after pruning/ leaf harvest). The pest is known to infest all the commercial mulberry varieties.

Period of occurrence: The infestation is observed on the onset of monsoon, i.e., from June and lasts upto February. However, the peak period of infestation is September–November.

Nature of damage and symptoms: The target area of the leaf roller is the apical portion of the mulberry shoot. The young caterpillar binds the leaflets together by silky secretion and settles inside and devours the soft green tissues of the leaf surface. Grown up caterpillars feed on tender leaves and their faecal matter can be seen on the leaves below the affected portions.

Management

As tackling of this pest is of immense importance the Pest Management Laboratory, Central Sericultural Research and Training Institute, Mysore has developed the following control package of Integrated Pest Management (IPM):-

1. **Mechanical:** Clipping off and destroying the pest infested parts of mulberry; Flood irrigation and deep ploughing to kill pupa; use of light traps to attract and kill adults are some of the mechanical methods.

2. **Chemical:**: Foliar application of 0.076% DDVP (Dichlorvos) 76% EC (one ml in one litre of water) 10 days after pruning/leaf harvest is found ideal. If infestation persists, repeated (2–3) sprays are required. The leaf can be utilised for silkworm rearing 7 days after DDVP application. Spraying of commercial Neem pesticide (0.03% AZ) 0.05% in 200 litre with safe period of 10 days is also recommended.

3. **Biological:** Release of *Trichogramma chilonis*—an egg parasitoid at the rate 25,000 acre/week and Tetrastichus *howardii*—a pupal parasitoid at the rate 25,000 acre/week has been recommended.

4. **Integrated Approach:** For effective management of the pest, an integrated method comprising of all the measures described as above has been developed and the schedule given below:

Sl. No.	Days*	Activity to be taken up	Qty. required/acre/crop
1	3	Release of pupal parasitoid *Tetrastichus howardii*	50,000
2	10	Release of egg parasitoid *Trichogramma chilonis* @ 18,000 eggs per card or per cc	5 cc
3	15	Spraying DDVP @ 0.076% in 200 lit.	200 ml
4	20	Release of *T. howardii* + *T. chilonis*	50,000 + 3 cc
5	25	Spraying Neem pesticide (0.03% AZ) 0.05% in 200 lt	1 liter
6	30	Release of *Trichogramma chilonis*	3 cc
Y	Top clipping (bud and next leaf) when the silkworms are in 4th molt * *after pruning*		

(Source: Disease and pests of Mulberry and their control, CSRTI Publication 1991, pp. 63)

Bihar Hairy Caterpillar

Among the leaf eating pests of mulberry, the Bihar hairy caterpillar, *Spilosoma obliqua,* assumes greater importance owing to its voracious feeding habit which causes extensive damage to mulberry crop.

Period of occurrence: In recent years, sporadic occurrence of this pest is known to cause considerable loss in the mulberry leaf yield in certain parts of Karnataka and Andhra Pradesh. Prevalence of the Bihar hairy caterpillar in mulberry gardens is reported throughout the year. However, its maximum occurrence is observed during August-February.

Type of damage and symptoms: The young caterpillars feed on chlorophyll layer of the leaf exposing the veins which impart dried/dead appearance to the leaves. The grown up larvae feed on the entire leaf rendering the branches without leaves.

Management

(1) Mulberry gardens are to be observed regularly for collection and destruction of egg masses and also gregarious young caterpillars. Ploughing the caterpillar infested gardens to a depth of about six inches to expose the pupae to scorching sun or to the sight of its natural enemies like birds. Besides, flood irrigation of the mulberry gardens will also help in killing the pupae. (2) Preparation of trenches (1 ft depth and 1 ft width) around the mulberry plot and placing poisonous baits (Preparation of baits: Dissolve 2 kg of Jaggery in 1 litre of water, add 20–25 kg of saw dust or wheat bran +3 litre water. To this add 250 ml Nuvacron, mix well and allow it to ferment for 1 day) evenly all around the mulberry plot inside the trenches, to check the migration of caterpillars from one garden to other. (3) Spray 0.15% DDVP to kill the caterpillars. Safe period: 15 days. (4) Release *Trichogramma chilonis* an egg parasitoid one trichocard per acre/week. Each trichocard should be cut into 12–16 pieces and staple them below mulberry leaf uniformly in entire mulberry garden for effective parasitisation of host egg. (5) Installation of light traps to attract adults and killing them using 0.5% soap solution in basin near the light source. Change the soap solution once in three days.

Thrips (Pseudodendrothrips mori) (Thysanoptera: Thripidae)

Though it is a major pest in Tamil Nadu and minor pest in Karnataka and Andhra Pradesh. It is popularly called as thunder flies and storm flies.

Period of occurrence: It is reported throughout the year but the infestation is severe during summer (Feb–April).

Type of damage and symptoms: Nymphs and adults of thrips lacerate the epidermal leaf tissues and suck the oozing cell sap leading to damage of guard cells and finally drying of the leaves. Affected leaves show streaks in the early stages and blotches in the advanced stage of attack.

Control: Sprinkler irrigation, and release of lady bird beetles *Menochilus sexmaculatus* and *Scymnus coccivora,* anthocrid *Orius* sp. and neuropterans were observed to feed on thrips in the field and laboratory.

Jassid (Empoasca flavescens F.) (Homoptera: Cicadellidae)

It is also a minor pest popularly called as leaf hopper. It is common during the summer months. Alternate host plants: Several malvaceous plants like lady's finger, cotton, castor, brinjal, green and black gram and cucurbits.

Type of damage and symptoms: The most prominent symptom is characteristic 'hopper burn' yellowing of leaves all along the leaf margin and reduces the plant vitality and yield. In the final stages of the attack the leaf becomes cup shaped and withers off prematurely.

Management

1. Physical/mechanical: Installation of Light trap and sticky traps are recommended.
2. Chemical: Spray of 0.1% Dimethoate or 0.05% DDVP has been recommended for its control.
3. Biological: Reduvid and Pentatomid predators are observed to feed on both nymphs and adults.

Spiraling White fly (Aleurodicus disperses) (Russel) (Homoptera: Aleyrodidae)

Though a minor pest, its severity has been of late reported from Tamilnadu and West Bengal areas. The spiraling of waxy material is the feature from which this whitefly derives its common name, the spiraling whitefly.

Period of occurrence: March-June; October-Dec. Prolonged dry spell followed by the hot humid weather favours the whitefly flare up.

Type of damage and symptoms: The majority of feeding damage is done by the first three nymphal stages. Infests upto the lower surface of leaves resulting in chlorosis, yellowing, upward curling of the leaves, leaf fall and retardation of growth. The nymphs and adults remain on the lower surface of the leaves and desap the plants. Continuous heavy feeding weakens the plant. The damaged leaf become unfit for rearing silkworm.

Management

1. Mechanical: Collection and destruction of infested leaves.
2. Cultural: Adoption of recommended spacing and fertilizer dose; removing alternate wild host plants like *Abutilon* and *Cassia auriculata* and other weeds from the field as well as neighboring areas, and reducing the activity and population buildup by fixing yellow sticky trap.

Chemical: Spray of 0.05% Dimethoate. Spray of 0.5% Neem oil mixed with 1 to 2 ml soap solution at 1: 2 ratio.

Scale insect (Saissetia nigra) (Homoptera Coccidae)

It is also a minor pest called as unarmored scales and soft scales insect.

Period of occurrence: It is reported throughout the year but becomes severe during summer. Alternate host plants: The known alternate host plants are Citrus, croton, *Chrysanthemum* and banana.

Type of damage and symptoms: Scales are similar in habit to mealy bugs and feed on plant sap. They are usually found on the underside of the leaves and stems.

Management

1. Physical: Swapping with a blunt edge wooden plate to dislodge the insect is recommended.
2. Chemical: Spraying of 0.05% Dimethoate with safe period of 10 days is recommended
3. Biological: Use of predators, such as the green lacewing, and parasitic encyrtid wasps.

Since leaves are fed to silkworm the application of pesticides in the mulberry has to be discouraged. In view of this, use of biopesticides is becoming more popular. Common biopesticides/control agents being recommended is given in Table 9.

Gaps and Way Forward

Even though mulberry once planted could yield continuously (five crops in an year) for nearly twenty years, the host plant cultivation for mulberry silkworm necessitates to throw attention on important issues as discussed by Dandin and Kumaresan (2003) and Kumaresan et al. (2012):

Table 9. Common biopesticides and their control agents.

No.	Mulberry Pest	Casual Organism	Bio-Control Agent
A. Animals (Animal products)			
1	Mealy bug	*Maconellicoccus hirsutus*	Coccinellid predators, *Cryptolaemus montrouzieri* Mulsant (Australian ladybird beetle), *Scymnus coccivora* Ayyar
2	Papaya mealy bug	*Paracoccusmar ginatus*	*Acerophagus papaya ,Cryptolaemus montrouzieri, Anagyrus loecki, Pseudleptomastix mexicana*
3	Bihar hairy caterpillar	*Paracoccusmar ginatus*	Pupal parasitoid *Tetrastichus howardii*, egg parasitoid *Trichogramma chilonis*
4	Leaf Webber or Leaf roller	*Diaphania pulverulentalis*	Pupal parasitoid *Tetrastichus howadrdii, Trichogramma chilonis, Apanteles* sp. and *Chelonus*, larval predator *Calosoma*
5	Spiralling white fly	*Aleurodicus disperses* Russel	Pupal parasitoid *Tetrastichus howardii, Trichogramma chilonis, Apanteles* sp. and *Chelonus*, larval predator *Calosoma*
6	Cut worm	*Spodoptera litura*	Spodolure pheromone trap.
7	Scale insect	*Saissetia nigra*	lace wing
B. Botanicals (Plant Products)			
8	Sucking pests	*Pseudodendro thripsmori*	*Citronella*
9	Mealy bugs	*Maconellicoccus hirsutus*	*Neem-Azadiractin*

(Source: Disease and pests of Mulberry and their control, CSRTI Publication 1991, pp. 63)

Gaps

➢ Urbanization leading to shrinking of cultivated land (area under cultivation is gradually declining due to population pressure, industrialization and urbanization, hence the availability of cultivable land for mulberry cultivation may be restricted for horizontal expansion).

➢ Slow spread of improved mulberry varieties (new high yielding mulberry varieties for irrigated and rainfed conditions grow slow due to the lack of proper mulberry multiplication system and its harnessing potential needs to be established through research).

➢ Deteriorating soil health (continuous and imbalanced application of chemical fertilizers and inadequate application of farmyard manure/compost, soil organic carbon, and phosphorus has considerably resulted in higher soil pH. This makes some of the soils develop salinity and alkalinity. Soils also are facing micronutrient deficiencies).

➢ Water stress (response to water input in terms of yield is substantial, but irrigation is proving to be a costly input due to water scarcity and depletion of ground water resources). There are indications that in many areas where sericulture is practiced soil becomes white if groundwater exploitation has been upto 65%, gray upto (6585%), and then black when more than 85%, respectively.

These issues need to be addressed to overcome the limitations in developing improved packages for productivity in mulberry.

Way Forward

➢ Popularization regarding package of practices for organic leaf production and certification mechanisms.

➢ Analysis of mulberry growth vigour and development of variety response to repeated shoot harvests.

➢ Techno-economic evaluation of new mulberry genotypes and cultivation technology.

> ➤ Identification of appropriate plant geometry for mechanization and quality mulberry leaf production.
> ➤ Development of cost effective water harvesting and moisture conservation practices for dry zones.
> ➤ Evolution of improved agronomical packages for sustenance of mulberry productivity.
> ➤ Metabolomics of mulberry to identify quality indices of mulberry and nutritional requirements of silkworm for improved silk productivity.
> ➤ Development of region-specific packages for high productivity.
> ➤ Precision agriculture—development of suitable practices for mulberry sericulture.
> ➤ Development of disease resistant and low water demanding mulberry varieties.
> ➤ Development of high yielding mulberry varieties with high input utilization efficiency.
> ➤ Development of mulberry varieties for biotic and abiotic stresses.
> ➤ Development of transgenic mulberry plants for biotic and abiotic stress.
> ➤ Production of triploid varieties with high leaf yield and better nutrition status.
> ➤ Establishment of seed complex for propagating mulberry through seeds in non-arable land.
> ➤ Genome editing (CRISPR/Cas9) for improving climate resilience in mulberry.
> ➤ Introduction of C4 mechanism into mulberry for improved photosynthetic efficiency and adaptation to increased heat radiation.

References

Anonymous. 1990. Report of the Committee for fixation of norms for maintenance of genes and genotypes in the germplasm bank. Central Silk Board, Bangalore.

Camerarius, R.J. 1664. *De sexu* plantarum epistola. (Ueber das Geschlecht der Pflanzen) Wilhem Engelmann, Liepzig.

Dandin, S.B., J. Jaiswal and K. Gridhar. 2000. Hand book of Sericultural Technologies, Central Silk Board, Bangalore.

Dandin, S.B. and P. Kumaresan. 2003. An empirical analysis of cost of cocoon production. Indian Silk 42(2): 5–10.

Dandin, S.B. and K. Giridhar. 2010. Hand book of Sericulture Technologies, Central Silk Board, Bangalore.

Kumaresan, P., S.B. Nagaraj, M.N. Morison, N.G. Selvaraju and P. Reddy. 2012. "Sericulture—An Ideal Enterprise for Sustainable Income". Lead Paper, National Workshop on Promotion of Sericulture for Sustainable Income, March 17–18, 2012 held at Annamalai University, Chidambaram, Tamil Nadu, pp. 22–26.

Linnaeus, Carl. "Species Plantarum" 1753. Library of the Linnean Society of London, LM Bot.

Mala, V. Rajan and D.D. Sharma. 2010. Mulberry diseases and pests-control measures—Mulberry diseases. pp. 153–172. *In*: Dandin, S.B. and K. Giridhar (eds.). Hand book of Sericulture Technologies. Central Silk Board, Bangalore.

Philip Tomy, Janardhan Latha, B.Gobindaiah, K.C. Mandal and A.K. Bajpai. 1995. Some observations on the incidence associated microflora and control of root rot disease of mulberry in South India 34(2): 137–139.

Philip, T., B. Gobindaiah and A.K. Bajpai, G. Nagabhushanma and N.R. Naidu. 1997. Aprliminary surbey on mulberry diseases in South India. Indian Journal of Sericulture 36: 128–132.

Rangaswami, G., M.N. Narasimhanna, K. Kashiviswanathan, C.R. Sastry and M.S. Jolly. 1976. Sericulture Manual 1. Mulberry Cultivation. FAO Agric. Bull., Rome.

Sengupta, K., Govindaiah and Pradeep Kumar. 1991. Diseases and pests of mulberry and their control, CSRTI, Mysore.ivfvcd.

Shekhar, M.A. and K. Sathyaprasad. 2010. Mulberry diseases and pests-control measures—Mulberry pests. pp. 153–192. *In:* Dandin, S.B. and K. Giridhar (eds.). Hand book of Sericulture Technologies. Central Silk Board, Bangalore.

Sivaprasad, V., M.T. Himananthraj, Satish Verma and Mogili. 2016. Commercial Chawki Rearing, Central Sericulture Research and Training Institute, Mysuru.

Sivaprasad, V. 2017. South Zone Mulberry Sericulture Technology Descriptor, Central Sericulture Research and Training Institute, Mysuru.

CHAPTER 16

Viroids, Viruses and Phytoplasmas of Mulberry

Keramatollah Izadpanah

INTRODUCTION

Viroids, viruses and phytoplasmas are among the most common pathogens infecting mulberry in all mulberry growing regions of the world. They are associated with symptoms ranging from none (i.e., symptomless infection) to severe decline of the trees. The common features of these agents are their total dependence on living host cells for their survival and multiplication and their evasion to be grown in artificial culture media. They are also graft transmissible, become systemic in plants and assume an intimate relationship with host cells. This is why until the discovery of phytoplamas (1967) and viroids (1970), these agents were regarded as viruses. In this chapter, a short description of these agents will be followed by the diseases they cause in mulberry plants.

The type of symptoms induced by viruses and viroids depends on the pathogen-host combination and environmental and growth conditions. However except for crypto and latent viruses, they all cause a degree of reduced growth and stunting. Other viroid/virus-like symptoms found in mulberry include vein clearing and flecking, mosaic, mottling, chlorotic or necrotic ringspots, paling of color in leaf sectors, unilateral growth retardation and bending of leaves, leaf puckering, reduction of leaf size, yellowing of leaves, die-back of shoots and dwarfing of the plant (Figure 1). Large chlorotic patches on leaves may become necrotic in late season (Figure 1G).

The most common symptoms of phytoplasma infection in woody plants include dwarfing, proliferation of tissues, witches' broom, yellowing and dieback. Herbaceous plants, in addition, may develop floral deformation.

Viroids and Viroid-Like RNAs

Viroids are small circular non-coding single-stranded RNA molecules capable of infecting, and multiplying in, susceptible plant hosts. With a size range of 246–401 nucleotides, viroids are the smallest known pathogenic agents. They replicate either in the nucleus or in the chloroplasts of the host cells via a rolling circle amplification mechanism using specific host polymerases. Inside the cells, they are believed to assume a rod-like secondary structure with a number of functional loops and motifs as a result of extensive base pairing.

Based on differences in structure, site of replication in the cell, and mode of rolling-circle amplification, the 32 viroids so far known are divided into two families of *Pospiviroidae* and *Avsunviroidae*, each with a number of genera and species. In the rod-like structure of pospiviroids, there are five loosely bordered domains called terminal left (TL), pathogenicity (P), central conserved region (CCR), variable (V) and terminal right (TR). Depending on the genus, there is also a "terminal conserved region" or a "terminal conserved loop" (Steger et al. 2017). Each viroid species consists of many variants which may differ in nucleotide sequence and/or biology.

Plant Virology Research Center, College of Agriculture, Shiraz University, Iran, Email: izadpana@shirazu.ac.ir

Figure 1. Various types of viroid/virus-like symptoms found in mulberry in Iran: (A) Vein clearing in terminal leaves. (B) Vein clearing and yellow speckling. (C) Chlorotic sectors and leaf deformation. (D) Mild mottling. (E) Unilateral growth retardation and bending of leaf. (F) Leaf puckering. (G) Mulberry shoot with small chlorotic leaf spots and necrosis developed on large patches of chlorotic tissues.

The type of symptoms induced by viroids depends not only on the viroid variant, but also on the species, cultivar and growth condition of the host plant. Plant stunting, leaf, flower and fruit deformation, color changes such as mottling and chlorosis, and plant decline are among the more common symptoms. Many viroids are also latent (i.e., induce no apparent symptoms) in at least some of their hosts.

Horizontal transmission of viroids takes place mainly through vegetative propagation and grafting in plants, and mechanically when contaminated tools are used in horticultural practices. Horizontal transmission by pollen has also been well documented (Yanagisawa and Matsushita 2018). In general, viroids have no biological vectors although aphids are reported as viroid vector in rare cases. Vertical transmission, i.e., through seed, first noticed in 1988 (Kryczynski et al. 1988), now seems to be common among viroids (Matsushita and Tsuda 2016).

Mulberry Viroids

Little information is available on viroid infection of mulberry. In fact, *Hop stunt viroid* (HSVd, genus *Hostuviroid*, family *Pospiviroidae*) is the only viroid so far known to naturally infect mulberry.

Hop Stunt Viroid in Mulberry

HSVd has a relatively wide host range among woody and herbaceous plants. The first report of mulberry infection by HSVd came from Lebanon and Italy where Elbeaino et al. (2012a) identified this viroid in less than 10% of surveyed mulberry trees. Later, HSVd infection of mulberry was reported from Iran (Amiri-Mazhar et al. 2013). So far, there is no report of mulberry infection by this viroid in other countries.

Symptoms. Elbeaino et al. (2012a) detected HSVd only in symptomless trees. Amiri-Mazhar et al. (2013), however, reported association of HSVd with vein clearing (Figure 1) and leaf deformation in Iran. They were able to transmit the viroid and reproduce the disease in healthy mulberry under controlled conditions. They were unable to detect the viroid in symptomless (apparently healthy) trees. The discrepancy between these results is not surprising as HSVd has many variants some of which may infect a particular plant without inducing symptoms.

In addition to vein clearing, HSVd has been isolated from trees exhibiting other types of symptoms such as yellow speckling and leaf puckering (Figure 1).

It should be noted that due to multiple infection of mulberry with viruses and virus-like agents, the cause-and-effect relationship of many symptoms (i.e., the Koch postulates) has remained uncertain. For example, Elbeaino et al. (2012a) using dsRNA assay failed to detect HSVd or any virus in mulberry leaves with severe vein clearing. On the other hand, Alishiri et al. (2016) detected fig badnavirus 1 and mulberry badnavirus 1 in mulberry plants with vein clearing and leaf puckering.

Transmission. HSVd is easily transmitted from plant to plant by budding and mechanical inoculation of infected sap (Amiri-Mazhar et al. 2013, Jahanshahi et al. 2015). Like other viroids, it is likely that HSVd can spread in the orchards by contaminated grafting knives and pruning shears. There is no information about seed or vector transmission of HSVd in mulberry.

Variation. HSVd is one of the most variable viroids so far studied. Amiri-Mazhar et al. (2013) recognized three sequence variants out of six isolates of HSVd in mulberry in the Fars province of Iran. They all were close to plum and almond isolates of HSVd but different from Lebanese and Italian isolates which, together with a fig isolate, were closer to citrus isolate of the viroid (Elbeaino et al. 2012b).

Jahanshahi et al. (2015) compared biological and structural features of three HSVd isolates from mulberry, fig and citrus in Iran. Most structural differences were found in the pathogenicity (P) domain of the secondary structure. However, in both primary and secondary structures, the mulberry isolate was closer to fig isolate rather than to citrus isolate (Figure 2). In greenhouse studies, the mulberry isolate induced more severe symptoms in sap inoculated cucumber, beet and tomato. Spinach was infected only by mulberry isolate.

Viroid-like RNAs

Viroid-like RNAs are small circular molecules similar to viroids but differ from the latter by lacking certain structural or biological features of viroids. The term is also used when, due to incomplete information, the molecule cannot be assigned to any of the known viroid groups.

Viroid-like RNAs of Mulberry

A disease of mulberry known as mosaic dwarf has been known in China for nearly 300 years (Fei et al. 2009). It seriously affects the Chinese silk industry. The disease has been transmitted to healthy mulberry by grafting and needle injection of total RNA extract from diseased trees (Fei et al. 2009, Wang et al. 2010). Electron microscopy of infected plant tissues did not reveal presence of a virus like entity. PCR with the use of viroid primers resulted in the isolation of a circular RNA entity of 356 nucleotides. However, common viroid structural features were absent in the depicted secondary structure of this molecule. They then designed specific primers to detect this RNA species in other mulberry trees. Using specific primer pair of Fei et al. (2009) in RT-PCR, an RNA species of about 300 nucleotides was frequently detected in the author's laboratory in mulberry with chlorotic speckling and fine

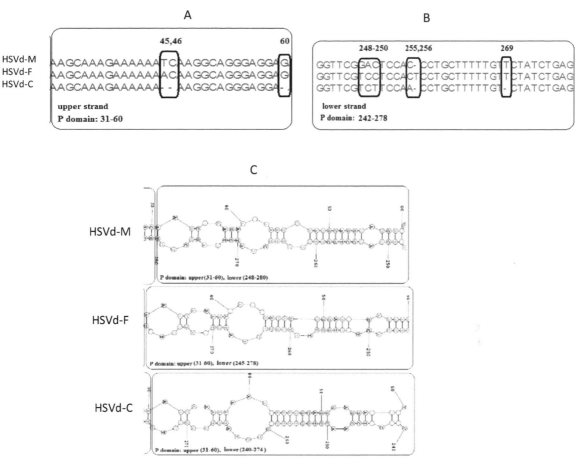

Figure 2. Primary (A, B) and depicted secondary (C) structures of P domain in mulberry (HSVd-M), fig (HSVd-F) and citrus (HSVd-C) variants of hop stunt viroid (adopted from Jahanshahi et al. 2015 by permission).

vein clearing (K. I., unpublished data). Further work is needed to establish the role of these viroid-like RNAs in the etiology of the disease. Ma et al. (2015) using next generation sequencing and rolling circle amplification found a monopartite geminivirus associated with mulberry mosaic dwarf in China. Other investigators have reported association of a nepovirus with this disease (Lu et al. 2015).

Viral Diseases

Detection and identification of viruses in mulberry has been based mainly on PCR analysis and occasionally enzyme-linked immunosorbent assay (ELISA) and electron microscopy of symptomatic or symptomless plants. Thus, several viruses from different viral families have been reported in various parts of the world. The following is a summarized account of these viruses.

Geminiviruses

Geminiviruses (family *Geminiviridae*) are small DNA viruses with geminate particles of about 20 × 32 nm. Their genome consists of one (in monopartites) or two (in bipartites) molecules of circular single-stranded DNA of about 3000 nucleotides housed in a protein shell of 110 subunits. These viruses replicate in the nucleus of host cells by rolling circle amplification using host DNA polymerase. Transcription of genome takes place in two directions, i.e., Open Reading Frames (ORFs) are located on both viral and its complementary molecules. These viruses are transmitted by insects (mostly leafhoppers and whiteflies) and often cause crop losses of economic importance. The family *Geminiviridae* is currently divided into 9 genera based on genome organization and expression strategy (Zebrini et al. 2017).

Mulberry mosaic dwarf-associated geminivirus. Mulberry mosaic dwarf-associated virus (MMDaV) was identified in China by high-throughput sequencing of small RNAs. Analysis of the virus genome showed that it was a unique monopartite geminivirus (Ma et al. 2015) that could not be assigned to any of the nine established geminivirus genera; it formed a distinct cluster with Citrus chlorotic dwarf-associated virus (Zebrini et al. 2017). Southern blot hybridization survey showed a close association (92.4%) of MMDaV with mosaic dwarf disease in Shaanxi province of China (Ma et al. 2015). The associated symptoms consisted of mottling and curling of leaves and dwarfing of the whole plant. The disease is of economic importance in silk industry in certain provinces of China. Other agents including a viroid-like RNA have also been associated with the disease (Fei et al. 2009). MMDaV is transmitted by grafting but there is no information on insect transmission of the virus.

Cassava mosaic virus. The second report of a mulberry infecting geminivirus came from Kerala, India (Shery 2016). In contrast to the Chinese MMDaV, the Indian geminivirus was bipartite and closely related to the Indian cassava mosaic virus and Sri Lankan Cassava mosaic virus which belong to the genus *Begomovirus*. In mulberry, it was associated with mosaic, leaf curling and deformation, enations and stunting. It was transmissible to herbaceous plants and from herbaceous plants back to mulberry by mechanical inoculation of sap. It was also transmitted efficiently by the cassava biotype of whitefly *Bemisia tabaci*. The coat protein (CP) gene of the virus was amplified from the whiteflies. The high identity of CP and movement protein genes suggests the mulberry virus is a variant of Cassava mosaic virus. Both mulberry and cassava may serve as the source for survival and spread of the virus (Shery 2016).

Badnaviruses

Badnaviruses are bacilliform dsDNA viruses of the genus *Badnavirus*, family *Caulimoviridae*. Virus particles are 30×130 nm in size although shorter and longer particles are also common. The genome consists of a circular dsDNA molecule of about 7.2–9.2 kbp. There are three gaps in the circular molecule. Three ORFs are present on the genome. Virus replication involves a step of reverse transcription facilitated by viral reverse transcriptase. Virus transmission takes place by mealy bugs or in certain viruses by aphids or lace bugs in a semi-persistent manner (Geering and Hull 2012). There is some evidence for integration of viral genome in the genome of host plants (Laney et al. 2012).

So far two badnaviruses have been identified in mulberry. Elbeaino et al. (2013) used electron microscopy of leaf dips and ultrathin sections of a sample from Lebanon to identify the first badnavirus in mulberry. The bacilliform particles of the virus (called mulberry badnavirus 1, MBV-1) were 30×150 nm in size. In phylogenetic studies the virus clustered with Cacao swollen shoot virus, Citrus yellow vein virus and Fig badnavirus 1. The virus differed from other badnaviruses in having a single ORF and two genomic sizes of 6 and 7 kbp (Chiumenti et al. 2016). A PCR survey for the virus in Italy, Turkey and Lebanon showed that 69% were infected with MBV-1.

Alishiri et al. (2016) used specific primers of Fig badnavirus 1 (FBV-1) and MBV-1 (Elbeaino et al. 2013) to survey for these viruses in mulberry in Iran. They found FBV-1 in 33% of the tested mulberry samples while MBV-1 was found in only 8.5% of the samples. Seven percent of the plants were infected by both viruses.

Alishiri et al. (2016) identified MBV-1 and FBV-1 in both asymptomatic trees and in trees showing mosaic and leaf deformation. In contrast all 38 MBV-1 infected mulberry trees tested by Elbeaino et al. (2013) except one were symptomless. In a different survey two of the 6 MBV-1 infected and 4 of 23 FBV-1 infected mulberry trees were asymptomatic (Alishiri et al. 2016). Thus based on the present data, association of MBV-1 and possibly FBV-1 with any specific disease is not constant.

Mulberry badnaviruses are transmitted by grafting but there is no information on their transmission by other means. There is also no evidence of the integration of MBV-1 genome in the genome of mulberry (Chiumenti et al. 2016) while FBV-1 genome is shown to integrate in the fig genome (Laney et al. 2012).

Tospoviruses

Members of the genus *Tospovirus* (family *Tospoviridae*, order *Bunyavirales*) are distinguished by their enveloped quasispherical virions of 80–120 nm diameter, tripartite circular single-stranded RNA genome, and negative or ambisense expression strategy. Known tospoviruses are transmitted mostly by species of *Thrips* and *Frankliniella* in a propagative manner (i.e., they replicate also in the body of insect vectors). The largest genomic RNA (designated L) has a negative polarity and is transcribed to an anti-genome with a single ORF encoding for the viral RNA-depended RNA polymerase (RdRP). The medium (M) and small (S) RNAs utilize an ambisense strategy to encode for other viral proteins.

The type species of the genus *Tospovirus* is *Tomato spotted wilt virus* (TSWV). In addition to the type species, several other species have been reported from ornamental plants and vegetable crops. They often cause chlorotic or necrotic ring spots, mottling, yellowing, various types of necrosis and wilt. So far the only report of mulberry tospovirus comes from China. Meng et al. (2013) used deep sequencing of mulberry plants with prominent yellow vein banding and ring spot symptoms to detect a tospovirus in the infected trees. Ring spots appeared in April while in autumn affected mulberries showed leaf deformation (including appearance of filiform leaves), vein necrosis, and necrotic spots. The virus was different from TSWV but closer to Watermelon silver mottle virus (WSMV) with 85.5% nucleotide sequence homology in RdRP gene and to Capsicum chlorosis virus with 74.4% nucleotide sequence homology in nucleocapsid protein gene. In thin sections of the leaves, the virus particles measured 80-100 nm in diameter. Serologically the virus was related to WSMV but not to TSWV or other serotypes (Meng et al. 2015). The virus is currently called mulberry veinbanding **associated** virus (MVBaV) because its cause-and-effect relationship with the disease symptoms is not established yet.

MVBaV was detected in 66.7% of symptomatic plants in Guangxi Province of China (Meng et al. 2013). A closely related variant of the virus was recently identified in *Pharbitis purpurea*, a wild herbaceous medicinal plant of the family Convolvulaceae in China (Gao et al. 2019). There is yet no report on the mode of natural spread of the virus. However, the mulberry thrips (*Pseudodendrothrips mori*), an important mulberry pest and a potential vector, has been found widespread in mulberry fields in China and other mulberry growing regions of the world (Etebari et al. 2004, Meng et al. 2015).

Nepoviruses

The genus *Nepovirus* is classified in the family *Secoviridae*, subfamily *Comovirinae* based on icosahedral symmetry of virions and genomic structure of the viruses in the group. Virus particles are 25–30 nm in diameter, made of 60 protein subunits of 52–60 kDa. Two molecules of RNA make the genome of the virus. Nepoviruses are divided into three subgroups of A, B and C based on their genome structure and sizes of genomic RNA molecules.

Many nepoviruses are soil borne and transmitted by longidorid nematodes of the genera *Xiphinema*, *Longidorus* and *Paralongidorus* (thus *ne*matode transmitted *po*lyhedral *virus*es). Quite a few nepoviruses are also transmitted through pollen and seed. One species is reported to be transmitted by an eriophyid mite (Susi 2004).

Nepoviruses infect a wide range of shrubs and herbaceous plants causing symptoms such as ring spots, mosaic, and leaf deformation depending on virus-host combination. Recovery, i.e., symptomless new growth, is also common.

So far two nepoviruses have been reported from mulberry. In Japan, Tsuchizaki et al. (1971) used electron microscopy to show presence of spherical virus particles in mulberry plants exhibiting ring spot, mosaic, filiform leaves and yellows. The virus was experimentally transmissible to a number of herbaceous plants by mechanical (sap) inoculation. The soil borne vector was the nematode *Longidorus martini*. The infested soil was still infected after 14–17 months storage in the absence of plants (Tsuchizaki 1975). The virus was also transmissible through soybean seeds. A low percentage of mulberry plants inoculated with purified virus developed ring spot symptoms after 3–6 months (Tsuchizaki et al. 1971).

The second nepovirus, the so-called mulberry mosaic leaf roll-associated virus (MMLRaV), was isolated from plants showing mosaic, upward rolling of leaves, reduction of leaf size, and severe stunting in China. Some affected trees also showed protuberances on leaf veins (Lu et al. 2015). According to Lu et al. (2015) the disease was previously called mulberry dwarf and described in an ancient book written around 1624. Lu et al. (2015)

purified the virus from mulberry leaves and studied some of its properties. The virus particles were 28–30 nm in diameter and like in other nepoviruses the viral genome consisted of two molecules of ssRNA (RNA1 and RNA2) of 7183 and 3742 nucleotides. MMLRaV was assigned to subgroup A, but distinct from other nepoviruses although closest to melon mild mottle virus. The cause-and-effect relationship of the virus with the disease remains to be established.

It is worth noting that in addition to MMLRaV, a filamentous virus and a viroid-like RNA were also found in some, but not all, affected trees.

Ilarviruses

The genus *Ilarvirus* comprises *i*sometric *la*bile *r*ingspot *virus*es in the family *Bromoviridae*. Virions are quasispherical of 26–36 nm diameter. In addition, some members may also have bacilliform particles of the same diameter and varying lengths. Like in other bromovirids, the genome in ilarviruses is tripartite (i.e., consists of three molecules of positive sense ssRNA). The genus is divided into several groups based on serological relationships. The viruses are unstable (labile) upon extraction from the plants.

Ilarviruses mostly infect the woody plants. Although some infections could be latent, most ilarviruses induce symptoms such as chlorotic/necrotic ringspots, line patterns, mosaic, leaf distortion, and leaf and shoot necrosis. Transmission of ilarviruses may occur by vegetative propagation of plants, and through pollen and seed. Insects visiting flowers, in particular honey bees and thrips, are instrumental in spread of the viruses if contaminated with infected pollens.

Two ilarviruses have been recently reported from mulberry. Rageshwari et al. (2019) reported infection of mulberry plants by Tobacco streak virus in Tamil Nadu, India. A 25% disease incidence was found in mulberry trees close to infected cotton fields. The symptom in mulberry consisted of cupping, curling and brittleness of young leaves, formation of dark brown necrotic rings and marginal leaf necrosis.

The virus was transmitted by sap inoculation from mulberry leaves to cowpea in which it produced necrotic local lesions. The virus was further transmitted from cowpea to healthy mulberry plants on which it produced leaf cupping and curling as in naturally infected plants.

The second ilarvirus was identified in symptomatic mulberry plants in Worcestershire, UK. The affected plants showed oak-leaf pattern in fully developed leaves. Skelton et al. (2018) used deep sequencing to identify Prunus necrotic ringspot virus (PNRSV) in infected plants. Presence of PNRSV in affected plants was subsequently verified by RT-PCR and sequencing. However, no infection was obtained by sap inoculation of test plants and no biological experiments were performed to verify the role of the virus in mulberry disease.

Other Mulberry Viruses

A filamentous virus (mulberry latent virus) was first detected in ring spot-affected mulberry plants in Japan but limited experimental work did not show its involvement in the ring spot or any other disease (Tsuchizaki et al. 1971). Later, Tsuchizaki (1976) was able to transmit a similar virus from mulberry to several herbaceous plants by sap inoculation followed by purification and characterization. The virus induced formation of indistinct local lesions and became systemic in *Chenopodium quinoa*. It was purified from *C. quinoa* by a combination of differential and density-gradient centrifugation. In purified preparations, the virus particles were 700 nm long. Serologically, the virus was distantly related to *Carnation latent virus*, the type species of the genus *Carlavirus*. Inoculation of healthy mulberry plants with purified virus preparation resulted in formation of indistinct local lesions and latent systemic infection (Tsuchizaki 1976). This is one of the few cases in which the Koch postulates have been fulfilled.

Mulberry carlaviruses are likely present in other mulberry growing countries as well. A filamentous virus of 11–13 nm diameter was also detected in mulberry plants affected with mulberry mosaic leaf roll in China (See Lu et al. 2015). It should be noted that although carlaviruses are generally known to be latent or induce only mild symptoms, in some virus-plant combinations they can cause severe disease symptoms. Also their possible interactions in co-infection with other viruses should be seriously taken into consideration.

While carlaviruses are naturally transmitted mostly by aphids and less frequently by whiteflies, there is no information regarding natural transmission of Mulberry latent virus.

Cryptoviruses are another group of temperate viruses which usually induce no symptoms in the infected plants. They are isometric viruses of 30–40 nm diameter with a bipartite or, less frequently, tripartite genome of dsRNA. They are classified in the family *Partitiviridae* (Sabanadzovic and Abou Ghanem-Sabanadzovic 2008). Cryptoviruses usually occur in low concentration in the infected cells, are not transmissible by mechanical inoculation or grafting, and have no vector. Seed and pollen transmission is their only means of survival. So far, Mulberry cryptovirus 1 is the only cryptovirus reported from mulberry. It was detected in mulberry plants affected by oak-leaf pattern in the UK by deep sequencing (Skelton et al. 2018). The plants were also infected with Prunus necrotic ringspot virus. The involvement of this cryptovirus in the mulberry disease is unlikely.

Another group of plant endogenous viruses are members of *Endornaviridae*. These are free single-stranded RNAs mostly present in the plant as stable replicative form (dsRNA). Plant endornaviruses are transmitted only through pollen and seed (Valverde et al. 2019). Several dsRNAs have been reported in mulberry (Nameth and Cheng 1994) but their relationship with endornaviruses is not known.

Reports of other viruses such as mulberry mosaic begomovirus (Marei et al. 2014), mulberry big spot virus and mulberry necrosis virus (Kuai 2010) have not been substantiated.

Phytoplasmas

Phytoplasmas are small wall-less pleomorphic bacteria infecting both plants and insects (vectors) as obligate parasites. Prior to 1967, these agents were regarded as "yellow" viruses because of their small size which was below the resolution of light microscope, their mode of transmission remained undetected causing and failure to isolate them in culture media. In 1967, Doi et al. (1967) used electron microscopy of ultrathin phloem sections to show the presence of mycoplasma-like organisms (MLOs) within the sieve cells of so-called yellows infected plants. This discovery made the foundation of recognizing a new category of plant pathogens distinct from previously known categories. The plant MLOs were renamed as phytoplasmas in the 10th meeting of International Organization of Mycoplasmology in 1994. Taxonomically, phytoplasmas belong to the family Acholeplasmataceae of the class Mollicutes (Lee et al. 2000). All phytoplasmas are considered as members of a single provisional genus, *Candidatus* Phytoplasma (IRPCM 2004). The genus is regarded as candidatus because phytoplasmas are presently un-culturable and have not been obtained in pure state for reinoculation to satisfy the Koch postulates.

Phytoplasmas are involved in hundreds of plant diseases of economic importance. They inhabit phloem cells of infected plants and disturb normal plant metabolism to induce various types of symptoms. The main symptoms are dwarfing, proliferation of various tissues, witches' broom, yellows, tissue necrosis and plant death. In herbaceous plants they also cause phyllody and virescence of flowers, i.e., transformation of flower parts into green leaf-like structures.

Natural transmission of phytoplasmas takes place by their phloem-feeder insect hosts. Most of these vectors are leafhoppers of the family Cicadellidae. Due to propagative nature of phytoplasmas in the vector tissues, such insects remain a source of inoculum throughout their life after initial acquisition of the pathogen. Phytoplasmas are also transmissible when infected vegetative propagation materials such as buds and grafts are used in horticulture. They are also perpetuated in plants from infected cuttings. It was generally believed that because of being limited to phloem, phytoplasmas are not transmitted through seeds of infected plants. However, recent reports suggest seed transmission of several phytoplasmas which may explain for unexpected epidemics in a region (Satta et al. 2019).

Experimental transmission of phytoplasmas may be facilitated by the use of parasitic plant dodder (*Cuscuta* spp.) particularly in research when transmission between taxonomically distant plants or between herbaceous and woody plants are required. In this process, dodder serves as a biological bridge connecting donor and recipient plants.

Within the candidatus genus Phytoplasma, grouping is primarily based on the nucleotide sequence homology of 16S ribosomal RNA genes (16Sr) followed by restriction fragment length polymorphism (RFLP). Thus, until now 33 16Sr groups (16SrI-16SrXXXIII) are recognized. Each 16Sr group may further be divided into subgroups (A, B, etc.) based on minor molecular or biological differences. Well characterized phytoplasmas and phylogenetically related strains assume the rank of *Candidatus* Phytoplasma species. There are now 38 recognized candidatus phytoplasma species (IRPCM 2004, Perez-Lopez et al. 2016).

Mulberry dwarf. The only phytoplasmal disease so far reported from mulberry is mulberry dwarf or yellow dwarf which has been known in Japan as early as 1603 (Okuda 1972). It is still a widespread disease of economic importance in Japan (Jiang et al. 2004). The symptoms of the disease are stunting of the trees accompanied by yellowing of leaves, dwarfing of shoots and witches' broom (Gai et al. 2014).

The phytoplasmal nature of the disease was first established by electron microscopy and sensitivity of the pathogen to tetracycline treatment, and further verified by molecular studies. In fact, mulberry dwarf was one of the two plant diseases shown to be of phytoplasmal nature by Doi et al. (1967). Mulberry dwarf phytoplasma belongs to the subgroup B of the 16SrI (aster yellows) group. The phytoplasma cells can be detected in root and reproductive organs of infected plants (Jiang et al. 2004).

Three leafhopper species, i.e., *Hishimonus sellatus*, *Hishimonoides sellatiformis*, and *Tautoneura mori* are reported to be vectors of mulberry dwarf phytoplasma (Jiang et al. 2005, Ishijima 1971). Phytoplasma cells have been identified in several organs and eggs of *H. sellatiformis* and *H. sellatus* (Kawakita et al. 2000).

Ji et al. (2010) attempted to purify mulberry dwarf phytoplasma from infected mulberry leaves and analyze the protein content of purified preparations. They reported presence of 209 experimentally verified and 63 hypothetical proteins by one-dimensional SDS-PAGE and nanocapillary liquid chromatography-mass spectrometry. Gai et al. (2014) studied micro RNA (miRNA) content of infected mulberry plants and concluded that differential expression of plant miRNAs and formation of novel miRNAs may be involved in symptom development by disturbing metabolism and hormone balance.

So far, mulberry dwarf phytoplasma has been reported only from Japan where it is associated with heavy losses to sericulture. Recently, however, mulberry trees with proliferation of yellow-dwarfed shoots (Figure 3) have also been observed in Iran. The associated phytoplasma belonged to the 16SrI-B, the same as Japanese

Figure 3. Proliferation of chlorotic dwarfed shoots in a mulberry tree from Yazd Province, Iran, infected with a phytoplasma strain of 16SrI subgroup B. (Photo courtesy of M. Salehi).

strain (M. Salehi, personal communication). Therefore, it is likely that distribution of mulberry dwarf is wider and includes mulberry growing regions outside Japan.

Detection and Control

Detection

Infection of Viroids, viruses, and phytoplasmas is often accompanied by the development of symptoms. While such symptoms may be useful in alarming the growers and plant pathologists for the seriousness of the problem, they are of limited value in diagnosis of the disease and identification of the causal agents. This is because similar symptoms may be induced by different biotic and abiotic agents. Type and severity of symptoms may be affected by pathogen strain, variety and cultivar of the host, growth conditions and environmental factors. Therefore, in most cases other methods must be employed for identification of disease agents.

There are numerous biological, serological, physical, chemical and molecular methods for diagnosis, detection and identification of viroids, viruses and phytoplasmas in plants (Hull 2014, Matthews 1993). The choice of methods largely depends on the nature of the pathogen, type and condition of the plant, the objective of the work and availability of facilities. In the absence of any information about the causal agent of a disease, a preliminary step in diagnosis is to show if the disease is transmissible. This will differentiate infectious diseases from those caused by abiotic factors such as environmental stress and genetic disorders. Generally, all viroids, viruses and phytoplasmas, except cryptoviruses and endornaviruses, are graft transmissible. Many viruses and viroids are also transmissible by mechanical (sap) inoculation. However, attempts to transmit these pathogens from woody plants often fail because of their low concentration and presence of inhibitors in the plant extract. Studies on tree pathogens will be greatly facilitated by their transmission to herbaceous plants. This can be achieved by bridging the woody and herbaceous plants with the parasitic plant dodder (*Cuscuta* spp.) which works best for phloem inhabiting pathogens. This method has been used to transmit many tree phytoplasmas to periwinkle and other herbaceous plants.

Methods which can be used for preliminary diagnosis of diseases and identification of possible pathogens include double stranded RNA (dsRNA) analysis, electron microscopy and deep sequencing. The latter has become increasingly popular in diagnosis of tree diseases (Maliogka et al. 2018). The method has been used to identify a monopartite geminivirus in mulberry mosaic dwarf, mulberry vein banding tospovirus, PNRSV and mulberry cryptovirus 1 in mulberry plants showing line pattern (Meng et al. 2013, Ma et al. 2015, Skelton et al. 2018). Deep sequencing can give a clue about involvement of pathogens in a disease. But due to high cost of the method, association of the candid pathogens with the disease must be verified by other molecular or serological methods such as polymerase chain reaction (PCR) or enzyme-linked immunosorbent assay (ELISA).

Serological methods can be used for detection of known viruses if their antisera are available. Several international companies provide serological reagents or kits as well as instruction for diagnosis of plant viruses. The specificity of the reactions is high. Popular methods applicable to woody plants include ELISA, dot blot immunoassay (DIBA), and tissue print immunoassay (TPIA). Details of procedures can be found in Hampton et al. (1990), Matthews (1993), and Huth (1999). TPIA (Huth 1999) is the method of choice for indexing large number of plants for phloem inhabiting pathogens. The sample membranes can easily be prepared in the field and sent to a laboratory for further process at appropriate time. Serological methods are not applicable for detection of viroids and most phytoplasmas.

PCR (including reverse transcription polymerase chain reaction, RT-PCR) has now become a routine method for detection and identification of plant pathogens. Most laboratories have equipment to carry out PCR assay. The desired primers based on target nucleotide sequences can be easily designed and ordered. They may be variant—or group-specific, i.e., designed to detect a certain variant or any member of a group. Multiplex PCR with combination of more than one set of primers can be used to detect several pathogens simultaneously. PCR is more sensitive than serological methods and becomes even more sensitive in nested PCR which is used when the concentration of target sequence is too low to be resolved by direct PCR. In nested PCR, the reaction is carried out twice using two sets of primers. The second set is designed to amplify the product of the first round of PCR. For phytoplasmas

Table 1. PCR primers used for detection of mulberry viruses and virus-like agents.

Pathogen	Primer sequence (5′-3′)	Product size (nt)	Reference
[1]Begomovirus	PCRV181: TAA TAT TAC CGG WTG GCC Bc: TGG ACY TTR CAW GGB CCT TCA CA	500	Rojas et al. 1993 Deng et al. 1994
[2]FBV-1	F: GCT GAT CAC AAG AGG CAT GA R: TCC TTG TTT CCA CGT TCC TT	214	Minafra et al. 2012
FBV-1	F: AAC GCC ACA GAT GAA GGA R: AGG CTA CCC ATC TCA CTC T	1192	Alishiri et al. 2016
FBV-1	F: ACC AGA CGG AGG GAA GAA AT R: TCC TTA CCA TCG GTT ATC TC	1070	Laney et al. 2012
HSVd	F: AAC CCG GGG CAA CTC TTC TC R: AAC CCG GGG CTC CTT TCT CA	303	Sano et al. 2001
MBV-1	F: TCA GGC GTA CAG AGA CAT CA R: AAC ATC ACT CGT GGG GCT AC	228	Elbeaino et al. 2013
MVBV	F: AAG CCA TCA ATG TGC CTC CGG A R: AAC ACC ATG TCT ACC GTC CGT C	1000	Meng et al. 2013
[3]Phytoplasmas	P1: AAG AGT TTG ATC CTG GCT CAG G P7: ATT CGT CCT TCA TCG GCT CTT R16F2: ACG ACT GCT AAG ACT GG R16R2: TGA CGG GCG GTG TGT ACA AAC CCC G	1800 1200	Schneider et al.1995 Lee et al.1993
PNRSV	F: GTT GCC ATG GTT TGC CGA ATT TAC R: CAT GGG GCC CTA GAT CTC AAG CAG GTC	480	Sanchez-Navarro et al. 1997
TSV	CPF: AGA TAA GTC GCT TCT CGG AC CPR: TGC TCG CAT GGA TCA TAG AC	900	Rageshwari et al. 2019
Viroid-like RNA	F: GTC CAG ACA CAC ATC T R: TGA TGA GTT CGA AAG AAC	356	Wang et al. 2010

[1]Degenerate primer pair for begomoviruses: B=C,G; R=A,G; W=A, T; Y=C,T.
[2]Acronyms: FBV-1=fig badnavirus-1; HSVd=hop stunt viroid; MBV-1=mulberry badnavirus-1; MVBV= mulberry vein banding virus; PNRSV= prunus necrotic ring spot virus; TSV= tobacco streak virus.
[3]Two pairs of universal primers used in nested PCR.

which have low concentration in the plant, the product of the p1/p7 universal primer pair is used as template for amplification by a second primer set such as R16F2n/R16R2 (Gundersen and Lee 1996). Table 1 shows several pairs of primers used in detection of mulberry viruses, viroids and phytoplasmas.

Agarose gel electrophoresis of PCR products and visualization of the bands by proper staining will show presence or absence of the target molecule and, if present, give an indication of product size. However, for more accurate identification and detecting possible variations, PCR product of selected samples must be sequenced and analyzed.

Electron microscopy has been used in detection and identification of many viruses, including mulberry nepoviruses (Tsuchizaki et al. 1971), latent virus (Tsuchizaki 1976, Lu et al. 2015), and vein banding tospovirus (Meng et al. 2015). It played a crucial role in the discovery of phytoplasmas as a distinct category of plant pathogens (Doi et al. 1967). However, exact identification of pathogen by electron microscopy is not possible unless combined with other methods such as immunology (Milne 1993).

Presence of dsRNA in diseased plants has been regarded as an indication of virus infection (Dodds 1993). In addition to the viruses with dsRNA genome, single stranded (ss) RNA viruses are also believed to form dsRNA, known as replicative form, during their replication. The isolated dsRNA can then be analyzed (for example by cloning and sequencing) for further identification.

A simple method for diagnosis of phytoplasmas in suspected plants is the use of the stain DAPI (diamidino-2-phenylindole). DAPI is a fluorescent dye which binds with DNA in the adenine-thymine rich region. It is used for staining sections of plant midribs or thin stems followed by observation under a light microscope. Stained phloem will be an indication of phytoplasma infection. It may also be used in combination with other methods such as PCR (Mero et al. 2016).

Management

Control of mulberry viruses, viroids and phytoplasmas follows the same principles used in the control of these agents in other perennial plants such as fruit trees. It is mainly based on preventive measures. Once a plant is infected with any of these agents, little can be done to free it from the pathogen without causing injury. This is because of the intimate relationship of the pathogens with host cells. Nevertheless, the type of management practices depends on the type of pathogen and economical importance of the disease, extent of pathogen incidence and distribution, type of vector if any, geographical location of plantation and other environmental factors. The following is a summary of major management practices against virus-like agents infecting mulberry. For further discussion, the readers are referred to Hadidi et al. (1998) and Hull (2014).

Preventing Pathogen Entrance

A common means of long distance spread of pathogens to other regions is movement of infected plant material facilitated by human activities. Most countries have a list of quarantine pests to prevent their entrance from outside sources. Some important mulberry pathogens are reported from certain countries but not others. For example, mulberry dwarf phytoplasma is reported only from Japan and mulberry vein banding tospovirus is not reported in countries other than China. Although lack of a pathogen report from a given country does not necessarily mean absence of that pathogen there, but before confirmation of its occurrence, it must remain in the quarantine list. In any case, accidental entrance and limited occurrence of a quarantined pathogen must be treated by eradication.

Some countries also enforce internal quarantine to bar the spread of pathogen from one region to other.

Past experiences show that quarantine regulations can at best delay the entrance of the pathogens into a country. Entrance of bacterial citrus canker into Florida is a good example. The first entrance of the bacterium into Florida took place in 1910. The disease was eradicated. The second canker introduction was in 1986 and eradication took place again. The third introduction was in 1995 which resulted in such a wide spread of the pathogen that eradication became impractical (Gottwald et al. 2002).

Eradication

Rogueing of mosaic dwarf-affected mulberry has been practiced in China since 17th century (Fei et al. 2009). In rogueing (eradication) the whole infected plant is removed and destroyed. The objective of this practice is to eliminate the pathogen source within an orchard. Rogueing of infected trees is practical only if the disease incidence is limited and the chances of pathogen spread from neighboring plantations is low. Badly affected un-economical plantations must be rogued completely and replaced with healthy, and if possible resistant, plants.

Use of Pathogen-free Propagating Material

When establishing a new plantation, it is most critical to use pathogen-free propagating material. When using vegetative propagating material such as cuttings and scions, it is essential to ascertain that the source plants are free from pathogens. This will be accomplished by visual inspection and by sensitive detection tests.

Heat treatment (thermotherapy) of source plants by allowing them to grow in growth chambers at 37–40°C for several weeks may result in elimination of many viruses from newly grown tissues, buds and meristems. Such tissues can be used as the source of virus-free material for grafting. Alternatively, meristem dome plus 1–2 leaf primordial (0.1×0.2–0.5 mm) may be cultured in artificial medium until rooting stage (reviewed by Tabler et al. 1998) or used in shoot tip grafting (reviewed by Faccioli and Marani 1998). By meristem tip culture and shoot-tip grafting it is expected to obtain a high percentage of virus-free plants.

Cross Protection

Each virus, viroid or phytoplasma is usually composed of a number of strains which differ from each other in minor properties including their effects on host plants. It has been known for many decades that a plant infected by a mild strain of a virus or phytoplasma is often protected from infection by severe strains of the same virus or phytoplasma (McKinney 1929, Kunkel 1955). This is called cross protection or mild strain protection.

The mechanism of cross protection has been the subject of many investigations and several hypotheses have been proposed to explain the phenomenon. Recent studies have pointed out post transcriptional genes silencing as a major mechanism of cross protection.

Cross protection has been used to control several plant viruses. A successful example is the control of citrus tristeza virus (CTV) in areas of high incidence of severe strains, such as Brazil where millions of new trees have been graft-inoculated with mild strains and used commercially against CTV severe strains (Garnsey et al. 1998, Moreno et al. 2008). Disadvantages of using mild strain protection include risk of genetic changes in the mild strains, possibility of synergistic interaction with other viruses and causing some losses to the crop. There is also the risk of mild strain forming a source for spread to other susceptible crop plants.

An alternative to mild strain protection is the so-called pathogen derived resistance (PDR) in which a pathogen gene or a part of it is introduced into the plant genome by biotechnology protocols. Such transgenic plants specifically resist the pathogen by the mechanism of gene silencing.

No cross protection or PDR studies have been reported for mulberry diseases. However, in the absence of natural resistance and for areas of high disease incidence, these control methods as well as other methods of gene transfer (Tabler et al. 1998) warrant investigation.

Vector Control

Most vectors of plant pathogens are among insects that feed on-and cause damage to plants. Confirmed vector-borne pathogens of mulberry are leafhopper transmitted mulberry yellow dwarf phytoplasma, whitefly transmitted cassava mosaic virus, thrips transmitted mulberry vein banding tospovirus, and nematode borne mulberry ringspot virus. By analogy, mulberry mosaic leaf roll associated virus, mulberry latent carlaviruses, mulberry badnavirus-1, and mulberry mosaic dwarf associated geminivirus are expected to be vectored by longidorid nematodes, aphids, mealy bugs, and whiteflies, respectively.

The aim of pest control is to reduce the insect populations to a level below the economic threshold. But this is not sufficient for preventing pathogen spread by insect vectors. While vector control could be beneficial for controlling phytoplasmas and persistent viruses in annual crops or tree nurseries, it would be more difficult to achieve satisfactory control of diseases in perennial plants. This is because perennial plants remain exposed to infection for several years and, on the other hand, only few insects remaining after pest control operations and those launched from neighboring plantations are enough to visit the crop plant and cause infection. Nevertheless, control of vector can prevent severe epidemics of diseases.

Use of Resistant Varieties

Use of good quality resistant varieties is the most economical and environmentally safest method of combatting plant pests and diseases. Natural resistance may be found among germplasms or horticulturally developed by crossing a good quality variety with a resistant variety. The latter method is quite laborious and time consuming and not suitable for mulberry species. An alternative is transferring of resistance genes by molecular biology techniques or taking advantage of gene silencing as discussed above.

There are many species in the genus *Morus* and hundreds of varieties in *Morus alba*. Although some mulberry screening has been made for resistance to bacterial and fungal pathogens (Banerjee et al. 2009, Pasha and Barman 1989), attempts to explore possible resistance among mulberry varieties against viruses and phytoplasmas have been limited (Machii et al. 2002).

Living with the Diseases

In the absence of resistant varieties and other satisfactory control measures in areas of high disease incidence, certain horticultural practices may help to obtain a reasonably better yield from the infected plants. Although mulberry is a hardy plant, proper irrigation and fertilizer applications as well as pest control may improve the growth of infected plants, especially for sericulture purposes. According to Tahama (1963), cutting back of plants to stubs in yellow dwarf affected mulberry plants **in winter** suppresses the disease symptoms for the following season. However, the phytoplasma are not eradicated from the trees. Cutting back these trees in summer caused a severe damage to plants.

Conclusions

A variety of viruses and virus-like agents, some of which associated with diseases of economic importance, have been detected in mulberry in different countries of the world. It is likely that still more of these agents will be discovered in the future. Like in other perennial plants, very often a combination of two or more agents are associated with a disease or a type of symptoms.

Present knowledge on virus-like-diseases of mulberry suffers from two important shortcomings. First, most studies have been confined to detection and have stopped at the "association" stage of Koch postulates. Thus the cause-and-effect relationship of various symptoms and associated agents has not been clearly established. This has caused a chaos in attributing various symptoms to the associated agents. For instance, HSVd has been reported from symptomless plants (Elbeaino et al. 2012a) and in association with vein clearing (Amiri-Mazhar et al. 2013). On the other hand, plants with vein clearing symptoms had also mulberry badnavirus-1 and fig badnavirus-1 (Alishiri et al. 2016). It is true that different agents cause similar effects or variants of the same agent may induce different symptoms. In addition, variety and growth conditions of plants can affect symptom expression. The second shortcoming that is somehow related to the first one is the scant of information available on mulberry resistance to the virus-like agents despite presence of hundreds of varieties. The future studies must be designed to focus on these shortcomings.

Acknowledgment

The author wishes to thank Dr. M.S. Sadeghi of the Plant Virology Research Center, Shiraz University, for technical assistance. Supported by Center of Excellence in Plant Virology.

References

Alishiri, A., F. Rakhshandehroo, M. Shams-Bakhsh and G.R. Salehi Jouzani. 2016. Incidence and distribution of fig badnavirus 1 and mulberry badnavirus 1 on mulberry trees in Iran. J. Plant Pathol. 98: 341–345.

Amiri-Mazhar, M., S.A.A. Bagherian and K. Izadpanah. 2013. Variants of Hop stunt viroid associated with mulberry vein clearing in Iran. J. Phytopathol. 162: 269–271.

Banerjee, R., N.K. Das, M. Maji, K. Mandal and A.K. Bajpai. 2009. Screening of mulberry genotypes for disease resistance in different seasons to bacterial leaf spot. Ind. J. Gen. Plant Breeding 69: 152–156.

Chiumenti, M., M. Morelli, A. De Stradis, T. Elbeaino, L. Stavolone and A. Minafra. 2016. Unusual genomic features of a badnavirus infecting mulberry. J. Gen. Virol. 97: 3073–3087.

Deng, D., P.F. McGrath, D.J. Robinson and B.D. Harrison. 1994. Detection and differentiation of whitefly-transmitted geminiviruses in plants and vector insects by the polymerase chain reaction with degenerate primers. Ann. Appl. Biol. 125: 327–336.

Dodds, J.A. 1993. dsRNA in diagnosis. pp. 273–294. *In*: Matthews, R.E.F. (ed.). Diagnosis of Plant Virus Diseases. CRC Press.

Doi, Y., M. Teranaka, K. Yora and H. Asayama. 1967. Mycoplasma or PLT group-like microorganisms found in the phloem elements of plants infected with mulberry dwarf and potato witches' broom. Ann. Phytopathol. Soc. Japan 33: 259–266.

Elbeaino, T., R. Abou Kubaa, E. Choueiri, M. Digiaro and B. Navarro. 2012a. Occurrence of Hop stunt viroid in mulberry (*Morus alba*) in Lebanon and Italy. J. Phytopathol. 160: 48–51.

Elbeaino, T., R. Abou Kubaa, F. Ismaeil, J. Mando and M. Digiaro. 2012b. Viruses and Hop stunt viroid of Fig trees in Syria. J. Plant Pathol. 94: 687–691.

Elbeaino, T., M. Chiumenti, A. De Stradis, M. Digiaro, A. Minafra and G.P. Martelli. 2013. Identification of a badnavirus infecting mulberry. J. Plant Pathol. 95: 207–210.

Etebari, K., L. Matindoost and R.N. Singh. 2004. Decision tools for mulberry thrips *Psudodendrothrips mori* (Niwa 1908) management in sericultural regions; an overview. Insect Sci. 11: 243–255.

Faccioli, G. and F. Marani. 1998. Virus elimination by meristem tip culture and tip micrografting. pp. 346–380. *In*: Hdidi, A., R.K. Khetarpal and H. Koganezawa (eds.). Plant Virus Disease Control. APS Press. St. Paul, MN, USA.

Fei, J., Y. Li, Y. Wu, X. Bai, G. Shi, F. Yu and Y. Kuai. 2009. Identification of a latent pathogen on mulberry tree with a disease of mosaic dwarf. Afr. J. Biotechnol. 8: 5358–5361.

Gai, Y.-P., Y.-Q. Li, F.-Y. Guo, C.-Z. yuan, Y.-Y. Mo, H.-L. Zhang, MH. Wang and X.-L. Ji. 2014. Analysis of phytopplasma-responsive sRNAs provide insight into the pathogenic mechanisms of mulberry dwarf disease. Scientific Reports 4: Article No. 5378.

Gao, X., Z.Q. Jia, B.J. Yang, X.L. Tan and Y.T. Liu. 2019. First report of mulberry vein banding virus infecting *Pharbitis purpurea* in Yunan, China. Plant Dis. 103. Doi: 10.1094/PDIS-12-18-2149-PDN.

Garnsey, S.M., T.R. Gottwald and R.K. Yokomi. 1998. Control strategies for citrus tristeza virus. pp. 639–658. *In*: Hdidi, A., R.K. Khetarpal and H. Koganezawa (eds.). Plant Virus Disease Control. APS Press. St. Paul, MN, USA.

Geering, A.D.W. and R. Hull. 2012. Family *Caulimoviridae*. pp. 428–443. *In*: King, A.M.Q., M.J. Adams, E.B. Carstens and E.J. Lefkowitz (eds.). Virus Taxonomy, Ninth Report of ICTV. Amsterdam, Elsevier Academic Press.

Gottwald, T.R., J.H. Graham and T.S. Schubert. 2002. Citrus canker: the pathogen and its impact. Plant Health Progress. Doi: 10.1094/PHP-2002-0812-01-RV.

Gundersen, D.E. and I.M. Lee. 1996. Ultrasensitive detection of phytoplasmas by nested PCR assay using two universal primer pairs. Phytopathol. Mediterr. 35: 144–151.

Hadidi, A., R.K. Khetarpal and H. Koganezawa (eds.). 1998. Plant Virus Disease Control. APS Press. St. Paul, MN, USA.

Hampton, R., E. Ball and S. De Boer (eds.). 1990. Serological Methods for Detection and Identification of Viral and Bacterial Plant Pathogens. APS Press, St. Paul, Min. USA.

Hull, R. 2014. Plant Virology. Fifth Ed. Academic Press.

Huth, W. 1999. Tissue print immunoassay—a rapid and reliable method for routinely detecting gramineae viruses. Plant Research and Development 49: 7–19.

IRPCM. 2004. Description of the genus '*Candidatus* Phytoplasma', a taxon for the wall-less non-helical prokaryotes that colonize plant phloem or insects. Int. J. System. Evolut. Microbiol. 54: 1243–1255.

Ishijima, T. 1971. Transmission of a possible pathogen of mulberry dwarf disease by a new leafhopper, *Hishimonoides sellatiformis* Ishihara. J. Seric. Sci. Jpn. 40: 136–140.

Jahanshahi, Z., K. Izadpanah, A.R. Afsharifar and S.A.A. Behjatnia. 2015. New hosts and comparison of biological and molecular characteristics of fig, mulberry, and citrus isolates of Hop stunt viroid in Iran. Iran. J. Virol. 9: 22–30.

Ji, X., Y. Gai, B. Lu, C. Zheng and Z. Mu. 2010. Shotgun proteomic analysis of mulberry dwarf phytoplasma. Proteome Sci. 8: 20.

Jiang, H., W. Wei, T. Saiki, H. Kawakita, K. Watanabe and M. Sato. 2004. Distribution pattern of mulberry dwarf phytoplasma in reproductive organs, winter buds, and roots of mulberry tree. J. Gen. Plant Pathol. 70: 168–173.

Jiang, H., T. Saiki, K. Watanabe, H. Kawakita and M. Sato. 2005. Possible vector insect of mulberry dwarf phytoplasma, Tautoneura mori Matsumura. J. Gen. Plant Pathol. 71: 370–372.

Kawakita, H., T. Saiki, W. Wei, W. Mitsuhashi, K. Watanabe and M. Sato. 2000. Identification of mulberry dwarf phytoplasma in the genital organs and eggs of leafhopper *Hishimonoides sellatiformis*. Phytopathology 90: 909–914.

Kryczynski, S., E. Paduch-Cichal and L.J. Skrzeczowski. 1988. Transmission of three viroids through seed and pollen of tomato plants. J. Phytopathol. 121: 51–57.

Kuai, Y.-Z. 2010. Research progress on mulberry viruses and viral diseases (I). Science of Sericulture 2010-05.

Kunkel, L.O. 1955. Cross protection between strains of aster yellows. Adv. Virus Res. 3: 251–273.

Laney, A.G., M. Hassan and T.E. Tzanetakis. 2012. An integrated badnavirus is prevalent in fig germplasm. Phytopathology 102: 1182–1188.

Lee, I.M., R.W. Hammond, R.E. Davis and D.E. Gunderson. 1993. Universal amplification and identification of mycoplasmalike organisms. Phytopathology 83: 834–842.

Lee, I.M., R.E. Davis and D.E. Gundersen-Rindal. 2000. Phytoplasma: Phytopathogenic mollicutes. Annu. Rev. Microbiol. 54: 221–255.

Lu, Q.-Y., Z.-J. Wu, Z.-S. Xia and L.-H. Xie. 2015. A new nepovirus identified in mulberry (*Morus alba* L.) in China. Arch. Virol. 160: 851–855.

Ma, Y., B. Navarro, Z. Zhang, M. Lu, X. Zhou, S. Chi, F. Di Serio and S. Li. 2015. Identification and molecular characterization of a novel monopartite geminivirus associated with mulberry mosaic dwarf disease. J. Gen. Virol. 96: 2421–2434.

Machii, H.A., A. Koyama and H. Yamanouchi. 2002. Mulberry breeding, cultivation and utilization in Japan. www.fao.org/3/X9895e05.htm.

Maliogka, V.I., A. Minafra, P. Saldarelli, A.B. Ruiz-Garcia, M. Glasa, N. Katis and A. Olmos. 2018. Recent advances on detection and characterization of fruit tree viruses using high-throughput sequencing technologies. Viruses: 10, 436: 1–23.

Marei, E.M., R.M. Elbaz, I. Elmaghraby and A. Sharaf. 2014. Occurrence of mulberry mosaic virus in Egypt. Internl. J. Microbiol. 6: 575–580.

Matsushita, Y. and S. Tsuda. 2016. Seed transmission of Potato spindle tuber viroid, Tomato apical stunt viroid, and Columea latent viroid in horticultural plants. Eur. J. Plant Pathol. 145: 1007–1011.

Matthews, R.E.F. (ed.). 1993. Diagnosis of Plant Virus Diseases. CRC Press.

McKinney, H.H. 1929. Mosaic diseases in the Canary Islands, West Africa and Gibraltar. J. Agric. Res. 39: 557–578.

Meng, J.R., P.P. Liu, C.W. Zou, Z.Q. Wang, Y.M. Liao, J. H. Cai and B.X. Qin. 2013. First report of a *Tospovirus* in mulberry. Plant Dis. 97: 1001.

Meng, J., P. Liu, L. Zhu, C. Zou, J. Li and B. Chen. 2015. Complete genome sequence of mulberry vein banding associated virus, a new tospovirus infecting mulberry. PLoS One 10: e0136196.

Mero, D., A. Bacu and K. Buzo. 2016. Phytoplasmas detection effectiveness based on symotomatology, DAPI staining and specific PCR at Plum plant material of different categories. Europ. J. Biotechnol. Genetic Engin. 3: 75–82.

Milne, R.G. 1993. Electron microscopy of in vitro preparations. pp. 215–251. *In*: Matthews, R.E.F. (ed.). Diagnosis of Plant Virus Diseases. CRC Press.

Minafra, A., M. Chiumenti, T. Elbeaino, M. Digiaro, G. Bottalico, V. Pantaleo and G.P. Martelli. 2012. Occurrence of fig badnavirus 1 in fig trees from different countries and in symptomless seedlings. J. Plant Pathol. 94: S4. 105.

Moreno, P., S. Ambros, M. Albiach-Marti, J. Guerri and L. Pena. 2008. Citrus tristeza virus: a pathogen that changed the course of the citrus industry. Mol. Plant Pathol. 9: 251–268.

Nameth, S.T. and S.L. Cheng. 1994. Identification and partial characterization of endogenous double-stranded ribonucleic acid in mulberry. J. Amer. Soc. Hort. Sci. 119: 859–861.

Okuda, S. 1972. Occurrence of diseases caused by mycoplasma-like organisms in Japan. Plant Protection 26: 180–183.

Pasha, K. and A.C. Barman. 1989. Selection of mulberry (*Morus alba* L.) variety resistant to leaf spot disease (1988). Agris 13: 103–108.

Perez-Lopez, E., C.Y. Olivier, M. Luna-Rodriguez and T.J. Dumonceaux. 2016. Phytoplasma classification and phylogeny based on *in silico* and *in vitro* RFLP analysis of *cpn60* universal target sequences. Int. J. Syst. Evol. Microbiol. 66: 5600–5613.

Rageshwari, S., P. Renukadevi, V.G. Malathi and S. Nakeeran. 2019. Occrrence of Tobacco streak virus in mulberry. Plant Disease 103. Doi: 10.1094/PDIS-08-18-1451-PDN.

Rojas, M.R., R.L. Gilbertson, D.R. Russell and D.P. Maxwell. 1993. Use of degenerate primers in the polymerase reaction to detect whitefly transmitted geminiviruses. Plant Dis. 77: 340–347.

Sabanadzovic, S. and N. Abou Ghanem-Sabanadzovic. 2008. Molecular characterization and detection of a tripartite cryptic virus from rose. J. Plant Pathol. 90: 287–293.

Sanchez-Navarro, J.A., C.B. Reusken, J.F. Bol and V. Pallas. 1997. Replication of alfalfa mosaic virus RNA 3 with movement and coat protein genes replaced by corresponding genes of Prunus necrotic ringspot ilarvirus. J. Gen. Virol. 78: 3171–3176.

Sano, T., R. Mimura and K. Oshima. 2001. Phylogenetic analysis of hop and grapevine isolates of hop stunt viroid supports a grapevine origin for hop stunt disease. Virus Genes 22: 53–59.

Satta, E., S. Paltrinieri and A. Bertaccini. 2019. Phytoplasma transmission by seed. *In*: Bertaccini,A. , P.G. Weintraub, G.P. Rao, and N. Mori (eds.). Phytoplasma: Plant Pathogenic Bacteria. II. Transmission and Management of Phytoplasma-Associated Diseases. Singapore, Springer.

Schneider, B., E. Seemüller, C.D. Smart and B.C. Kirkpatrick. 1995. Phylogenetic classification of plant pathogenic mycoplasmalike organisms or phytoplasmas. pp. 369–380. *In*: Razin, S. and J.G. Tully (eds.). Molecular and Diagnostic Procedures in Mycoplasmalogy. Vol. I. Academic Press.

Shery, A.V.M.J. 2016. A new variant of *Cassava mosaic virus* causes mulberry mosaic disease in India. Int. J. Plant, Animal and Environmental Sci. 6: 83–92.

Skelton, A., A. Fowkes, I. Adams A. Baxton-Kirk, V. Harju, S. Forde, R. Ward, M. Kelly, P. Barber and A. Fox. 2018. First report of Prunus necrotic ringspot virus and mulberry cryptic virus 1 in mulberry (*Morus alba*) in the United Kingdom. New Disease Reports 37: 23.

Steger, G., D. Riesner, M.-C. Maurel and J.-P. Perreault. 2017. Viroid structure. pp. 63–70. *In:* Hdidi, A., R. Flores, J.W. Randles and P. Palukaitis (eds.). Viroids and Satellites. Academic Press.

Susi, P. 2004. Black currant reversion virus, a mite-transmitted nepovirus. Mol. Plant Pathol. 5: 167–173.

Tabler, M., M. Tsagris and J. Hammond. 1998. Antisense RNA and Ribozyme-mediated resistance to plant viruses. pp. 79–93. *In*: Hdidi, A., R.K. Khetarpal and H. Koganezawa (eds.). Plant Virus Disease Control. APS Press. St. Paul, MN, USA.

Tahama, Y. 1963. Studies on the dwarf disease of mulberry tree. (V) Suppression of symptoms in diseased trees cut back in winter. Japan. J. Phytopathol. 28: 53–57.

Tsuchizaki, T., H. Hibino and Y. Saito. 1971. Mulberry ringspot virus isolated from mulberry showing ringspot symptom. Japanese J. Phytopathol. 37: 266–271.

Tsuchizaki, T. 1975. Mulberry ring spot virus. CMI/AAB Descriptions of Plant Viruses. No. 142, 4 pp.

Tsuchizaki, T. 1976. Mulberry latent virus isolated from mulberry (*Morus alba*). Japanese J. Phytopathol. 42: 304–309.

Valverde, R.A., M.E. Khalifa, R. Okada, T. Fukuhara, S. Sabanadzovic and ICTV Consortium. 2019. ICTV Taxonomy Profile *Endornaviridae*. J. Gen. Virol. 100: 1204–1205.

Wang, W., J. Fei, Y. Wu, X. Bai, F. Yu, G. Shi, Y. Li and Z. Kuai. 2010. A new report of a mosaic dwarf viroid-like disease on mulberry tree in China. Polish J. Microbiol. 59: 33–36.

Yanagisawa, H. and Y. Matsushita. 2018. Differences in dynamics of horizontal transmission of *Tomato planta macho viroid* and *Potato spindle tuber viroid*—infected pollen. Virology 516: 258–264.

Zebrini, F.M., R. Briddon, A. Idris, D.P. Martin, E. Moriones, J. Navas-Castillo, R. Rivera-Bustamante, P. Roumagnac and A. Varsani. 2017. ICTV virus taxonomy profiles: Geminiviridae. J. Gen. Virol. 98: 131–133.

CHAPTER 17

Conservation of Mulberry Genetic Resources to Sustain Sericulture under Climate Change

Jhansilakshmi K[1],* and Ananada Rao, A[2]

INTRODUCTION

Plant Genetic Resources are essential raw materials for genetic improvement of the crop to meet present needs and also to address future problems. The significant threats for loss of diversity are because of replacement of local cultivars and landraces by a few genetically uniform modern varieties, habitat destruction, deforestation, fragmentation/degradation, the spread of invasive alien species, climate change, urbanization, pollution, over-grazing and changes in the land-use pattern (Kaviani 2011). Crop species that are endemic and without alternative habitats will be vulnerable to extinction (Jarvis et al. 2008). It is estimated that 15 to 37%, including the wild relatives of many crop species, will be threatened with extinction due to climate change by 2050 (Thomas et al. 2004). Mulberry is the only food plant for mulberry silkworm, *Bombyx mori* L. and hence, development of new mulberry varieties utilizing diverse mulberry genetic resources is crucial for increased productivity, input use efficiency and to withstand various abiotic (drought, heat, cold, alkalinity) and biotic stresses (root-rot, root-knot, mealybug, thrips, mites, whitefly) under the present context of climate change. Therefore, conservation and utilization of the genetic variability in mulberry have immense value to cope up with less predictable and extreme weather events for sustainable sericulture.

Uniqueness of Mulberry

Mulberry is a fast-growing deciduous tree and grows well in a variety of soils as well as climatic conditions (upto 250 mm minimum rainfall and from mean sea level to high altitudes of 4000 m). It can be cultivated in bush plantation, dwarf tree plantation, grown as an avenue tree and can withstand repeated pruning.

Although mulberry is the only food plant for silkworm, *Bombyx mori* L. and widely cultivated for its foliage, every part of mulberry has economic importance. Leaves can be feed to cattle, goats, sheep and rabbits (Sanchez 2000, Singh and Makkar 2000). It provides excellent timber which can be used for sports goods (Limaye 1952) and the local needs of furniture. The fuel value of stems is much superior to most of the agricultural residues (Chinnaswamy and Hariprasad 1995). Thin twigs are used for basket making. Branches can be used as raw material for paper production and medium for mushroom production. Fruits can be eaten fresh, preserved, venified or dried for winter use. Leaves, root, bark, fruit and latex have medicinal value (Yamatake et al. 1976). Mulberry leaf is

[1] Scientist-D, REC, Central Silk Board, Krishnagiri and Tamil Nadu.
[2] Scientist-D (Retd.), Central Sericultural Germplasm Resources Centre, Hosur, Tamil Nadu.
* Corresponding author: jhansimulberry@gmail.com

effective against high blood pressure because they are rich in gamma-aminobutyric acid (Chen et al. 2016). Also, abundant in deoxynojirimycin, which is reported to have an effect to lower the blood-sugar level closely related to diabetes (Tian et al. 2016). Hence, mulberry tea is popular as a health food now-a-days (Machii et al. 2000).

Mulberry fruit is a powerhouse of nutrients. Mulberry fruit is reported to nourish the blood, benefit the kidneys and treat weakness, fatigue, anemia and premature graying of hair. It is also used to treat urinary incontinence, tinnitus, dizziness and constipation in the elderly and the anemic. Mulberry fruit is a potential source of minerals and vitamins *viz.*, iron, calcium, vitamin A, C, E and K, folate, thiamine, niacin, pyridoxine and also fiber. Antioxidants found in mulberry repair free radicals. Resveratrol, an antioxidant, is found plentiful in mulberries and helps to promote the health of heart and overall vitality. Mulberry fruit is also rich in phytonutrients like anthocyanin, flavonoids, lutein, zeaxanthin, B carotene and A carotene. The abundant anthocyanin that is found in the fruits has curative properties for certain diseases and it is also used as a natural food colorant. It also useful in clearing and rejuvenating skin, makes skin soft and radiant. It is reported to rejuvenate hair follicles and promotes hair growth (Lee et al. 2004, Ozgen et al. 2009, Zhang et al. 2018).

Importance of Mulberry Genetic Resources in the context of Climate Change

Provides Parental Material to Develop New Climate Resilient Mulberry Varieties

Mulberry Genetic resources are invaluable assets and are the reservoirs of genes with adaptive value. They can play vital role in improving the adaptability and resilience of mulberry (capacity to continue functioning and producing in the face of changes and shocks) in different production systems. The prerequisite for use of these resources are, identification of mulberry accessions with specific functional traits and development of screening tools for identifying the desired genotype in large populations.

Prevents the Risk Associated with Narrowing Genetic Base

Genetic vulnerability is caused when a few cultivars derived from limited genepools are extensively grown in larger areas, susceptible to a pest, pathogen or environmental hazards due to its genetic make-up. This results in considerable crop losses particularly when a new type of pest or disease emerges as a consequence of climate change. The utilization of diverse mulberry genetic resources for the development of new mulberry varieties leads to broad genetic base and hence avoids vulnerability to pests and diseases. Incorporation of novel traits that are often found in wild relatives of crop into the cultivars reduces genetic vulnerability (Lane and Jarvis 2007, Maxted et al. 2008). Growing "a genetically diverse portfolio of improved crop varieties, suited to a range of agro-ecosystems and farming practices, and resilient to climate change" is a validated means for enhancing the resilience of production systems (FAO 2011).

Potential Parental Material for Increasing Productivity

Mulberry is a C_3 plant and a higher concentration of carbon dioxide in the atmosphere will have positive effect if all other factors remain favourable since the photosynthesis depends on the concentration of carbon dioxide that is available in the atmosphere. Adverse moisture conditions during crop growth, insufficient available nitrogen or temperatures above the optimum range may offset this effect. Utilizing genetic resources with high physiological efficiency and having adaptive features to withstand adverse weather conditions helps to increase the productivity in future cultivars.

Stress Tolerance

It is predicted that the tropics, will suffer the most from the effects of increased and sporadic temperature spikes as well as decreased and sporadic precipitation in general (IPPC 2007, Gornall et al. 2010). In many areas, the crop varieties currently grown by farmers cannot tolerate these stresses which results in loss in mulberry productivity.

Mulberry genetic resources consist of a vast diversity of heritable traits that enable the crop to adapt to abiotic and biotic stresses, e.g., drought (Susheelamma et al. 1990, Guha et al. 2010, Jhansilaskhmi and Dandin 2011, Jhansilakshmi et al. 2014, Jhansilakshmi and Gargi 2018), heat (Chaitanya et al. 2001, Ananda Rao et al. 2011), salinity and alkanity (Agastian and Vivekananda 1997, Vijayan et al. 2003, Vijayan et al. 2009, Jhansilakshmi et al. 2016, Jhansilakshmi et al. 2018), cold (Ananda Rao et al. 2009, Sukla et al. 2016) and water logging (Ghosh et al. 2012). Potential sources for tolerance to different biotic stresses were identified by several researchers by screening mulberry germplasm (Benerjee et al. 2009, Maji et al. 2009). Harnessing this genetic diversity to develop stress tolerant mulberry varieties helps for adapting to the consequences of climate change.

Improve Input use Efficiency

The major challenge for sustainable sericulture will be to increase crop production with limited water and other inputs particularly Nitrogen. Decreasing N fertilizer utilization is also required to save fossil fuel and to prevent detrimental environmental consequences of excess N fertilizer application. Exploiting genetic variation in the uptake and assimilation of N is a potential route for generating gains in NUE. Introgression of water mining (root) and water use efficiency traits helps to develop new varieties with increased drought tolerance and water productivity. The mulberry genetic resources that are identified (Vinoda et al. 2016, Jhansilakshmi et al. 2018) can be utilized in development of new cultivars.

Provides Novel Genes / Alleles

Wild genetic resources collected from natural habitat, and centers of diversity, possess genes tolerant to abiotic and biotic stresses and transfer of these genes into the agronomically superior varieties helps to increase the mulberry productivity and adaptability. Prebreeding efforts show that the wild species *M. serrata*, the natural Himalayan mulberry confined to North West India, and *M. laevigata* can contribute to crop improvement (Tikader and Kamble 2008). The potential of proteomics and metabolomics is yet to be exploited in mulberry. An integrated analysis of genomic, transcriptomic, proteomic, and metabolomic analysis is expected to yield desirable results for understanding mechanisms of enhancing the productivity of mulberry (Khurana and Checker 2011). High-quality genomic and transcriptomic data generated through Next Generation Sequencing (NGS) platforms provided information of novel candidate genes and comprehensive molecular markers. Thus, the scope for trait-specific genetic improvement in mulberry in terms of productivity, quality and climate resilience is enhanced (Vijayan 2010).

Improve Livelihood Options through Non-Sericultural uses of Mulberry

Mulberry is a hardy multipurpose tree (Singh and Makkar 2000, Sharma and Zote 2010) and has rich diversity. Planting of multipurpose tree species in non-forest land categories serves a dual purpose, i.e., conservation of diversity and carbon sequestration. Tree cultivation in agricultural land improves biomass productivity and also helps in the utilization of nutrients from different soil layers. There is a lot of potential for mulberry tree plantation in non-forest lands of village ecosystem as it provides multiple benefits and increases livelihood options in rural areas. Mulberry genetic resources with traits of economic importance and tolerance to abiotic and biotic stresses were identified for agroforestry and ecological restoration (Jhansilakshmi and Gargi 2018).

Germplasm Collection/Augmentation

The first step in conservation of genetic resources is collection of germplasm. This activity is done by:

A. Survey, Exploration and collection from farmers' fields and wild habitats, particularly in areas known as centers of diversity, and

B. Procuring materials of interest from other institutes or organizations

As per the standards provided by FAO (2014), all germplasm accessions added to the gene bank should be legally acquired, with relevant technical documentation. All material should be accompanied by at least a minimum of associated data as detailed in crop passport descriptors and samples acquired from other regions within country or other countries should pass through the relevant quarantine process.

Need for Collecting Germplasm

➢ To avoid loss of genetic diversity—The replacement of land races and local cultivars with a few high yielding mulberry varieties has resulted in reduced genetic diversity. The variability available in the natural habitats particularly in centres of diversity may be lost permanently due to increased deforestation, urbanization, natural hazards, and abnormal weather conditions in context of climate change.

➢ Incase the genotypic diversity is missing or insufficiently represented in an existing collection.

➢ Requirement for collecting material with attributes such as adaptation to high temperature, drought stress, salinity, alkalinity, cold, low soil fertility or wild relatives.

➢ For opportunistic reasons if germplasm contains striking features or found under unusual circumstances.

Germplasm Collection by Survey and Exploration

The fundamental objective of collecting plant genetic resources is to capture the maximum amount of useful genetic variation (Marshall and Brown 1975) available in nature. The basic parameter for measuring variation in a given population is allelic richness: the number of distinct alleles at a single locus (Brown and Marshall 1995). Germplasm collection requires theoretical knowledge of sampling, practical knowhow of plant diversity, and environmental including the socio-economic and cultural aspects of the farming. Tactics, logistics, preparations and procedures have been elaborately dealt with by Hawkes (1980) and Arora (1981).

Owing to the origin of the genus *Morus* in sub-Himalayan belt of Indo-China, mulberry is available in natural form in this region. In India, four species of mulberry viz., *Morus indica, M. alba, M. laevigata* and *M. serrata* were reported to be available in nature. *M. serrata* is confined to higher altitude of North Western Himalayas and sparsely distributed in North East India. It is known as Himalayan mulberry and found up to an altitude of 2200 mts above MSL particularly in Salna, Urgam valley, Joshimath (approximately 1220 years old mulberry tree protected as sacred grove), Chakrata (Ravindran et al. 1997). Dandin et al. (1993) reported the occurrence of mulberry in Jammu division, Kangra, Chamba, Nahan and Kullu districts of Himachal pradesh and Garwal Himalayas of Uttaranchal.

M. alba and *M. indica* are available throughout India mostly in cultivated forms. However, occurrence of natural distribution of *M. alba* was also observed in Surari Garhal of Himalayan region up to an elevation of 1365 mts above MSL and in Ladakh Himalayan region up to an altitude of 3300 m above MSL (Ananda Rao et al. 2006). The cultivated forms of *M. indica* and *M. alba* are available in Gujarat, Rajasthan, Uttar pradesh, West Bengal, Sikkim, Assam, Meghalaya, Arunachal Pradesh (Chandrasekar 2001, Saraswat 2002).

M. laevigata is naturally distributed in many parts of the country particularly in sub-himalayan regions of India and Andaman Islands. Besides, it has been developed as gene pool reserves in Joara, Dhar, Shivapuri, Gwalior, Bilaspur, Pachmarhi reserve forests of Central India (Dandin et al. 1995, Jain and Kumar 1989, Ravindran et al. 1998), and in isolated pockets at Shevoroy hills of Salem district of Tamilnadu (Yadav and Pavan Kumar 1996).

Apart from naturally growing mulberry, mulberry genetic resources are maintained as avenue trees on road side, as shade tree, sacred grooves, on-farm conservation mainly in coffee and tea states, farm bunds and back yards. The future collections need to be focussed on climate stressed regions since there is scope for collecting the samples with adaptive traits to make mulberry field gene banks 'climate ready' to meet the challenges of future.

The species status of important countries engaged in sericulture are presented in the Table 1 and Table 2.

Germplasm Conservation

The Convention on Biological Diversity (CBD) recognized two ways of conserving genetic resources: *in situ* (in the place of origin) and *ex situ* (outside the place of origin).

Table 1. Mulberry species distribution in China, Japan and Korea.

China		Japan		Korea	
Species	Accessions	Species	Accessions	Species	Accessions
M. bombycis Koidz.	22	*M. bombycis* Koidz.	583	*M. bombycis* Koidz.	97
M. multicaulis Perr.	750	*M. latifolia* Poir.	349	*M. latifolia* Poir.	128
M. alba L.	762	*M. alba* L.	259	*M. alba* L.	105
M. wittorium Hand-Mazz.	8	*M. acidosa* Griff.	44	*M. acidosa* Griff.	1
M. mizuho Hotta	17	*M. indica* L.	30	*M. indica* L.	5
M. rotundiloba Koidz.	4	*M. rotundiloba* Koidz.	24	*M. rotundiloba* Koidz.	-
M. australis Poir.	37	*M. kagayamae* Koidz.	23	*M. kagayamae* Koidz.	1
M. mongolica Schneider	55	*M. notabilis* C.K.Schn.	14	*M. mongolica* Schneider	1
M. bionensis Koidz.	-	*M. bionensis* Koidz.	11	*M. bionensis* Koidz.	-
M. nigriformis Koidz.	-	*M. nigriformis* Koidz.	3	*M. nigriformis* Koidz.	-
M. atropurpurea Roxb.	120	*M. atropurpurea* Roxb.	3	*M. atropurpurea* Roxb.	-
M. serrata Roxb.	-	*M. serrata* Roxb.	3	*M. serrata* Roxb.	-
M. laevigata Wall.	19	*M. laevigata* Wall.	3	*M. laevigata* Wall.	1
M. nigra L.	1	*M. nigra* L.	2	*M. nigra* L.	3
M. formosensis Hotta.	-	*M. formosensis* Hotta.	2	*M. formosensis* Hotta.	-
M. rubra L.	-	*M. rubra* L.	1	*M. rubra* L.	-
M. mesozygia Stapf.	-	*M. mesozygia* Stapf.	1	*M. mesozygia* Stapf.	-
M. celtifolia Kunth.	-	*M. celtifolia* Kunth.	1	*M. celtifolia* Kunth.	-
M. cathayana Hemsl.	65	*M. cathayana* Hemsl.	1	*M. cathayana* Hemsl.	-
M. tiliaefolia Makino	-	*M. tiliaefolia* Makino	1	*M. tiliaefolia* Makino	14
M. microphylla Bickl.	-	*M. microphylla* Bickl.	1	*M. microphylla* Bickl.	-
M. macroura Miq.	-	*M. macroura* Miq.	1	*M. macroura* Miq.	-
		Morus spp.	15	*Morus* spp.	259

Source: Tikader and Vijayan (2010)

In situ Conservation

The most appropriate way of conserving biodiversity is to conserve species in their natural habitats. *In situ* conservation is often accomplished as (a) '*in situ* conservation' of genetic resources in their wild native habitats and (b) 'on-farm management' of genetic resources in the agricultural systems. The aim of *in situ* conservation, thus, is to allow the populations to maintain/perpetuate itself within the adapted environment so that it has the potential for continued evolution. UNEP (1992) extended the CBD's definition of *in situ* conservation as follows: 'the conservation of ecosystem and natural habitats, the maintenance and recovery of viable populations of species in their natural surroundings and, in the case of domesticated and cultivated species, in the surroundings where they have developed their distinctive properties'. The genetic resources are thus preserved by protecting the ecosystem in which it occurs naturally. The Indian government took initiative and established 18 Biosphere Reserves in India. Among them, Nandadevi, Manas, Nokrek, and Great Nicobar are the potential sites for *in situ* conservation of mulberry (Naik and Mukharjee 1997). Namdapha of Arunachal Pradesh and North Andaman are important areas also amongst the selected potential sites for Biosphere Reserves choosen by the Ministry of Forests and Environment (Govt. of India). Similar efforts were also made in Canada to conserve red mulberry in Hamilton's Royal Botanical Gardens, Ball's Falls Conservation Area, Niagara Glen, Rondeau Provincial Park, Point Pelee National Park, Fish Point Provincial Nature Reserve, Pelee, Middle and East Sister Islands (Tikader and Vijayan 2017).

Table 2. Details of germplasm collected from major survey programmes conducted by CSGRC, Hosur in India.

S. No.	State	No. of surveys	No. of districts covered	Mulberry species Collected
1	Andhra Pradesh	1	1	*M. indica*
2	Arunachal Pradesh	2	6	*M. laevigata, M.indica*
3	Assam	3	2	*M. laevigata, M.indica*
4	Bihar	1	2	*M. indica, M. laevigata*
5	Chhattisgarh	2	3	*M. laevigata, M.indica*
6	Goa	2	1	*M. indica, M. latifolia*
7	Gujarat	1	6	*M. indica*
8	Haryana	1	6	*M. laevigata*
9	Himachal Pradesh	5	9	*M. indica, M. serrata, M. alba, M. laevigata*
10	Jammu and Kashmir	3	6	*M. indica, M. alba, M. serrata, M. laevigata*
11	Jharkhand	1	3	*M. indica, M. laevigata, M. alba*
12	Karnataka	2	2	*M. indica, M. alba*
13	Kerala	4	5	*M. indica, M. laevigata*
14	Madhya Pradesh	5	18	*M. laevigata, M .indica, M. alba*
15	Maharashtra	4	12	*M. indica, M. laevigata, M. alba*
16	Manipur	1	1	*M. laevigata*
17	Meghalaya	6	2	*M. laevigata, M. indica*
18	Mizoram	1	3	*M. laevigata, M. indica*
19	Nagaland	1	3	*M. indica*
20	Odisha	1	1	*M. indica*
21	Punjab	3	5	*M. indica, M. alba, M. laevigata*
22	Rajasthan	4	15	*M. indica, M. laevigata, M. alba*
23	Sikkim	3	2	*M. laevigata M .indica*
24	Tamil Nadu	5	4	*M. indica, M. serrata, M. alba, M. laevigata*
25	Tripura	1	1	*M. indica*
26	Uttar Pradesh	14	20	*M. indica, M. laevigata, M. alba*
27	Uttarakhand	2	4	*M. serrata, M. indica, M. alba, M. laevigata*
28	West Bengal	4	4	*M. laevigata, M. indica, M. alba*
29	Andaman and Nicobar Islands	3		*M. laevigata*
30	New Delhi	3		*M. indica, M. laevigata*
31	Puducherry	1		*M. indica, M. laevigata*

Source: Jhansilakshmi and Gargi (2018)

The objective of on-farm conservation is to maintain crop evolution in farmers' field, home gardens and land scapes (Bellon et al. 2014). Farmers' efforts to select new traits and exchange selected materials with friends and relatives are processes that allow the new genetic variability to evolve and change over time (Jarvis and Hodgkin 2000). Owing to highly heterozygous and cross pollinated nature of mulberry, rich *Morus* diversity exists under managed habitats, i.e., in the horticultural gardens, agricultural lands and roadside plantations, farmhouses, backyards, kitchen gardens, etc., in India. These are the first-hand selections of the farmers and tribals for diverse utilizations. The genetic resources conserved by Rajasthan Armed Constables at their farm, Bikaner (Rao et al. 2011) and propagated samples of *M. indica* and *M. Laevigata* by Govt. U.P. School, Bithiya, Pali may be utilized for on-farm conservation. In addition, mulberry plants are used as shade trees in Tea and coffee estates of Yercaud, Tamilnadu and fruit yielding mulberry trees grown on the farm bunds in fruit orchards of Maharastra need to be encouraged for on-farm conservation.

An important type of traditional nature of conservation practiced as part of the religion-based conservation ethics of ancient people in many parts of the world is the protection of small areas of forests as sacred grooves or of particular tree specimens as sacred trees. For ages, mulberry is being worshiped by tribals of Uttaranchal in the Himalayan region and conserved as "Sacred grooves". The sacred mulberry tree (*M. serrata*) at Joshimath (2000 above MSL) worshipped by the pilgrims of Badrinath, is said to be 1200 years old and is the largest mulberry tree (21.6 mts circumference) in the world (Rau 1967). The great Indian sage Adiguru Shri Shankarcharya is said to have meditated under this tree. Due to this fact, *M. serrata* trees are protected at several places namely, Shirmoli near Almora, Ulkadevi temple at Pithoragarh, Gwaldam, Garhwal and Kumaon regions of Himalayas.

However, *in situ* conservation experiences some risks due to demographic uncertainty resulting from random population migration events, environmental uncertainty due to unpredictable weather conditions, competition with other plant species, natural catastrophes such as floods, fires, or droughts and genetic uncertainty or random changes in genetic make-up due to genetic drift (Shaffer 1981).

Ex situ Conservation in Field Gene Bank

Ex situ conservation is the preservation of components of biological diversity outside their natural habitats. This involves conservation of genetic resources collected from natural wild habitats, climate stressed regions and cultivated lands at a designated site. It may include part of the plant or organism from which the concerned plant can be reproduced and preserved or whole plant where its stock of individuals is kept outside its natural habitat either in a plantation area, botanical garden, or gene bank, etc. *Ex situ* conservation has been practiced by several nations and considerable number of mulberry genetic resources that belong to different species have been conserved in field gene banks in silk producing countries. Among these countries, China is holding the largest mulberry germplasm collections (2,600) followed by Japan (1,375), India (1,291), Korea (615) and Bulgaria (140). Different *Morus* species conserved in different countries were presented in Table 1 and Table 3. Since mulberry is highly heterozygous cross pollinated plant and can be clonally propagated through stems cuttings or grafting, it is conserved as whole plant in the field or through preservation of plant parts like winter dormant buds/shoot tips *in vitro* that can be regenerated as whole plant.

In India, CSGRC being the Nodal Agency and National Active Germplasm Site (NAGS) for mulberry genetic resources, conducted more than 80 surveys and exploration trips in India which include the cold deserts of Ladakh Himalayan region, arid and semiarid regions of Rajasthan, saline regions of Andaman Islands and managed habitats of south India and collected 1291 (1006 indigenous and 285 exotic) mulberry accessions (Table 3) from 30 countries located in different geographical areas of the world which are being maintained as dwarf trees in the field gene bank that represents 13 *Morus* species viz., *Morus indica, M. alba, M. laevigata, M. serrata, M. rubra, M. cathayana, M. nigra, M. australis, M. bombycis, M. sinensis, M multicaulis, M. rotundiloba* and *M. tiliaefolia*.

Establishment and Management of Ex situ Field Gene Bank

The key principles at the centre of gene bank operation are the preservation of germplasm identity, maintenance of genetic integrity, and the promotion of utilization of accessions. This includes providing associated information to facilitate use of conserved plant material (FAO 2014).

Identity of Accessions

This process begins with recording passport data and collecting donor information if applicable. As soon as any new genetic resource is added in the field gene bank, it should be given number to identify that accession. To avoid maintenance and multiplication of same mulberry accession under different names by different institutes and also

Table 3. *Ex situ* conservation of mulberry germplasm in the field gene bank of CSGRC, Hosur.

Sl. No.	State	No of accessions	Sl. No.	Country	No of accessions
1	Andaman & Nicobar	15	1	Afghanistan	3
2	Andhra Pradesh	4	2	Australia	2
3	Arunachal Pradesh	9	3	Bangladesh	5
4	Assam	11	4	China	55
5	Bihar	9	5	Cyprus	1
6	Chhattisgarh	4	6	Egypt	3
7	Goa	11	7	France	32
8	Gujarat	16	8	Hungary	1
9	Haryana	13	9	India	1006
10	Himachal Pradesh	36	10	Indonesia	8
11	Jammu & Kashmir	41	11	Italy	7
12	Jharkhand	17	12	Japan	72
13	Karnataka	159	13	Myanmar	7
14	Kerala	71	14	Nepal	1
15	Madhya Pradesh	12	15	Pakistan	8
16	Maharastra	32	16	Papua New Guinea	1
17	Manipur	12	17	Paraguay	4
18	Meghalaya	23	18	Philippines	1
19	Mizoram	8	19	Portugal	1
20	Nagaland	9	20	Russia	1
21	New Delhi	3	21	South Korea	6
22	Orissa	1	22	Spain	2
23	Pondicherry	4	23	Sri Lanka	2
24	Punjab	18	24	Thailand	11
25	Rajasthan	60	25	Turkey	1
26	Sikkim	15	26	USA	4
27	Tamil Nadu	86	27	Venezuela	1
28	Tripura	2	28	Venosa	1
29	Uttar Pradesh	146	29	Vietnam	5
30	Uttaranchal	8	30	Zimbabwe	11
31	West Bengal	151		Unidentified	28
	Total	1006		Total	1291

for avoiding duplicity of the material, NBPGR provides national accession number upon the request by the national active germplasm sites along with the supportive information. After obtaining the national accession number, it is to be used while distribution of germplasm for utilization. If the accession is received from other institutes, the number should be informed to them for using the same in their publications and documentation of data, etc.

Herbarium specimen often plays an important role in the correct identification samples. Field layout plans should be properly documented to ensure proper identification of accessions in field gene banks. Field labels are prone to be damaged due to various external factors such as bad weather conditions. Modern techniques like Radio-Frequency Identification (RFID) tags and molecular markers, accession labels with printed barcodes, can greatly facilitate germplasm management by reducing the possibility of error.

Maintenance of Genetic Integrity and Health of Germplasm

Gene banks should ensure that collected germplasm is genetically representative of the original population. As field gene banks are vulnerable to the impacts of environmental factors, such as weather conditions, incidence of pests, various molecular techniques are needed to assess whether genomic stability has been maintained. Monitoring of genetic stability is equally important for germplasm conserved *in vitro*, especially in view of the risk of somaclonal variation.

Physical Security of Collections

An underlying principle of germplasm conservation is that the infra structure of the gene bank facilities are of adequate standard to secure the materials from any external factors, including natural disasters, and damage caused by human-beings. Adequate security systems are required to ensure that gene bank cooling equipment, as well as backup generators are in good running condition. Furthermore, it is vital to maintain the adequate levels of LN in cryovessels.

Availability and use of Germplasm

Conserved material must be made available for current and future use. Hence it is important that all processes in gene bank operations and management contribute to this goal. Although there are a few individuals per accession in field gene banks with a limited capacity for distribution to users, the genebank should have a strategy in place for multiplying quickly any required germplasm for distribution.

Documentation and Sharing of Information

Essential, detailed, accurate, and up-to-date information should be recorded in electronic databases to ensure communication of information while supplying germplasm. Accessibility, availability and sharing of this information should be treated with high priority as it leads to better utilization of germplasm. Interactive databases with search-query can assist germplasm clients in choosing the germplasm as per their requirement. Feedback information collected from users and incorporating the data on further evaluation adds to the value and utility of the collection. If information on the conserved germplasm is easily available and accessible, it will enhance germplasm usage.

Proactive Management of Gene Banks

Effective and sustainable conservation of genetic resources depends on active management of conserved germplasm. Proactive management is crucial for ensuring that germplasm is efficiently conserved and made timely available and in adequate quantity for further use by plant breeders, farmers, researchers and other users. Standards were defined by FAO (2014) for acquisition of material to field gene bank, establishment of collections, field management, regeneration, propagation and for safety duplication.

Safety Back ups

Every field gene bank accession should be safety duplicated at least in one more sites and/or backed up by an alternative conservation method/strategy such as *in vitro* cryopreservation, where ever, possible to prevent the loss due to extreme weather conditions. In India, safety back up for temperate and tropical mulberry accessions were established under CSRTI, Pampore and CSRTI, Mysore respectively.

Common Problems in Mulberry Field Gene Bank and Measures Taken for their Management

Poor Establishment of Mulberry Genetic Resources in the Field Gene Bank

Young saplings should not be used for plantation in the field gene bank. Healthy 10–12 months old are to be used for planting. In case of weak stems, support should be given until tree attains sufficient girth and strength.

Soil Erosion and Maintenance of Soil Health and Fertility

Mulberry is a perennial tree and hence, maintained as a dwarf tree in field gene bank. Because of wider spacing in the field gene bank, there is every possibility of loss in top soil during heavy rains. Hence, mulching and minimum tillage are to be practiced. In addition, organic content of the soil is to be maintained by addition of sufficient manures, raising of green manure crops in order to facilitate the growth of favourable micro-organisms and in turn soil health.

Integrated Pest Management

Continuous monitoring and plant protection measures are required in mulberry field gene bank particularly with regard to root diseases and stem borer to avoid loss of germplasm. Special attention is needed for old plantations.

(i) Termites control

Continuous drought years and low organic content in the soil will aggravate termite attack. By the time the symptoms appear on the above ground, there may be severe damage in the root zone with basal portion of the trunk becoming hallow. Hence, much attention has to be paid on maintenance of organic content of the soil. If any termite attack is observed, the necessary control measures are to be followed.

Extracts of *Agave americana* have been found to be effective to control termites as well as stem borer. About 25 kg of Agave is crushed, soaked in 100 litres of water for over night and sprayed on the trunk. This will not have any harmful effect on the plant or person engaged for this work. In severe cases, chemical control measures may be practiced.

(ii) Damage by ant hills

If neglected, the ant hills in the basins of the plant can damage the tree. Ant hills after rains form big holes near trunk of the tree which expose roots and tree loses the support which in extreme cases may die.

(iii) Removal of parasitic plants

Parasitic plants need be removed, if observed, otherwise tree becomes weak and parasitic plant penetrate deep into the host plant and cause severe damage.

Formation of Deadwood and Poor Sprouting

Pruning of the trees in the field gene bank should be carried out by skilled workers. Some mulberry accessions cannot withstand repeated pruning or harsh pruning. The plants showing poor growth should not be pruned every year. For regeneration of such old trees, deadwood should be removed and that should be followed by full and frequent irrigations to allow latent buds to sprout in addition to application of sufficient farm yard manure.

Mixed Branches from Rootstock and Scion

Removal of basal branches at frequent intervals is necessary in mulberry accessions that are established in the field gene bank by grafting to avoid growth and mixing of shoots from the root stock.

In vitro Conservation

The required germplasm can be conserved *in vitro* by the addition of growth inhibitors in the culture medium to reduce the number of cycles of regeneration per year. The identification of optimal storage conditions for *in vitro* cultures must be determined according to the species (Gupta et al. 2009). A regular monitoring is required for checking the quality of the *in vitro* cultures in slow-growth storage against possible contamination. Monitoring genetic stability becomes an integral activity to ensure avoidance of somaclonal variation (Agrawal et al. 2019).

Cryopreservation of Mulberry for Long term Conservation

Storage of plant germplasm in liquid nitrogen (LN) is a relatively new practice (Engelmann 2004). Cryobanks provide the only option for long-term storage of vegetatively propagated plants and for seeds that can not be dried for storage. In cryopreservation, the plant materials are stored at −196°C in LN. At this temperature, cell division and metabolic activities remain suspended and the material can be stored without any genetic changes for long periods. Controlled rate cooling, vitrification, encapsulation dehydration, dormant bud preservation, and combinations of these techniques are now directly applicable for plant genotypes representing hundreds of species. Practical considerations and ready to use protocols are available (Reed 2008). Different types of plant cells, tissues and organs can be used for cryopreservation of plant genetic resources. Nonetheless, the most common organs used for cryopreservation of vegetatively propagated plants are shoot tips or dormant axillary buds (Towill and Forsline 1999). In mulberry, the most appropriate material for cryopreservation was found to be winter buds, though embryonic axes, pollen, synthetic seeds have also been used (Niino and Sakai 1992, Niino et al. 1992, 1993, Niino et al. 1995).

Protocol for Cryopreservation of Winter-dormant Buds by Dehydration and Slow Freezing

Axillary winter-dormant buds are collected from 8–10 old shoots after pruning trees during January–February and the buds are cryopreserved in liquid nitrogen (−196°C) after following dehydration in silica gel and slow freezing (Niino et al. 1995). For mulberry the following procedure (Anandarao et al., 2007, 2009) is followed:

➢ Dormant buds along with bark tissue are dissected from stem cuttings.

➢ Bark tissue along with the outer five to seven bud scales are removed without damaging the bud.

➢ The moisture content of the dormant buds of fresh samples is determined and buds desiccated in silica gel. For desiccation (500 g/150 mm internal diameter) buds are kept for 4–6 h to attain the moisture content of 12–15%.

➢ The slow freezing is achieved by treating the buds at sequentially lower temperatures of −5°C/day up to the temperature of −30°C (Niino et al. 1995). For this buds are placed in 2.5 ml polypropylene cryo-vials and stored in the cryotank containing liquid nitrogen (−196°C).

Regeneration from Cryopreserved Winter Buds

➢ For recovery, the cryopreserved axillary buds are taken out from cryotanks and thawed rapidly at 38°C in a water bath for 5 min.

➢ A rehydration treatment lasting 24 h is given by keeping the buds in sterile filter papers in the desiccators filled with the sterile soil-rite mixture.

➤ The rehydrated buds are surface-sterilized with liquid detergent Tween-20, disinfected with 70% ethanol for 25 s followed by 0.1% mercuric chloride for 6 to 8 min depending on the size of the bud. Rehygrated buds are given repeated three sterile water washes, and placed in sterile culture tubes (25 × 150 mm, Borosil make, fitted with polypropylene caps) containing MS (Murashige and Skoog 1962) medium with 3% (w/v) sucrose gelled with 0.7% (w/v) agar. The medium is supplemented with 1.0 mg l^{-1} 6-benzyl amino purine (BAP) initially.

➤ After bud break and the emergence of three to four young leaves, a dark treatment of 1 wk is given to initiate stem elongation.

➤ The sprouted buds which did not elongate were subcultured in the MS medium with 1.0 mg l^{-1} BAP, 0.2 mg l^{-1} gibberlic acid.

➤ Initially, the cultures were kept under dark for 4–7 d and then transferred to normal growing conditions at 25 ± 1°C with 55–60% relative humidity and light intensity of 65 ± 10 µmol m^{-2} s^{-1} with 16 h photoperiod in vitro culture room.

➤ *In vitro* sprouted buds were subcultured after 30 d of inoculation to the fresh MS medium with cytokinins and auxins.

➤ The accessions, which have high phenolic exudation, were repeatedly subcultured in the medium with activated charcoal 400–500 mg l^{-1}.

➤ The well-developed micro shoots are transplanted to 1/2 MS medium with indole butyric acid (IBA) 0.5 mg l^{-1} and activated charcoal (AC) 100 mg l^{-1} and incubated at 25 ± 1°C with 55–60% relative humidity and light intensity of 65 ± 10 µmol m^{-2} s^{-1} for inducing rooting.

➤ The *in vitro* rooted plants were hardened in plastic pots using autoclaved soilrite mixture covered with polyethylene bags and kept in the poly house with 75% relative humidity.

➤ The plants were watered once in 2 d with 1/2 MS liquid medium.

Protocol for Cryopreservation of Shoot-tips by Encapsulation and Dehydration

For this purpose, the *in vitro* grown shoot tips are used. Shoot cultures are hardened at 5°C at 8 h/day photoperiod for 3 weeks. The gel matrix should be prepared by adding 4% (w/v) sodium alginate, and 3% sucrose, 100 mg l^{-1}, myoinositol in MS medium without CaCl$_2$. A solution of CaCl$_2$.H$_2$O (100 mM) in sterilized distilled water should be prepared separately. The *in vitro* shoot tips/shoot buds surface sterilized should be dropped in the sodium alginate gel matrix solution, mixed thoroughly, and kept for 5 minutes. The explants kept in sodium alginate gel matrix be picked up with tweezers and dropped into sterile solution of calcium chloride ensuring one bead to contain one bud and allowed 30 minutes for complete polymerization. The alginate beads with the buds should be treated with sucrose (0.5 M to 1 M) solution for 24 hours, air diried in the laminar flow for 4 hours, and then cryopreserved in liquid nitrogen. The procedure for *in vitro* regeneration of cryopreserved encapsulated buds after thawing in water bath at 37°C is similar to already described for regeneration of winter buds.

Protocol for Cryopreservation of Apical Meristems by Vitrification

The apical meristems of *in vitro* grown shoot tips can be used in this process. The cultures are hardenend at 5°C at 8 hours/day photoperiod for 3 weeks. The meristematic tissue with two leaf primorida has to be dissected carefully using microscope under aseptic conditions. The explants should be pretreated in loading solution with 0.7 M sucrose for one day before vitrification. Apical meristem explants are treated with PVS2 (comprising 1 ml of standard growth liquid medium containing 0.4M sucrose, 30% (w/v) glycerol, 15% (w/v) ethylene glycol, 15% (w/v) dimethyl sulfoxide in 2.5 ml polypropylene cryovials) vitrification solution for 90 minutes. Then explants along with fresh PVS2 solution in the cryovials are preserved in LN. For recovery, the explants are removed from the cryotank, thawed immediately at 37°C, treated with unloading solution (1.25 M sucrose), and after decanting PVS2 solution subjected to plant regeneration using *in vitro* tissue culture techniques.

Core Collection of Mulberry Genetic Resources

Conservation efforts by the countries involved in sericulture have resulted in assembling and conserving a large number of mulberry accessions in the national field gene banks. However, a large size collection may lead to difficulties in germplasm management, adding to the cost of conservation, and the use of the collections may be restricted due to lack of knowledge about genetic diversity among the accessions. It is also difficult to identify the gaps while adding new material to the large collection. Frankel (1984) proposed establishment of "core collection" from the existing collection for effective conservation and utilization of genetic resources.

For a gene bank, a core collection with a limited number of the accessions is chosen to represent the genetic spectrum of the whole collection with minimum repetitiveness and maximum genetic diversity (Brown 1995). As the core collection is developed more and more information becomes available regarding the entries. The demand for core collection entries is likely to be substantially greater than for other accessions in the whole collection. The core provides evidence of gaps in collections (groups with very small number of accessions or substantial discontinuities between groups) which may indicate that further collection is required.

van Hintum et al. (2000) described a general procedure for developing a core collection in five steps: (i) Identifying the material that will be represented; (ii) Deciding the size of core collection; (iii) Dividing all accessions in whole collection into distinct groups; (iv) Deciding the number of entries per group; (v) Choosing the entries from each group that will be included in the core. Many researchers proposed other methods for development of a core collection such as PowerCore (Kim et al. 2007), Mstrat (Gousenard et al. 2001), stepwise clustering (Hu et al. 2000), and least distance stepwise clustering (Wang et al. 2007). These methods depend on some factors like genetic diversity of species, size of the whole collection, grouping of the whole collection, and data type (i.e., phenotypic or molecular data).

There are many suggestions regarding the size of a core collection. Based on the theory of neutral alleles and simulated scenes with different numbers and allelic frequencies for the loci of several populations, the core collection must contain at least 10% of the accessions of the total collection (Brown 1989). A core collection with high percentages (20–30%) is proposed especially when the objective is to retain the genetic diversity of quantitative traits (Noirot et al. 1996).

An efficient strategy using advanced molecular tools and morphological data are required to establish the genetic identity and establishing the diversity of core collections for enhancing the genetic resources utilization. Establishment of such a collection is a logical first step of intensive screening for desirable alleles. The core collection should serve as a working collection that could be intensively screened for the desired genotypes and the accessions excluded from the core collection would be retained as the reserve collection.

In India, Tikader and Kamble (2009) developed a core collection of 135 accessions that represent different countries and geographical regions which exhibited variability for different qualitative and quantitative characters out of a total of 628 mulberry accessions. Guruprasad et al. (2014) analyzed 850 mulberry accessions from 23 countries using molecular and phenotypic markers and selected a core collection with 100 diverse entries. In China, Chen et al. (2008) selected 11 accessions as core collection of *Morus multicaulis* Perr. from 46 accessions that originated in Shandong and Hebei province, while Zhang et al. (2011) selected 16 entries as core collection from 73 Gelu ecotype mulberry accessions. Yanfang et al. (2019) used 560 accessions with 40 morphological descriptors and stratified sampling strategies for a core collection. The core collection consisted of 28 accessions, accounting for 5% of the whole collection. Of these seven accessions belonged to *Morus alba*, one to *M. alba* var. macrophylla, four of *M. atropurprea*, one of *M. nigra*, three of *M. australis*, seven of *M. multicaulis*, two of *M. wittorum* and three of *M. bombycis*, respectively. The core collection can be considered as a preferential collection for conservation and characterization of mulberry.

Characterization and Evaluation of Mulberry Genetic Resources

The genetic resources without accompanying information on its characteristics are of limited value. The information generated through characterization and evaluation on each of the mulberry genetic resources and identifying important specific traits in them will be of immense value particularly in mulberry that is having wider adaptability with multiple uses.

Characterization is the description of plant germplasm based on the expression of highly heritable characters ranging from morphological, reproductive, anatomical to molecular markers, whereas evaluation is the recording those characteristics whose expression is often influenced by environmental factors. Evaluation involves the systematic collection of data on agronomic and quality traits through appropriately designed experimental trials. It frequently includes insect pest resistance, disease resistance, environmental traits (drought/cold tolerance, heat tolerance, salt tolerance, etc.) and quality evaluations (e.g., leaf protein content). Adding this type of information allows more focused identification of germplasm to develop climate resilient mulberry varieties with adaptive traits. Descriptors and methodology for characterization and evaluation of mulberry genetic resources are available (Ananda Rao 2002 and Alok Sahay et al. 2016)). DUS guidelines to assess distinctiveness, uniformity and stability (DUS) of different characters and their states of expression were developed at Central Sericultural Research and Training Institute, Mysuru under the aegis of Protection of Plant Varieties and Farmers' Right Authority, Govt. of India, New Delhi (http://plantauthority.gov.in/ pdf/Mulberry.pdf). In India, Central Sericultural Germplasm Resources Centre (CSGRC), Hosur which is recognized as National Active Germplasm Site (NAGS) for mulberry by National Bureau of Plant Genetic Resources (NBPGR), New Delhi presently holding 1291 accessions and documented the data generated through characterization and evaluation in the form of catalogues. Also, screened mulberry genetic resources for different abiotic and biotic stresses and identified the following promising accessions.

Promising Mulberry Accessions based on Characterization and Evaluation in the Field Gene Bank at CSGRC, Hosur

Eighteen exotic promising accessions viz., ME-0253, ME-0173, ME-226, ME-0188, ME-0251, ME-0245, ME-0214, ME-0244, ME-0170, ME-208, ME-0256, ME-0221, ME-0178, ME-0143, ME-0211, ME-0199, ME-0142, ME-0016 were identified for crop improvement in temperate regions.

Thirty one indigenous promising accessions MI-0657, MI-0458, MI-0665, MI-0670, MI-0827, MI-0246, MI-0568, , MI-0643, MI-0486, MI-0470, MI-0677, MI-0675, MI-0226, MI-0499, MI-0310, MI-0286, MI-0491, MI-0812, MI-0828, MI-0828, MI-0437, MI-0548, MI-0834, MI-0640, MI-0288, MI-0376, MI-0829, MI-0324, MI-0376, MI-0326, MI-0439 were identified based on morphological, anatomical, growth, yield, biochemical and propagation and any of these parameters of desirable trait can be applied for crop improvement.

Promising Mulberry Accessions Identified by Multilocation Trails in Different Environments

CSGRC, Hosur: MI-0437, MI-0376.

CSRTI, Berhampore: MI-0310, MI-0324, MI-0376.

CSRTI, Pampore: ME-0167, ME-0130, ME-0173 and ME-0168.

CSRTI, Mysure: MI-0310, MI-0326.

RSRS, Coonoor: ME-0007, ME-0033, ME-0130 ME-0169.

RSRS, Jorhat: MI-0154, MI-0369, MI-0416, MI-0349, MI-0388.

RSRS, Miransahib: MI-0324 , MI-0252.

Sahaspur: MI-0439, MI-0416, MI-0431.

Promising Mulberry Accessions for Abiotic Stress Tolerance

RSRS, Ananthapur (Tropical semi arid): MI-0463, MI-0458, MI-461, MI-0244, MI-0573, MI-0469, MI-0173.

RSRS, Chamarajnagar (Alkaline rainfed): MI-0204, MI-0173, MI-0024, ME-0006, ME-0052.

RSRS, Jorhat (Tropical hot and humid): MI-0549, MI-0576, MI-0469, MI-0587.

RSRS-Salem (Tropical alkaline irrigated): ME-0052, MI-0204, ME-0006, MI-0369, MI-0437, MI-0439, ME-0065, MI-0211, MI-0172.

Pampore (Temperate cold): ME-0006, ME-0191, ME-0201, ME-0210.

Mulberry Genetic Resources Identified for Biotic Stress Resistance

Nine accessions viz., MI-0055, MI-0555, MI-0206, MI-0096 MI-0208, ME-0165, ME-0239, Acc-199 & Acc-355 found resistant to tukra.

22 accessions viz., MI-0095, MI-0006, MI-0054, MI-0167, MI-0061, MI-0048, MI-0013, MI-0098, MI-0138, ME-0104, ME-0182, ME-0189, ME-0045, ME-0039, ME-0095, ME-0178, ME-0030, ME-0062, ME-0019, ME-0173, ME-0129 observed resistant to powdery mildew.

Promising Mulberry Accessions Identified based on Physiological Characterization in response to Water and Nitrogen used, Efficiency (Deficiency)

17 accessions viz., MI-0214, MI-0768, ME-0016, MI-0025, MI-0332, ME-0244, ME-0107, MI-0699, MI-0026, MI-0256, MI-0477, ME-0125, MI-0298, MI-0762, MI-0437, MI-0763, MI-0314 were found to be superior based on multiple traits under water-limited conditions.

Two accessions viz., MI-0685 and MI-0683 showed superior performance only under low N input (Efficient genotypes).

Six accessions viz., MI-0139, MI-0178, MI-0573, MI-0416, MI-0193, MI-0533 showed superior performance only under high N input conditions (Inefficient Responders).

17 accessions viz., MI-0256, MI-0332, MI-0768, MI-0762, MI-0477, MI-0622, MI-0226, MI-0657, MI-0763, MI-0346, MI-0025, MI-0699, MI-0314, MI-0214, MI-0670, MI-0827, MI-0161 performed well both under low and high N input (Efficient and Responders).

Promising Mulberry Germplasm Accessions for Tolerance to Alkalinity/Salinity

Screened 102 mulberry accessions with 2 levels of salinity (EC 6 and 8 dS/m) using C.776 as check. Identified 20 mulberry salinity tolerant accessions viz., MI-0437, MI-0376, MI-0327, MI-0670, MI-0657, MI-0012, MI-0476, MI-0242, MI-0129, MI-0245, MI-0161.

MI-0763, MI-0716, MI-0310, MI-0145, MI-0497, MI-0499, MI-0027, MI-0139 & MI-0764 at EC 8 ads/m (Jhansilakshmi et al. 2016).

Screened 100 accessions at 2 levels of alkalinity (pH 8.5 and 9) in micro plots using AR-12 as check. Eighteen alkaline tolerant accessions viz., MI-0226, MI-0670, MI-0836, MI-0652, MI-0762, MI-0449, MI-0764, MI-0437, MI-0716, MI-0822, MI-0310, MI-0248, MI-0702, MI-0190, MI-0643, MI-0499, MI-0788 & MI-0466 were identified as alkaline tolerant at pH 9.0.

Promising Mulberry Accessions Suitable for Social Forestry and Marginal Lands/Ecological Restoration

The accessions were identified based on multiple trait approach (Giannini et al. 2016) for characterization and evaluation of mulberry genetic resources in the field gene bank (Jhansilakshmi and Gargi 2018). The traits which help for survival and regeneration of the plants on barren marginal lands are selected on parameters such as survival % (rooting ability), number of branches, traits with tolerance to biotic and abiotic stress and traits having economic importance to human beings and animals. The accessions that have a high total score, cumulative index value, and regeneration capacity after repeated prunings in the field gene bank, are suitable for social forestry or marginal lands/ecological restoration.

Twenty one exotic mulberry genetic resources viz., ME-0217, ME-0218, ME-0253, ME-0220, ME-0251, ME-0245, ME-0214, ME-0226, ME-0246, ME-0182, ME-0215, ME-0027, ME-0174, ME-0040, ME-0232, ME-0041, ME-0149, ME-0189, ME-0194, ME-0129 and ME-0170 were identified for social forestry in temperate regions. Among these, ME-0253, ME-0220, ME-0251, ME-0245, ME-0226, ME-0246, ME-0027 and ME-0232 have high rooting and regeneration ability and hence can be used in marginal lands/ecological restoration.

Among the indigenous accessions, 35 mulberry genetic resources *viz.,* MI-0243, MI-0512, MI-0218, MI-0322, MI-0268, MI-0275, MI-0555, MI-0280, MI-0499, MI-0576, MI-0792, MI-0827, MI-0229, MI-0804, MI-0326, MI-0214, MI-0225, MI-0267, MI-0269, MI-0270, MI-0276, MI-0486, MI-0657 MI-0665, MI-0675, MI-0681 MI-0836 MI-0304 MI-0310 MI-0786 MI-0789 MI-0292 MI-0294 MI-0790 MI-0791 were identified for social forestry. All these promising genetic resources are being maintained in the mulberry field gene bank at CSGRC, Hosur.

Future Strategies for Meeting Mulberry Germplasm Requirements in Context of Climate Change

Collection of Germplasm from Climate Stress Regions

Integrated geoinformatics technology which is integrated approach to use geospatial technologies (remote sensing, geographic information system, global positioning system and information system) can be used in gap analysis, planning and execution of future exploration programme at National level (Semwal and Ahlawat 2015). Enrichment of germplasm need to be taken up with new collections from climate stress regions and breeding lines selected for specific traits from the screening programmes in hotspots.

Identification of New Traits or Alleles that Improves Adaptability or Increases Tolerance to Abiotic and Biotic Stress

The earlier breeding programs in India were aimed at increasing yield and leaf quality of mulberry which directly effect the profitability in sericulture. In the present context of climate change, the superior plant habit, improved adaptation for increased tolerances to environmental stresses (Reynolds et al. 2015), and higher efficiency in the utilization of limited soil nutrients (Hirel et al. 2011) are considered more important. Resistance to fungal diseases (e.g., powdery mildew and root rot), root-knot nematodes (*Meloidogyne* spp.), and insects (e.g., Mealy bug, thrips, mites), particularly those that cause significant yield reductions assume more importance. As climate change influences the occurrence, prevalence, and severity of plant pests and diseases, screening programmes to identify the resistance sources need to be taken up as per the need. More emphasis need to be given for identifying trait specific mulberry genotypes.

The nutritional content of leaves, stems, roots, fruits and tubers of C_3 plants grown at areas having elevated carbon dioxide levels is expected to lower the levels of protein, minerals and trace elements, such as zinc and iron (Taub et al. 2008, Loladze 2014). Hence, identification and utilization of mulberry genetic resources with high leaf quality and nutrient use efficiency helps to reduce the cost of mulberry production, maintain the leaf quality which in turn results in profitability in sericulture without adversely effecting the environment.

Though, nutrient use efficiency can be achieved by adopting more efficient crop management practices and breeding more nutrient use efficient cultivars, genetic improvement is more important and economical for perennial vegetatively propagated crops like mulberry. Further, in case of micronutrient deficiencies induced by high pH (i.e., Zn, Fe, Mn deficiencies), agronomic solution is not always successful, and genetic solution is necessary.

For the optimum growth of mulberry and good sprouting of the buds, the mean atmospheric temperature should be in the range of 13°C to 37.7°C. The ideal temperature is 24–28°C (Patnaik 2008). Mulberry has been shown to grow abnormally at 35°C (Fukui and Naoi 1996). Hence, the temperature above 37°C is considered as high temperature, below 13°C as low temperature for mulberry growth. Raising temperatures in recent times necessitates including thermo-tolerance as one of the objectives in tropical sericulture areas.

Utilization of Wild Relatives for Incorporation of Novel Genes that Contribute to Adaptability

Wild species are especially vulnerable to climate change because they do not receive management interventions that could help them adapt to changing conditions. Hence, collection and conservation of wild species is utmost important. There has been concern that the genetic variation within cultivated mulberry varieties has become too restricted (Tikader and Kamble 2007) and attempts were made to develop genetically broad-based plants through inter-specific hybridization. Pre-breeding (i.e., the generation of intermediate materials for use as parents in plant breeding) is a means to introgress novel alleles from non-adapted materials into crop varieties (Nass and Paterniani 2000). Pre-breeding efforts were initiated to transfer desirable genes from wild mulberry genetic resources (*Morus laevigata* and *Morus serrata*) into the cultivated varieties to make use of unadapted mulberry germplasm (Tikader and Dandin 2005, 2007). At present, 17 mulberry germplasm resources derived through prebreeding are available in the *ex situ* field gene bank of CSGRC, Hosur. However, there is a greater need to enhanced utilization of unadapted mulberry germplasm through novel prebreeding strategies to identify and make use of new genes/alleles and pyramiding of the desired genes for their effective utilization for developing climate resilient mulberry varieties.

Development of New Phenotyping and Genomic Tools for Rapid and Accurate Screening of a Large Number of Populations

Accurate and rapid screening tools need to be developed and used for identifying the desired genotype to make use of germplasm more effectively. Development and use of new genomic tools and utilization of new tools like spectral reflectance of plant canopy, infrared thermography, image analysis techniques, etc., for large scale screening and identifying the desired genotype more accurately, is one of the priority area. Understanding mulberry genetic resources more thoroughly in network mode particularly under different environmental challenges and utilization of identified genetic resources in target environments helps in sustainable sericulture.

Exploration of Genetic Diversity for Non-Sericultural Purposes for Improving Livelihood of Farmers

There is a lot of scope to make use of rich diversity in mulberry germplasm for non-sericultural purposes as mulberry is a multipurpose tree. Collection of feedback information and incorporating the new information developed through further evaluations by different germplasm users add value to the germplasm and increases livelihood options of rural poor.

Integration of Genomics and Genetic Resources for Crop Improvement

Association mapping helps for identification of marker tags associated with trait of interest. Phenotyping the diverse panel of mulberry genetic resources for different abiotic and biotic stress, genotyping with molecular markers (e.g., SSRs, SNPs), quantification of LD using molecular marker data and correlation of phenotypic and genotypic data reveals marker tags. Marker tags are the most effective tools for crop improvement and help in incorporation of the genes of interest from donor lines to the breeding material through marker-assisted selection (MAS).

Availability and Accessibility of Mulberry Genetic Resources as per the Requirement

Conservation of mulberry genetic resources must be linked to its sustainable use by unlocking genetic potential stored in field gene bank. Encouraging exchange of germplasm between different states, organizations within the country and other countries helps to enhance utilization of mulberry germplasm for meeting the new challenges due to climate change. New techniques/treatments need to be employed for detection of pests and diseases in the newly procured germplasm.

Safety Backups

As field gene banks are vulnerable of extreme weather conditions, the coreset of mulberry genetic resources are to be maintained in other centres as safety backups. Cryo-preservation can be used as an alternative long-term conservation strategy.

Participation of Private Organizations and Society

Participation of NGOs and local people need be encouraged for On-farm and *In situ* conservation of mulberry genetic resources.

References

Agastian, S.T.P. and M. Vivekananda. 1997. Effect of induced salt stress on growth and uptake of mineral nutrients in mulberry (*Morus alba*) genotypes. Indian J. Agricult. Sci. 67: 469–472.

Alok, S., K. Jhansilakshmi, P. Saraswathi and S. Sekar. 2016. Manual on Mulberry Gene Bank Operations and Procedures. Technical Manual, CSGRC, Hosur.

Ananda Rao, A. 2002. Conservation status of mulberry genetic resources in India. *In*: 19th Int. Seric Congres Expert Consultation on Promotion of Global Exchange of Sericulture Germplasm Satellite Session. September 21–25, Czech J. Genet. Plant Breed., 48, 2012 (4): 147–156 Bangkok. Available at http://www.fao.org/DOCREP/005/ AD107E/ad107 e0m.htm.

Ananda Rao, A., K. Thangavelu and K.R. Sharma. 2006. Distribution and variation of mulberry genetic resources in high altitude and cold desserts of Ladakh Himalayan region. Indian Jour. Pl. Gen. Res. 18(20): 255–259.

Ananda Rao, A., Rekha Chaudhury, Susheel Kumar, D. Velu, R.P. Saraswat and C.K. Kamble. 2007. Cryopreservation of mulberry germplasm core collection and assessment of genetic stability through ISSR markers. Int. Indust. Entomol. 5: 23–33.

Ananda Rao, A., R. Chaudhury, S.K. Malik, S. Kumar, R. Ramachandran and S.M.H. Qadri. 2009. Mulberry biodiversity conservation through cryopreservation. *In Vitro* Cell Dev. Biol. Plant. 45: 639–649.

Ananda Rao, A., S.S. Chauhan, R. Radhakrishnan, A. Tikader, M.M. Borpuzari and C.K. Kamble. 2011. Distribution, variation and conservation of mulberry (*Morus* spp.) genetic resources in the arid zone of Rajasthan, India. Bio Biodiver. Bioavailab. 5(1): 52–62.

Agrawal, A., A. Shivani Singh, Era Vaidya Malhotra, D.P.S. Meena and R.K. Tyagi. 2019. *In vitro* conservation and cryopreservation of clonally propagated horticultural species. pp. 529–578. *In*: Tyagi, R.K., P.E. Rajasekharan and V.R. Rao (eds.). Conservation and Utilization of Horticultural Genetic Resources. Springer, Singapore.

Arora, R.K. 1981. Plant genetic resources exploration and collection: planning and logistics. pp. 46–54. *In*: Mehra, K.L. et al. (eds.). Plant Exploration and Collection. Sci. Monogr. 3. National Bureau of Plant Genetic Resources, New Delhi, India.

Banerjee, R., M.D. Maji, P. Ghosh and A. Sarkar. 2009. Genetic analysis of disease resistance against *Xanthomonas campestris pv. mori* in mulberry (*Morus* spp.) and identification of germplasm with high resistance. Archives of Phytopathology and Plant Protection 42(3): 291–297.

Bellon, M.R., E. Gotor and F. Caracciola. 2014. Conserving land races and improving livelihoods. How to assess the success of conservation projects. International Journal of Agriculture Sustainability. http://dx.doi.org/10.1080/14735903.2014.98 6363.

Brown, A.H.D. 1989. Core collections: a practical approach to genetic resources management. Genome 31: 818–24.

Brown, A.H.D. 1995. The core collection at the crossroads. pp. 3–19. *In*: Hodgkin,T., A.H.D. Brown, T.J.L van Hintum and E.A.V. Morales (eds.). Core Collections of Plant Genetic Resources, John Wiley and Sons, New York, USA.

Brown, A.H.D. and D.R. Marshall. 1995. A basic sampling strategy: Theory and practice. pp. 75–91. *In*: Guarino, L., V.R. Rao and R. Reid (eds.). Collecting Plant Genetic Diversity Technical Guidelines. CAB International, Oxon, UK.

Chaitanya, K.V., D. Sundar and A. Ramachandra Reddy. 2001. Mulberry leaf metabolism under high temperature stress. Biol. Plant. 44(3): 379–384.

Chandrasekar, M. 2001. Technical Report on Survey and Exploration of mulberry genetic resources in Gujarat, CSGRC, Hosur.

Chen, J.B., Y. Huang, L. Zhang, W.G. Zhao and Y.L. Pan. 2008. Construction of the core collection of *Morus multicaulis* Perr-germplasm resources from Shandong and Hebei based on ISSR molecular markers. Science of Sericulture 34: 587–592.

Chen, H., X. He, Y. Liu, J. Li, Q. He, C. Zhang and B. Wei. 2016. Extraction, purification and anti-fatigue activity of gamma amino butyric acid from mulberry (*Morus abla* L.) leaves. Sci. Rep. 6. doi:10.1038/srep 18933.

Chinnaswamy, K.P. and K.B. Hariprasad. 1995. Fuel Energy Potentiality of Mulberry. Indian Silk 34: 15–18.

Dandin, S.B., Basavaiah, R. Kumar and H.V. Venkateshaiah. 1993. Phytogeographical studies in the genus *Morus* L. II. Geographical distribution and natural variation of *M. serrata* Roxb. Indian Journal of Plant Genetic Resources 7: 223–226.

Dandin, S.B., Basavaiah, R. Kumar and R.S. Mallikarjunappa. 1995. Phytogeographical studies in the genus *Morus* L. II. Geographical distribution and natural variation of *M. laevigata* Wall Ex. Brandis. Indian Journal of Plant Genetic Resources 8(1): 129–131.

Engelmann, F. 2004. Plant cryopreservation: progress and prospects. *In Vitro* Cellular and Developmental Biology 40(5): 427–433.

FAO. 2011. Save and grow, a policymaker's guide to the sustainable intensification of smallholder crop production. Rome.

FAO. 2014. Genebank Standards for Plant Genetic Resources for Food and Agriculture. Rev. ed. Rome.

Frankel, O.H. 1984. Genetic perspectives of germplasm conservation. pp. 161–170. *In*: Arber, W.K., K. Llimensee, W.J. Peacock and P. Starlinger (eds.). Genetic Manipulation: Impact on Man and Society Cambridge University Press, Cambridge, UK.

Fukui, K. and T. Naoi. 1996. Growth injury and morphological characteristics of mulberry under high temperature conditions. J. Seric. Sci. Jpn. 65: 165–169.

Ghosh, M.K., A.K. Misra, Shivnath, C. Das, T. Senguptha, P.K. Ghosh, M.K. Singh and B.B. Bindroo. 2012. Morpho-physiological evaluation of some mulberry genotypes under high moisture stress condition. Journal of Crop and Weed 8(1): 167–170.

Giannini, T.C., A.M. Giulietti, R.M. Harley, P.L. Viana, R. Jaffe, R. Alves, C.E. Pinto, N.F.O. Mota, C. F. Caldeira, V. Imperatriz-Fonseca, A.E. Furtini and J.O. Siqueira. 2016. Selecting plant species for practical restoration of degraded lands using a multiple-trait approach. Austral. Ecol. 42: 510–521.

Gornall, J., R. Betts, E. Burke, R. Clark, J. Camp, K. Willett and A. Wiltshire. 2010. Implications of climate change for agricultural productivity in the early twenty-first century. Philosophical Transactions of the Royal Society B: Biological Sciences 365(1554): 2973–2989.

Gouesnard, B., T. Bataillon, G. Decoux, C. Rozale, D.J. Schoen and J.L. David. 2001. MSTRAT: An algorithm for building germplasm core collections by maximizing allelic or phenotypic richness. J. Hered. 92: 93–94.

Guha, A., G.K. Rasineni and A. Ramachandra Reddy. 2010. Drought tolerance in mulberry (*Morus* spp.): a physiological approach with insights into growth dynamics and leaf yield production. Expl. Agric. 46: 471–488.

Gupta, S., P. Singh and A.A. Rao. 2009. Micropropagation of mulberry (*Morus indica* L.) using explants from mature tree: Effects of plant growth regulators on shoot multiplication and rooting. Progressive Horticulture 41(2): 136–144.

Guruprasad, R.R. Krishnan, S.B. Dandin and V.G. Naik. 2014. Groupwise sampling: a strategy to sample core entries from RAPD marker data with application to mulberry. Trees 28: 723–731.

Hawkes, J.G. 1980. Crop genetic resources—A field collection manual, IBPGR/EUCARPIA, Univ. of Birmingham, UK.

Hirel, B., T. Tetu, P.J. Lea and F. Dubois. 2011. Improving nitrogen use efficiency in crops for sustainable agriculture Sustainability 3: 1452–1485

Hu, J., J. Zhu and H.M. Xu. 2000. Methods of constructing core collections by stepwise clustering with three sampling strategies based on the genotypic values of crops. Theor. Appl. Genet. 101: 264–268.

IPCC. 2007. Summary for policymakers. *In*: Solomon, S., D. Qin, M. Manning, Z. Chen, M. Marquis, K.B. Averyt, M. Tignor and H.L. Miller (eds.). Climate Change 2007: The Physical Science Basis. Contribution of Working Group I to the Fourth Assessment Report of the Intergovernmental Panel on Climate Change Cambridge University Press, Cambridge, United Kingdom and New York, NY, USA.

Jain, A.K. and R. Kumar. 1989. Mulberry species and their distribution in North Eastern India. pp. 37–42. *In*: Sengupta, K. and S.B. Dandin (eds.). Genetic Resources of Mulberry and Utilisation, CSR & TI, Mysore.

Jarvis, D and T. Hodgkin. 2000. Farmer decision making and genetic diversity: linking multidisciplinary research to implementation on-farm. pp. 261–278. *In*: Brush, S.B. (ed.). Genes in the Field: On-Farm Conservation of Crop. Genes in the Field: On-Farm Conservation of Crop Diversity. Lewis Publishers, Boca Raton, FL, USA.

Jarvis, A., A. Lane and R. Hijmans. 2008. The effect of climate change on crop wild relatives. Agr. Ecosyst. Environ. 126: 13–23.

Jhansilakshmi, K. and S.B. Dandin. 2011. Genetic diversity of selected mulberry (*Morus* spp.) genetic resources for drought resistance characters. Sericologia 53(4): 545–556.

Jhansilakshmi, K., A. Ananda Rao and M.M. Borpuzari. 2014. Potential mulberry genetic resources for drought tolerance. Paper presented in 23rd International Congress on Sericulture and Silk Industry held on 24–27th Nov., 2014, Bangalore.

Jhansilakshmi, K., M.M. Borpuzari, A. Ananda Rao and P.K. Mishra. 2016. Differential response of mulberry (*Morus* spp.) accessions for salinity stress. J. Appl. Biosci. 42(1): 30–35.

Jhansilakshmi, K. and Gargi. 2018. Utilization of Mulberry Germplasm in Breeding Programmes—Current Status and Future Prospects. pp. 40–45. *In*: Status papers Seri-Breeders' Meet. Central Silk Board, Bangalore.

Kaviani, B. 2011. Conservation of plant genetic resources by cryopreservation. Australian Journal of Crop Science 5: 778–800.

Khurana, P. and V.G. Checker. 2011. The advent of genomics in mulberry and perspectives for productivity enhancement. Plant Cell Rep. 30: 825.

Kim, K.W., H.K. Chung, G.T. Cho, K.H. Ma, D. Chandrabalan, J.G. Gwag, T.S. Kim, E.G. Cho and Y.J. Park. 2007. PowerCore: A program applying the advanced M strategy with a heuristic search for establishing core sets. Bioinformatics 23: 2155–2162.

Lane, A. and A. Jarvis. 2007. Changes in climate will modify the geography of crop suitability: Agricultural biodiversity can help with adaptation. Journal of Semi-arid Tropical Agricultural Research 4: 1–12.

Lee, J.Y., S.O. Moon, Y.J. Kwon, S.J. Rhee, H.R. Park and S.W. Choi. 2004. Identification and quantification of anthocyanins and flavonoids in mulberry (*Morus* sp.) cultivars. Food Sci. Biotechnol. 13: 176–184.

Limaye, V.D. 1952. Timber for sports goods. The Indian Forester 78: 371–378.

Loladze, I. 2014. Hidden shift of the ionome of plants exposed to elevated CO_2 depletes minerals at the base of human nutrition. *eLife* 3:e02245. doi: 10.7554/eLife.02245.

Machii, H., A. Koyama and H. Yamanouchi. Mulberry breeding, cultivation and utilization in Japan. *In*: Proceedings of the Mulberry for Animal Production: FAO Electronic Conference, Feed Resources Group (AGAP), FAO, Rome, Italy, May–June 2000.

Maji, M., H. Sau and B.K. Das. 2009. Screening of mulberry germplasm lines against Powdery mildew, *Myrothecium* leaf spot and *Pseudocercospora* leaf spot disease complex. Arch. Phytopathol. Plant Protect. 42(9): 805–811.

Marshall, D.R. and A.H.D. Brown. Optimum sampling strategies in genetic conservation. pp. 53–80. *In*: Frankel, O.H. and J.G. Hawkes (eds.). 1975. Genetic resources for today and tomorrow (eds). Cambridge Univ. Press, Cambridge.

Maxted, N., B.V. Ford-Lloyd and S.P. Kell. 2008. Crop wild relatives: Establishing the context. pp. 3–30. *In*: Maxted, N., B.V. Ford-Lloyd, S.P. Kell, J.M. Iriondo, M.E. Dulloo and J. Turok (eds.). Crop Wild Relative Conservation and Use. CABI Publishing.

Murashige, T. and F.A. Skoog. 1962. A revised medium for rapid growth and bioassay with tobacco tissue culture. Physiol. Plant 15: 473–497.

Naik, V.G. and P. Mukherjee. 1997. An exploration to Andaman and Nicobar Islands for wild mulberry germplasm. Journal of Environmental Resources 5(1-4): 17–19.

Nass, L.L. and E. Paterniani. 2000. Pre-breeding: a link between genetic resources and maize breeding. Sci. Agric. 57: 581–587. doi: 10.1590/S0103-90162000000300035.

Niino, T. and A. Sakai. 1992. Cryopreservation of alginate—coated *in vitro* grown shoot tips of apple, pear and mulberry. Plant Sci. 87: 199–206.

Niino T., A. Sakai, S. Enomoto, J. Megoshi and S. Kat. 1992. Cryopreservation of *in vitro* grown shoot tips of mulberry by vitrification. CryoLetter 13: 303–312.

Niino T., K. Shirata and S. Oka. 1995. Viability of mulberry winter buds cryopreserved for 5 y at −135°C. Jour. Seric. Sci. Jap. 64: 370–374.

Noirot, M., S. Hamon and F. Anthony. 1996. The principal component scoring: A new method of constituting a core collection using quantitative data. Genet. Resour. Crop Evol. 43: 1–6. https://doi.org/10.1007/BF00126934.

Ozgen, M., S. Serçe and C. Kaya. 2009. Phytochemical and antioxidant properties of anthocyanin-rich *Morus nigra* and *Morus rubra* fruits. Sci. Hortic. 119: 275–279.

Patnaik, R.K. 2008. Mulberry Cultivation. Bioteck books, New Delhi.

Rau, M.A. 1967. The sacred mulberry tree of Joshimath, UP. Indian Forester 93(8): 333–335.

Ravindran, S., A.A. Rao, V.G. Naik, A. Tikader, P. Mukherjee and Thangavelu, K. 1997. Distribution and variation in mulberry germplasm. Indian J. Plant Genet. Resour. 10(2): 233–242.

Ravindran, S., A. Tikader, V.G. Naik, A.A. Rao and P. Mukherjee. 1998. Distribution of mulberry species in India and its utilization. Indian J. Plant Genet. Resour. 2: 163–168.

Reed, B.M. 2008. Plant Cryopreservation: A Practical Guide. Springer, NewYork, 511p.

Reynolds, M., G. Molero, M. Tattaris, C.M. Cossani, P. Alderman and S. Sukumaran. Improving crop adaptation to climate change through strategic crossing of stress adaptive traits. *In* Agriculture and Climate Change - Adapting Crops to Increased Uncertainty (AGRI 2015) Procedia Environmental Sciences 29(2015) 298–299. doi: 10.1016/j.proenv.2015.07.272.

Sanchez, M.D. 2000. World distribution and utilization of mulberry, potential for animal feeding. FAO Electronic Conference on mulberry for animal production, World Anim. Rev. 93:1–11, FAO, Rome.

Saraswat, R.P. 2002. Technical Report on Survey, Exploration and Collection of mulberry genetic resources in Arunachal Pradesh and Meghalaya, CSGRC, Hosur.

Semwal, D.P. and S.P. Ahlawat. 2015. Application of geoinformatics in PGR studies. E-Publication (NBP-16-03), ICAR-National Bureau of Plant Genetic Resources, New Delhi.

Shaffer, M.L. 1981. Minimum population sizes for species conservation. Bioscience 31: 131–134.

Sharma, S.K. and K.K. Zote. 2010. MULBERRY—A multi purpose tree species for varied climate Range Management and Agroforestry 31(2): 97–101.

Shukla, P.K.R. Gulab, A.S. Aftab and S.P. Sharma. 2016. Prospect of cold tolerant genes and its utilization in mulberry improvement. Ind. Hortic. J. 6(Special): 127–129.

Singh, B. and H.P.S. Makkar. 2000. The potential of mulberry foliage as feed supplement in India. pp. 139–153. *In*: Sanchez, M.D. (ed.). Mulberry for Animal Production. Animal Health and Production Paper No. 147. FAO, Rome, Italy.

Susheelamma, B.N., M.S. Jolly, K. Giridhar and K. Sengupta. 1990. Evaluation of germplasm genotypes for the drought resistance in mulberry. Sericologia 30: 327–340.

Taub, D.R., B. Miller and H. Allen. 2008. Effects of elevated CO_2 on the protein concentration of food crops: a meta-analysis. *Glob.* Change Biol. 14: 565–575. doi: 10.1111/j.1365-2486.2007.01511.x.

Thomas, C.D., A. Cameron, R.E. Green, M. Bakkenes, L.J. Beaumont, Y.C. Collingham, B.F.N. Erasmus, M. Ferreira De Siqeira, A. Grainger, L. Hannah, L. Hughes, B. Huntley, A.S. Van Jaarsveld, G.F. Midgley, L. Miles, M.A. Ortega-Huertas, A.T. Peterson, O.L. Phillips and S.E. Williams. 2004. Extinction risk from climate change. Nature 427: 145–148.

Tian, S., M. Tang and B. Zhao. 2016. Journal of Traditional Chines Medical Sciences 3(1): 3–8.

Tikader, A. and S.B. Dandin. 2005. Biodiversity, geographical distribution, utilization and conservation of wild mulberry *Morus serrata* Roxb. Caspian J. Envir. Sci. 3(2): 179–186.

Tikader A. and S.B. Dandin. 2007a. Prebreeding efforts to utilize two wild Morus species. Current Science 92(12): 1729–1733.

Tikader, A. and S.B. Dandin. 2007b. Pre-breeding efforts to utilize two wild *Morus* species. Curr. Sci. 92: 1072–1076.

Tikader, A. and C.K. Kamble. 2007. Mulberry breeding in India: a critical review. Sericologia 47(4): 367–390.

Tikader, A. and C.K. Kamble. 2008. Mulberry wild species in India and their use in crop improvement—a review. Aust. J. Crop. Sci. 2: 64–72.

Tikader, A. and C.K. Kamble. 2009a. Monograph on mulberry, CSGRC, Central Silk Board.

Tikader, A. and C.K. Kamble. 2009b. Development of core collection for perennial mulberry (*Morus* spp.) germplasm. Pertanika J. Sci. Tech. 17: 43–51.

Tikader, A. and K. Vijayan. 2010. Assessment of biodiversity and strategies for conservation of genetic resources in Mulberry (*Morus* spp.). Bioremediation, Biodiversity and Bioavailability 4: 15–17.

Tikader, A. and K. Vijayan. 2017. Mulberry (*Morus* Spp.) Genetic Diversity, Conservation and Management. pp. Ahuja, R. and S.M. Jain (eds.). Biodiversity and Conservation of Woody Plants, Sustainable Development and Biodiversity. Springer International Publishing AG 2017 M.

Towill, L.E. and P.L. Forsline. 1999. Cryopreservation of sour cherry (*Prunus cerasus* L.) using a dormant vegetative bud method. Cryo-letters 4 (20): 215–222.

UNEP. 1992. Global Biodiversity Strategy. World Resources Institute, The World Conservation Union and United Nations Environment Programme, Washington, U.S.A.

Van Hintum, Th. J.L., A.H.D. Brown, C. Spillane and T. Hodgkin. 2000. Core collections of plant genetic resources. IPGRI Technical Bulletin No.3. International Plant Genetic Resources Institute, Rome Italy.

Vijayan, K., S.P. Chakraborti and P.D. Ghosh. 2003. *In vitro* screening of axillary buds for salinity tolerance in mulberry genotypes. Plant Cell Rep. 22: 350–357.

Vijayan, K., S.G. Doss, S.P. Chakraborti and P.D. Ghosh. 2009. Breeding for salinity resistance in mulberry (*Morus* spp.). Euphytica 169: 403–411.

Vijayan, K. 2010. The emerging role of genomic tools in mulberry (*Morus*) genetic improvement. Tree Genet. Genomes 6: 613–625.

Vijayan, K., S.P. Chakraborti and P.D. Ghosh. 2003. *In vitro* screening of axillary buds for salinity tolerance in mulberry genotypes. Plant Cell Rep. 22: 350–357.

Vinoda, K.S., B. Madhura, S. Sheshshayee, K.C. Narayanaswamy, Udayakumar, Madan. 2016. Assessing genetic variability in water use efficiency of *Morus* species based on stable isotopes (δ13 C) and RAPD profiling. 12. 7–16.

Wang J.C., J. Hu, H.M. Xu and S. Zhang. 2007. A strategy on constructing core collections by least distance stepwise sampling. Theoretical and Applied Genetics 115: 1–8.

Yadav, B.R.D. and T. Pavan Kumar. 1996. Occurrence of new distributional ranges of *M. laevigata* Wall. and its variants in south India. Ind.J. Seric. 35: 163–164.

Yadav, D.B.R., J. Sukumar and K.V. Prasad. 1993. Screening of potential resistance in the mulberry to leaf spot (*Cercospora moricola*) disease. Sericologia 33: 81–90.

Yamatake, Y., M. Shibata and M. Nagai. 1976. Pharmacological studies on Root Bark of Mulberry Tree (*Morus alba*), Japan. J. Pharmacol. 26: 461–9.

Yanfang Zhang, Dechang Hu, Jincheng Zuo, Ping Zhang, Zhaohong Wang, Chuanjie Chen. 2019. Development of a mulberry core collection originated in China to enhance germplasm conservation. Crop Breed. Appl. Biotechnol. [Internet] 19(1): 55–61.

Zhang, L., J.B. Chen, Y. Huang, X.J. Shen, L. Liu, W.G. Zhao and S. Qiang. 2011. Screening of core germplasms of Gelu ecotype mulberry based on ISSR marker. Science of Scriculture 37: 380–388.

Zhang, H., Z.F. Ma, X. Luo and X. Li. 2018. Effects of Mulberry Fruit (*Morus alba* L.) Consumption on Health Outcomes: A Mini-Review. Antioxidants (Basel). 21; 7(5): 69.

Index